Life as a Hunt

Life as a Hunt

Thresholds of Identities and Illusions on an African Landscape

Stuart A. Marks

berghahn
NEW YORK • OXFORD
www.berghahnbooks.com

Published in 2016 by

Berghahn Books

www.berghahnbooks.com

Library of Congress Cataloging-in-Publication Data

Names: Marks, Stuart A., 1939– author.
Title: Life as a hunt : thresholds of identities and illusions on an African landscape / Stuart A. Marks.
Description: New York : Berghahn Books, [2016] | Includes bibliographical references and index.
Identifiers: LCCN 2015047945 | ISBN 9781785331572 (hardback : alk. paper) | ISBN 9781785331589 (ebook)
Subjects: LCSH: Bisa (Zambian people)—Hunting. | Subsistence hunting—Luangwa River Valley (Zambia and Mozambique) | Wildlife conservation—Social aspects—Luangwa River Valley (Zambia and Mozambique) | Luangwa River Valley (Zambia and Mozambique)—Environmental conditions.
Classification: LCC DT3058.B58 M373 2016 | DDC 305.896/391—dc23
LC record available at http://lccn.loc.gov/2015047945

British Library Cataloguing in Publication Data

A catalogue record for this book is available from the British Library

ISBN 978-1-78533-157-2 hardback
ISBN 978-1-78533-158-9 ebook

Contents

Figures and Tables

Figures

Tables

Preface and Acknowledgments

Since Aristotle (384–322 BCE) and Pliny the Elder (23–79 CE) used an ancient Greek proverb that "something new was always expected from Africa" to depict the bizarre arrays of unknown animals emerging from its interior, Northern Hemisphere inhabitants have organized these species according to contemporary knowledge and used them imaginatively in imperial schemes. Aristotle's hypothesis about Africa's tantalizing arrays of similar animals (antelopes, cats) was that they were the products of promiscuous matings around scarce watering holes within an arid land (Feinburg and Solodow 2014). During Europe's ages of discovery and colonization, Linnean classification provided a "planetary consciousness" for making sense, a descriptive mechanism, by integrating unknown with known species, and for mastering new worlds. Distant peoples and different cultures also fit within this classification (based upon descriptions, reproductive structures, and hierarchies) as readily as new species of fauna and flora (Pratt 1992). British enthusiasts of natural history extracted specimens from tropical habitats and from contexts within other economic and cultural systems to establish their own visions and schemes (Ritvo 1987). This acquisitiveness of material and cultural collections situated within evolutionary and revolutionary arrangements bore witness to European military and intellectual superiorities over Africa and elsewhere (MacKenzie 1988). Today large landscapes within Africa remain, at least on paper, off limits to their former inhabitants, if not their original creators and erstwhile stewards, as these national parks or protected areas are sustained ineffectively through the resources of outside interests and tourists as well as the dispositions of scientists (Leach and Mearns 1996; Brockington, Duffy, and Igoe 2008).

The conquest, subjugation, and division of tropical Africa among Europeans during the latter part of the nineteenth century defined the area, and much of Sub-Saharan Africa was converted into "an imperial laboratory where political, economic, and scientific experiments could be pursued with relative impunity. These experiments had a lasting impact both within and beyond the continent" (Tilley 2011: 313). As Helen Tilley demonstrates for the colonial world, these projects and actors were not simply one-way vectors of influences, for Africa and its peoples had profound effects on shaping the "key elements of the modern world" as well as the structures and concepts of "disciplines, theories, institutions, and even laws" (2011: 314). She shows that African experiences and precedents were prominent in defining disciplinary boundaries and concepts within anthropology (prehistory, linguistics, biological and sociocultural anthropology), ecology (plant, animal, community), and also the history of international conservation (Hayden 1942; Adams 2004; Tilley 2011).

This book is the product of interdisciplinary studies spread over six decades through the lens of anthropology, animal ecology, development studies and history. It constitutes both a personal quest for understanding as well as a portrayal of global intrusions and progressively imposed state regulations on village lives, identities, and livelihoods as well as wildlife abundance within a prime game management area (GMA) in Zambia's central Luangwa Valley. It is also a history of local and individual responses to these interventions. In this valley, marginal for hoe agriculture and inimical to the keeping of domestic stock, wildlife has been a historical as well as significant cultural resource, a definite threat yet an anticipated source of nurturance at all times, especially during droughts. The proximity of dangerous wildlife to human cultivators presents a persistent risk to lives and properties that acknowledges the need for both restraint and protection.

This book is composed of two accounts, two histories, two stories, neither of which is fictitious. My first exposures of fourteen months in the 1960s, motivated by intellectual curiosity and adventure, took additional several years to assimilate and to structure as a dissertation and a book; the second contacts and studies, built upon protracted readings, differing experiences and interactive reflections within the same and different locales globally, occurred over the successive decades. The initial studies provided the baselines and additional guideposts for subsequent accounts about this particular place while the latter inquiries led me to pay closer attention to what was happening to individuals within my host Zambian community and to their more distant connections elsewhere. Both narratives seek to bring order out of what appeared to be, at first blush, chaos and uncertainty. The original explorations provoked the confusions of a personal emersion within another's organizational culture, which later became the different toils to interpret these local frameworks within conventional professional and disciplinary bounds. These earlier experiences resulted in obtaining an interdisciplinary PhD degree and with publications to counter the "crisis narratives" about Africans as proverbial destructive actors then in vogue among biologists, conservationists, and administrators during the 1960s and later decades (Roe 1999).

The connections between rural societies and their adjacent natural environmental assets are rarely exclusively local or immediately visible, yet these more distant demands of regional, national, and global scales also impact village lives, relationships, and surrounding "properties." I initially dismissed these intruding threads, subtle remote influences, as I strained to contain this valley world within the confines of a normative equilibrium ecology and structural-functional anthropology. At least superficially, the position of the Munyamadzi Corridor on the maps of the 1950s, surrounded by "natural features" (game reserves, steep escarpment, and a river) appeared as a convenient "petri-dish" frame for such an "objective" study. Yet the artificial boundaries I contrived around villages, habitats, and wildlife to satisfy time and degree requirements did not persist during subsequent visits. Neither did the ethnographic present tense or that of a homogeneous "culture."

The second account of village lives, livelihoods, hunting, and wildlife is about processes that have taken longer to accumulate, sort out, comprehend, and inscribe. There are additional reasons for the delays between the gathering of these materials, their analyses, and their appearance in this book. First, as customary wildlife entitlements and rural environmental interactions have become "criminalized" and violent in many African countries, residents of rural communities are increasingly culpable to the prosecutorial gaze of outside and inside agents, as well as to beatings, arrests, imprisonments, and other callous charges. As my studies sought understanding and depended upon rapport as well as the hospitality of others, I have sought to protect my sources by refusing to contribute unilaterally to their ongoing vulnerabilities by prematurely publishing liable information. As the passage of time, the passing of individuals, and the use of pseudonyms (should legally) serve as personal shields from such prosecution, I have used all of these buffers to protect my sources.

Secondly, active studies in the Luangwa Valley absorbed most of my time while on site, leaving until later the more protracted periods to analyze and assemble its information for publications. Upon leaving the valley after each visit, I returned to academia or other employments, which likewise demanded most of my time and energy. During these other engagements, I assembled bits and pieces of ongoing processes, which I anticipated folding later into a larger work when they became instructive about continuing activities on the ground and were no longer legally liable. This volume on local hunters and their changing worlds was left to my retirement, when their stories were history and I could assemble and integrate the processes into the current text.

While serving on the faculty of the United States Agency for International Development's (USAID) Development Studies Program on a year's IPA contract (1978), I learned government "speak," consultancies and development work as well as the global reach of bureaucratic and commercial tentacles through international programs and projects. Beyond bouts of teaching in Washington, DC, the program's staff participated in field exercises in rural North Carolina counties and international consultancies. After this assignment in 1979 and the 1980s, I undertook extended excursions through my neighbors' backyards in rural

Scotland County, North Carolina. *Southern Hunting in Black and White: Nature, History, and Ritual in a Carolina Community* (Marks 1991) is a reflection of these diversions in order to make sense of my own cultural region's storied past of colonization and cultural and political-economic upheavals. I contended also with what I was hearing about what local people considered as normative recreation was disconnected from that learned from my reflections on history, readings in the archives, and other studies about hunting. The "beast" of violence, whether directed destructively by colonizing Europeans toward other humans (Indians, African Americans, ethnic minorities) or other forms of life (wildlife), had become largely "silenced" on this southern US landscape but still remained residual, visible on occasions, and capable of rising again in difficult political or economic moments.

Later, as a rural sociologist within the Botswana Natural Resources Management Project (1990–92), I had further involvement with bureaucratic "cultures" within international aid organizations and a national government as well as with career agents representing personal, national, and international commercial interests. Later still (1995–2002) were lessons as a science advisor at national and global conventions, overseeing global conservation projects while vetting the hunting interests of Safari Club International Foundation (1995–2002) members and their clients for access to global lands.

Thirdly, there are differences between the orders (governance, regulations, management, labor, or salvation) that those, such as government officials, missionaries, employers or even teachers, who consider themselves "above," at least in their knowledge or claimed authority, seem compelled to enforce upon others, whom these agents considered "below" them. Those supposedly subordinate persons also seek to retain their own different creative orders within their lives out of and outside of these compulsory commands or restraints. These local constructions are the responses to the imposed administrative rules, allegedly for "their welfare," but which may cause dislocations from locally accepted practices. Such binary hierarchical oppositions between top and bottom, traditional and modern, developed and developing imply an understandable critical discourse, yet they are limited often to just calibrations between two opposing frames of obvious difference. Such assessments are about power and relations between groups, and are typically organized along lines of superiority, demand, or cause and effect.

If "developed" or "modern" societies are not monolithic, neither are those at the opposite end, whether designated as "traditional" or "developing" composites. If one begins with this disaggregated proposition to describe the natures of smaller groups within larger-scale societies, then disparate historic connections and events among individuals and their links elsewhere have differentially influenced their welfares and continue to do so at different rates and times. Gerald Sider (2014) argues that such discontinuities and differences constitute the everyday lives of people made vulnerable and chaotic through links with dominant individuals and governing groups. He contends that what is typically referred to as a "traditional society" is "always fully modern," as local traditions of "an ultra-low-cost supply of goods and labor" create dependencies that would "be more realistically labeled 'impoverished'" (Sider 2014: 164). Studies under such conditions demand a loosening of earlier methods that use notions of preconceived rationality, questions, or interpretations and a movement more toward those means of unplanned listening, exposures to the discontinuities and disjunctions within lives, and extended/suspended observations on the diversity of interactions within communities, particularly those on cultural "frontiers."

Imposed developments and modernization schemes are rarely uniform or one-way processes for all actors and groups, as donors, intermediaries, or recipients are influenced and changed throughout the duration of these exchanges. In her superb book, *Africa as a Living Laboratory*, Helen Tilley (2011) clarifies these interactional processes and reviews the instructive roles that ecological and social scientists played in developing as well as destabilizing imperial designs within British colonial Africa. Some of her prime examples are of anthropologists, agriculturalists, and ecologists who, through their groundbreaking work, challenged and helped to undermined the imperial order within former Northern Rhodesia, a protectorate that obtained political independence as Zambia in 1964.

Prominent European biologists and many expats wrote and talked about the African world of the late 1950s and 1960s as "biospheres in crisis" rapidly passing into presumably untrained, political, African hands. Among biologists, Sir Julian Huxley was an influential activist and promoter of conservation strategies within British African colonies. In 1960, he was sent by the United Nations Educational, Scientific, and Cultural Organization (UNESCO), an international entity he had helped establish and served as its first director-general, to review and make recommendations on the status of wildlife and natural habitats throughout Central and Southern Africa (Huxley 1961; Marks 1994b). Supported by the British government, Huxley's consultancy is worth reviewing as his summaries reveal a European professional and administrative consensus about Africans and their presumed adverse resource relationships. His "crisis narrative" promoted continued European (British) expertise and hegemony over the management of the African environment. This UNESCO mission led to global conservation projects within Africa under the aegis of the International Union for the Conservation of Nature and Natural Resources (Riney 1967; Riney and Hill 1967) and the Food and Agricultural Organization of the United Nations (Dodds and Patton 1968). One of these projects, including others that would come later, had direct application within Zambia's Luangwa Valley and would impact its landscapes of it and the Valley Bisa, the subjects of this book.

As Huxley's terms of reference were to provide the way forward, he framed the African situation in ecological terms. "The ecological problem is fundamentally one of balancing resources against human needs, both in the short and in the long term. It must be related to a proper evaluation of human needs and it must be based on resource use, including optimal land use and conservation of habitat" (Huxley 1961: 13). For Huxley, his balance was oriented toward the past (an "original ecosystem"), against indigenous knowledge and practices, and within a new synthesis about how the natural world worked. The original ecosystem of two main variables (habitats and wildlife) was seriously "degraded" and transformed millennia earlier by African migrants through their nomadic livestock, shifting cultivation, and passions for fire. Within the past century, European explorers, settlers, scientists, missionaries, medical doctors, and administrators had greatly accelerated these indigenous processes.

Beyond habitats, Huxley (1961) reported that African wildlife, despite its uniqueness and potential value, was "gravely diminished," placing its future at risk. The worse contributors to this crisis were endemic within that continent and included "alarming increases in organized poaching" and "the equally alarming increase in population." Colonial policies, including the wholesale shooting of wildlife to control tsetse flies (the intermediate hosts of a deadly parasite of domestic livestock), administrative orientations toward technological and agricultural developments rather than the use of wild animals in their natural habitats, the monetization of tribal economies, and the rapid transitions to all African governments, exacerbated these prime factors.

Huxley's recommendations for saving Africa's wildlife were as follows: First, wildlife must be studied by scientists to establish appropriate land use policies and preserved "as a spectacle" for the enjoyment of global tourists (1961: 16). Second, scientific management of wild land could supply a large wild animal and plant harvests. Besides these commercially profitable harvests, this "meat crop" would diminish the continents' meat appetite reflective of "the region's marked protein deficiency," and thereby reduce poaching. Poaching was the extensive "illegal trade in slaughtered wild animals," an act where only the "trophies" were taken and the meat was left. The immediate elimination of this "shocking trade" was necessary just as the prior elimination of the slave trade earlier had been for the continent's future. According to Huxley, "Like the slave trade, [wildlife poaching] is profitable, highly organized, extremely cruel and quite ruthless" (1961: 16). He attributed wildlife massacres and poaching to many causes, including the European indulgence for African needs for protein and their demands to possess muzzle-loading guns. Allowing Africans to retain these weapons for protecting their crops from wildlife damages had not worked in practice, as most hunting was "for money" and the external sale of "trophies" (1961: 17). These muzzle-loaders were "cheap," "cruel" and "inefficient," and their possession provided a "temptation to break the [game] law[s]"; their availability could become a public liability during "tribal or political disputes" (1961: 19).

Given the European diaspora, the world had become ecologically "a singular interconnected whole," rendering the fates of

Africa's wildlife and environments "no longer a local matter" but of world concern (1961: 14). The work ahead was for ecological scientists and conservationists to convince Africans and their impending governments that the ecological approach served their best interests in the short and long terms. In Huxley's view, African wildlife was for "Profit, Protein, Pride and Prestige, with enjoyment and scientific interest thrown in" (1961: 23). If ecology and appropriate land uses were to become significant in the modernization of these underdeveloped lands, then international finance would support the initiating projects, and funds from world opinion and international tourism would sustain them.

Beyond the 1960s, I have found many of Huxley's themes about northern science and outside expertise resolving Africa's problems, the horrific poaching epidemics, Africans' destructiveness of wild plant and animal life, and the lack of African political will to appropriate sufficient funding incorporated repeatedly within conservation "crisis narratives" and in the northern media. Global conservation enriched and extended itself by learning how to stretch its imperial hands and wildlife designs into the lower southern latitudes jointly with the media, commerce, and with the consent of (in)dependent national governments. Global propositions repeatedly claimed the wisdom and methods to save Africa, particularly from its rural peoples, by cultivating the same ground that germinated anticolonial responses. The validity and expectations of conservation's outside assertions about its successes are difficult to assess, as the employment, pedigrees, and careerism of its professionals work toward a commercial consensus rather than as constructive critics (Blaikie 2006; Buscher 2013). Furthermore, within the consumer's dominated economy of the North, celebrity, language, and tourism rarely provide its participants occasions to glimpse what's really happening on the ground (Brockington 2009). Yet statistics of increasing poverty, income inequalities, and disparities in basic needs on the world stage tell another, more compelling story.

Changes are inevitable, occurring faster at times more so than at others. In many ways this book is about a path not taken, about the map of conservative managerial constraints conscripted and imposed to continue imperial territorial designs and to colonize another's cultural landscape. This book's cover depicts images of a local hunter and wild animals, with their names in ciBisa, drawn by children attending Nabwalya primary school in 1966. These images expose the imaginative creativity and novelty of indigenous, largely untutored local kids, whose sketches originally adorned the jacket of my initial book (Marks 1976). Some of these drawings, reappear now as white ghostly figures suspended over a Luangwa Valley landscape, reminiscent of lives that once existed in more animated and workable relationships and perhaps still inhabit the minds of a few youthful and aging residents. My thesis is that northerners and members of their worldwide diaspora might learn much from the inhabitants of this valley, and perhaps from others like them, who have lived intimately with wild animals for generations and know about environmental limits. The appeal is not for a return to an improbable past, but for understanding and for the restoration of local initiatives that might benefit the "sustainability" of environments and of us all. Such rebuilding of local resourcefulness requires the time to listen, to learn the appropriate questions to ask, and to reflect upon the nature of the employment, deployment, or pleasure processes that bring northerners to these foreign "scapes."

During the Cold War, the global world was typically discussed in terms of three units: the first and second blocks (the political economies of the United States versus the Soviet Union), with the less "developed" dependents and nonaligned states constituting a pressured but inchoate "third world." Since then, civilization has become more complex, slippery, diverse, and uncertain. In some ways, our globe appears to have become bipolar in nature, with the northern hemisphere states dependent upon the inexpensive labor and raw resources of those states located in the south. I simplify and characterize the "worlds of livelihoods and the appeals of wildlife" in these two spheres: that of the "North," inhabited by a changing blend of "northerners" (decreasingly Euro-Americans) and recent migrants, an industrial and service economy; and that of the "South," which mainly encompasses Africa below the Sahara, whose rural and expanding urban populations also have settlers and displaced persons from other parts of the world. The South, through its struggles to develop by exporting goods and labor to the North, has transformed the latter into a more diverse if uncertain world (Comaroff and Comaroff 2012). For my purposes, the contrasts are between the urban northern peoples, who are mainly attracted to wild animals for reasons that are largely based in transcendental values such as "freedom," "wilderness," "trophies," "adventure" and aesthetic images and who can afford touristic visits with those still living on the rural southern African landscapes whose experiences and values are of these animals in more practical, pragmatic terms, and exist beside them. There is also the shadowy flows of rural bushmeat from Africa's "heartlands" into its burgeoning urban (and global) centers, primal subsidies in the earlier transformations of northern economies. that conservationists (as "status quo" conservative actors in the South) seek to destroy or silence.

These two spheres carry a heavy load of imagined and actual baggage from the past, yet its structures are prevalent in the hardball political-economic realities of donor funding (structural adjustments), influences, and credits within international conventions and institutions, which may obscure rather than reveal. I wonder why development has to come out of an exclusive export economy rather than from a more utilitarian enhancement of regional growth and diversification at smaller scales (Guyer 2004)? How might one begin to frame a search for an "ex-centric" view of place outside the currency of the North that provides "an angle of vision . . . from which to estrange the history of the present in order to better understand it" (Comaroff and Comaroff 2012: 47). My "hunt," in reality a scholarly quest, builds upon extensive and diverse experiences in both spheres, founded in the procedures of field observations within the biological sciences and as a participant observer of social processes from cultural anthropology, presented here as an extended southern case study.

As personal experiences and reflections have important bearings on these protracted engagements, I explore them below. My initiations into biology and anthropology began in my teens when I lived on an interior frontier within a former colony, the Belgian Congo (1948–57). When not in the secondary school classroom on a small mission station, I actively collected zoological specimens for the Musee Royale d'Afrique Centrale in Belgium. When my family spent months in Brussels to study French (1951–52), two taxonomists at this museum taught me to collect and preserve samples of tropical mammal and bird life. These scientists sought to classify the biodiversity within the relatively "unknown" (at least for these taxonomists) frontier of the Kasai Province to which my family was returning. Until my university education began in the United States, I curated and sent study specimens to these scientists for identification, the accumulations of weeks and weekend encampments in the bush with a few Congolese hunters. During the daylight hours, these adults were tutors, mentors. as well as fellow pursuers of creatures, for which they had names and knew where to find the "prey." Their questions and stories around the evening campfires in Luba, entertained and sensitized me to other strengths and struggles in their lives. These were the stories of their grandparents and parents under colonial and missionary dominance, about different worlds and worldviews, even about the natures and names of the creatures we were pursuing collectively. I learned to listen, and, with humor as well as their indulgence, my tutors persisted in their entertaining but disturbing stories.

Upon finding two "previously unknown" species of shrews in my collections, my museum mentors responded by naming each species after themselves (Walker 1964: 153, 158) and complimented me on preserving them. Even as my distant mentors got the professional awards, I could at least, as an "amateur" collector, boast about these accomplishments to my classmates and parents. The campfire stories were not ones that I could discuss with my parents or anyone else, but they traveled with me as memories on my journey across the Atlantic Ocean for university studies.

I emerged from the Congo in late 1957 with a strong penchant toward biology (bringing order to a biodiverse world) intertwined with unexamined reflections upon my exposures in different cultural worlds, which took decades to order meaningfully. The undergraduate academies of the 1950s presented these options as different professional careers geared to similarly dichotomized employments. More flexibility appeared possibly later in graduate school, if I could obtain financial support, draft an acceptable proposal, and identity supporting faculty to allow such an interdisciplinary quest. In 1962, a six-week sojourn as a lone biologist monitoring wildlife relationships among the Inuit of St. Lawrence Island, Alaska, led to additional insights about wildlife as well as local anxieties over foreign-dominated restraints on essential local resource relationships. Reflections on these experiences along with those earlier in the Congo became materials for a proposed interdisciplinary dissertation.

Upon my return to Michigan State University the next year, I found a group of willing graduate advisors to serve on my PhD committee. This committee—chaired by an animal ecologist, George A. Petrides, and included John E. Cantlon (plant ecology), Moreau Maxwell (arctic anthropology), and Charles G. Hughes (African Studies Center and anthropologist)—helped me to negotiate the shoals and academic waves of that era's graduate school. A Foreign Area Fellowship (1965–68) funded a year of academic immersion within the University of London, and was followed by research in Zambia under the aegis of the Institute for African Studies (formerly the Rhodes-Livingstone Institute) and a few months to write a dissertation.

While studying in London, the new Zambian Ministry of Lands and Natural Resources suggested that I study the Valley Bisa, inhabiting the Munyamadzi Corridor as my proposed study was for a village group and its uses of environmental assets. Upon my arrival in Zambia in 1966, I learned that the colonial staff at the Department of Game had a history of ongoing conflicts with these residents over wildlife and that the ministry planned to host an international conservation and development program. That year, the department hosted an advisory team consisting of two North American biologists, both professional novices in the colonial and tropical worlds, working from Huxley's "conservation crisis narrative" and who were employed to write a proposal for a United Nations Development Project for the Luangwa Valley. This program was to develop a land-use and management scheme that would demonstrate the profitability of wildlife as a revenue generator for the new state of Zambia. From these biologists and others, I learned that the department was hoping that my studies would help them build a consensus among government ministries to evict the Valley Bisa from between the two extensive Luangwa Valley Game Reserves. Their resettlement would rid the department of a perpetual "contentious indigenous menace" and grant it a vast game estate as an experimental space to demonstrate the economic worth of (re)centralized wildlife management (Marks 2005). In 1968, the year I completed the requirements for my doctorate, administrators, ecologists, and resource managers, along with specialist consultants, were engaged within Zambia's Luangwa Valley. They followed Huxley's prescriptions, but the dilemmas they faced were responded to within the narrower pedigree of earlier colonial designs (Darling 1960; Anker 2001; Tilley 2011).

By the beginning of my next round of studies in Zambia during 1973, the multiyear Luangwa Valley Conservation and Development Project was ending after running its tumultuous course. The project published a number of surveys and volumes but failed to deliver on many of its objectives, including the resettlement of the Valley Bisa as well as demonstrating that wildlife cropping by the state was a profitable land-use enterprise. The mission itself experienced a succession of administrative changes, yet trained abroad a few Zambians as biologists, who eventually inherited a "modernized" department with a new title, National Parks and Wildlife Service, with insufficient funding (Astle 1999). Rather than become managed according to ecological principles, Zambian wildlife was destined for national political patronage and scrambling to control the intermittent resource streams of external funding (Gibson 1999). Successive decades of external funding and media cycles continued to stage images of anecdotal conservation successes while overlooking the shambles of growing human poverty, decreasing wildlife numbers, and degrading habitats.

My next year of residence at Nabwalya in 1988–89 built upon a conviction that the effective integration and development of rural societies must begin with a clear vision and understanding of who rural people were, what they did, and what they wanted. It would involve more creative analysis, grounded imagination, open agendas, and often "surprises," beyond that of contemporary disciplinary studies. Global conservationists allied with funding from donors and commercial interests were initiating new programs in rural areas throughout southern Africa.

My studies coincided with the appointment of the new chief at Nabwalya, an important visit by government dignitaries there including a ranking member of the reigning political party's Central Committee, and the beginning of ADMADE, Zambia's "community-based" wildlife program. This program's deployment of local wildlife scouts as enforcement along with the inflating national economy had changed significantly environmental relations within these local communities and would continue to impact them progressively in the future. These circumstance required active participation by residents to write records and observations so their voices could be heard and become the base line in future studies. Once I found ways to implement these objectives, my role became to encourage and monitor their observational and recording processes so that, in my unencumbered moments, I might have time to listen, to notice gaps and follow opportunities as they occurred.

For local participation we established two basic surveys, one on village activities and one on the agendas of local hunters and wildlife counts in the bush. These surveys describe important sectors and their intersections within village life, hopefully revealing some paradoxes. The village survey began with the new chief, the formation of a local committee, and a household survey of neighboring households and the choice of a small sample of households, divided by gender for wealth, age and work categories, sampled for a whole day once week by s recorder (Marks 2014). The observations on household members incorporated both quantitative (timed) and qualitative (notes) information similar to Else Skjonsberg's village surveys in Zambia's Eastern Province (Skjonsberg 1981; 1989). Summaries of these records became a baseline for that agricultural year, have served as benchmarks for tracking individuals' welfare during each of my subsequent visits (Marks 2014: 7–8, 77–102). Selected brief notes of four individuals' daily activities that portray some normative rounds in village life during 1988–89 appear in chapter 2.

I typically met with these assistants late each afternoon as soon as they completed their recordings. This allowed us the time to go over their records, to talk about any problems or new observations, and to acquaint me with the work pulses and gossip rounds of village lives. More importantly, they allowed time for me to examine what was happening or might be slipping beyond the frames of other inquiries.

The second survey focused on local hunters, their forays, and activities in the nearby bush. These surveys and interviews extended and expanded earlier studies to include the uses, distributions, and management of these local assets, particularly wildlife, within households. The initial wildlife surveys began in 1966 and were repeated during each of my subsequent visits (Marks 1973b, 1977a, b). Always comparatively few, the local hunters around the chief's village permitted my frequent recordings of their life histories and accomplishments (Marks 1979).

Local hunters are the main actors, their life histories and frequent tragic stories contributing significantly to the themes and arguments of this book. At this juncture I sought a different perspective on hunting involvements afield and within villages that went beyond that of an outsider's accounts. In conjunction with four local hunters of different ages in 1988, we devised a method in which they would record and time their activities in the bush (Marks 1994a). Rather than following a fixed transect or road (as is the common method among grounded wildlife professionals), hunters proceeded toward sites where they expected to encounter wildlife and timed other activities they engaged in during each foray, besides searching for game. To make their forays statistically viable, I encouraged each to conduct up to ten counts per month, each lasting several hours (figure3.1). Hunters kept wildlife counts during all months during 1989–90. After than they wished to take their counts only during the dry seasons so they could better accomplish other household requirements during the rains.

One or two hunters have made these counts almost every year since 1989; two have since died. Another left Nabwalya after graduating from the local school. Their records contribute a valuable database for determining wildlife fluctuations on the Nabwalya Study Area as they reference human predation, increasing human habitat uses and disturbances around villages, and the vagaries of local rainfall. Significantly, these counts offer a different, separate record from the interests of safari firms and government scouts, who also take intermittent notes on observations of valuable "trophies" and whose "anti-poaching" patrols outweigh any positive benefit that most villagers receive under the "community-based" wildlife program. From these records, readers may detect the crafty and clever hands of some who persistently applied their customary identities by their assertion of rights and entitlements to wildlife. Such studies would be impossible without the active assistance and detailed recordings of those involved. In an ideal world, these participants would be credited and named, but it appears more prudent to protect them from any unanticipated reprisals. Ironically, under current conditions most users of wildlife, including some local hunters, derive more benefits (short term) by hunting entrepreneurially outside current legal restrictions than by conforming to them.

During the 1990s and 2000s, I made seven subsequent visits to Nabwalya. Each of these trips built upon previous studies and materials sent from Valley Bisa residents during the interim to keep me current with events affecting their lives and welfare. The longest stay of four months occurred in 2006. By then the "community-based" wildlife program was eighteen years old and the wildlife agency had been "modernized" again to become the Zambia Wildlife Authority (ZAWA). Yet the administrative authority remained underfunded, dependent upon outside donors and expertise while still claiming to search for its "national vision." A small grant allowed me to return and evaluate village-level responses to this wildlife program within the Munyamadzi GMA. At Nabwalya, I discussed these plans with local advisors and gained their enthusiasm along with that of the local Community Resource Board (CRB). Several recorders engaged in earlier studies brought their energies and skills to this survey. We named our team and its advisors "Mipashi Associates," based on the vernacular meaning of *mipazhi*, good spirits whose beneficial activities are readily known and visible. In four months, this local team revised my questionnaire and administered it to a broad range of over five hundred residents throughout the GMA. We also interviewed employed individuals, scouts, and those representing other interests. Oversight of this survey in Zambia, as well as analyzing and publishing its information, remained my responsibility.

The assessments of government management policies on local welfare and on wildlife were printed initially as Marks and Mipashi Associates (2008), and were later expanded as a book (Marks 2014). Both reports were distributed to selected residents. These and other published materials appear minimally in the current book and are cited as background and cultural contexts in the interactive worlds of local hunters within the Nabwalya Study Area. Figure 6.3 (see p. 272) and some materials in chapter 6 were published earlier as "Buffalo Movements and Accessibility to a Community of Hunters in Zambia" in *Journal of the East African*

Wildlife Journal (1977: 251–61). Tables 1.1, 7.1, and 9.2 appeared respectively as tables 5.3, 9.1, and 9.3 in *Discordant Village Voices: A Zambian Community-Based Wildlife Program* (Marks 2014) as did figure 2. Table 9.1 was revised from table 7, and it and figure 1 were published initially in *Large Mammals and a Brave People: Subsistence Hunters in Zambia* (Marks 1976) and republished as a paperback (Transaction Publishers 2015). These references are textually integrated into the current narrative in new ways.

In many ways this book is an unfinished and reciprocal product, as many of those known during my recurrent visits have taught me more about learning and unlearning about life and persistence in ways that I am still musing over on how to distill and communicate. In this sense it is as personal as any intellectual work of value in the sciences or arts that demands such subjective commitments and sustained learning from and with others. At its core are the intermittent conversations, contacts, and observations made with three successive generations of local hunters, most of whom resided in the villages (the Nabwalya Study Area) and practiced their crafts in the bush on both sides of the Munyamadzi River (the Nabwalya Study Area) during the last half of the twentieth century. The book is about these hunters' life histories, about their beliefs, motivations, and changes in their villages and in their interactions with wild animals in the nearby bush, as well as with wildlife scouts in both spheres. The book chronicles their accomplishments and disappointments as important actors engaged in "gathering people" together within this particular place, and registers the shifts in their knowledge and in relationships with the local fauna. Their environmental space also bears the progressive interactive impacts of a chancy climate, uncertain consequences affecting the quests of all those living there, the radiating effects of increasing human demands for environmental products, and imposed regulations by national and foreign agents seeking to reconfigure this natural world in ways that destroy these hunters' customary identities as providers and protectors. The latter intrusions include officials from diverse international and national ministries, conservation organizations with complex global links, tourist and hunting concessionaires with diverse international clients, and service organizations pursing souls and clients.

During the decades taken to assemble and interpret these materials, I have accumulated many profound debts, which I gratefully acknowledge. Although I have earlier recognized some of these individuals in former publications, here I wish to pay homage to the enduring personal friendships and the contributions that many have made directly and indirectly during the *longue duree* of this project. Within Lusaka, Robert and Namposya Serpell, Jacob and Ilse Mwanza, Guy Scott, and Charlotte Harlan Scott provided lodging and delicious meals, or briefed me on current events during my passage through the capital city, and they often helped in obtaining the requisite clearance for research in the hinterland. The late Norman Carr, Adrian Carr, Judy Carr, Vic Guhrs, Pamela Guhrs, Athol Frylinck, Jeremy Pope, the late Richard Bell, and Phil Berry have been friends almost from the beginning, kindly offering housing, meals and thought-provoking conversation on my passages through or brief stays in Mfuwe. They often suggested or provided the connecting links northward up the eastern side of the Luangwa Valley or close by where I could walk to Nabwalya. On the western side of the valley, above the Muchinga escarpment, another group of gracious friends, Barbara Collinson and Dutch Gibson with British Aid at Mpika and Mike and Lari Merrett at Mutinondo Wilderness, often provided temporary quarters and revitalizing boarding after my egresses from the valley. Mike and Paddy Fisher took me back to Kitwe from Mutinondo Wilderness, where I obtained treatment for an ailment on my way back to Lusaka in 2006. Since I possessed a vehicle only during my initial stays (1966–67, 1973), most travels at Nabwalya have been limited largely to my own legs or a bicycle. I am grateful to Luawata Conservation, Leopard Ridge Safari, and professional hunters during the dry seasons for their assistance with occasional transportation and other needs.

My debts to all of the residents at Nabwalya are more extensive and profound. I am indebted in many ways, as they have been my hosts, helpers, caretakers, instructors, and neighbors for extended periods over many decades. I begin by acknowledging the contributions and kindnesses of the late Chief Nabwalya (Kabuswe Mbuluma). He was a cordial mentor during the earlier studies, always helpful and solicitous when my wife and I arrived unexpected in his courtyard. He was generous with his time, enjoyed telling us about local history and interpreting events, establishing and suggesting contacts, and had a keen grasp of how we would mutually assist one another. His successor, His Royal Highness Chief Nabwalya (Blackson Nsomo), whom I first met in 1973 and next in 1987, welcomed me again and was likewise a caring and attentive host. Both chiefs facilitated these studies and always willingly complied with requests for assistance and interviews, and they kept me informed of important happenings in the interims between visits.

Special gratitude goes to the hunters, farmers, traders, friends, and research assistants, and all those whose paths I crossed while in the central Luangwa Valley. They have been the best of teachers on the interlacing of cultural and environmental processes through their words and actions, while allowing me a privileged measure of access into their worlds. I trust that the mention of their names here will serve as appropriate "respect" and acknowledgment to associations together. In no particular order: Jason Ngulube, Kapumba Jona, Funo Kalonda, Chizola Zakaria and Zolani, Hapi Luben, Chibale Chinzambe, Chopu Chop, London Tembo, Mumbi Chalwe, Seluka Chinzambe, Mumba Mulungwe, Luisa Kailobwele, Kameko Chisenga, Mutombo, Robinson Simosokwe, Robert Mukupa, Peter Tembo, Luci, Katabole Taibo, Donati, Chibindi, Taulo Chigone, Frank Mbewe, Chitumbi Muzenga, Mulenga Morgan, Achim Mbuluma, Frank Mukosha, Julius Mwale, Philip Ndaba, Elvis Kampamba, Andrew Chulu, July Musa, Hellen Kapito, Jackson Katongola, Luben Kafupi, Morgan Kabongo, Florence Somo, Pelembe, Blackwell Banda, Kabuswe Musumba, Lightwel Banda, Doris Kanyunga, Joseph Chanda, Feleson Komechi, Fanwell Zulu, Kangwa Samson, Rhoderick Mbuluma, Joseph Farmwell, Moses Ngulube, Stephen Ngulube, Rose Chiombo, Nixon Mwale, and Dyson Mwamba. I add to this list Fr. Waldemar Potrapeluk, who as the resident Fidei Donum priest in residence constructed Our Lady of the Rosary Parish at Nabwalya. Both the priest and the chief kindly invited me to attend the parish's consecration in 2011. I did so knowing that its construction was a cherished dream of the latter, who believed that this institution would provide additional development options for the community. Given this host of knowledgeable instructors and mentors, I accept the responsibility for inaccuracies, omissions, limitations, or imprecise translations of their voices.

Over the decades, I have met or corresponded with a broad range of scholars and researchers whose studies overlapped or impinged on my own, and I have visited or read about a number of wildlife projects throughout southern Africa. These following individuals have generously shared their thoughts in conversations with me as well as in their published and unpublished materials at various stages. These include Barbara Ward, George Kay, Robin Fox, Francois Bouliere, Mary Douglas, Andrew Roberts, Michael Mann, Audrey Richards, Rollin Baker, Harm DeBlij, Wilton Dillon, William Beinart, J. Desmond Clark, Glynn Isaac, Hans Kruuk, the late Keith Eltringham, the late A. J. Jobaert, Thane Riney, Marshall Murphree, Lucy Emerton, the late Ivan Bond, Russell Taylor, Ian Manning, Terence Ranger, Kate Showers, George Treichel, David Cumming, Ian Parker, John Sinton, H. Jachmann, Hugo van der Westhuisen, the late Larry D. Harris, Clive Spinage, Mwelwa Muambachime, the late Alec Cambell, Robert Hitchcock, the late Victor Turner, Thomas Beidelman, the late Richard Bell, the late Jon Barnes, Ed Steinhart, Gary Haynes, Richard Blue, Clark Gibson, the late James Teer, Rob Gordon, Kate Crehan, Robert Bates, Norman Rigava, the late Ivan Bond, the late Frank Ansell, the late Bill Astle, and Harry Chabwela.

As I assembled and reworked the drafts and edits of the sections of this book over the past twenty-five years, I have shared copies with other scholars. Among these readers are Gary Haynes, Andrew Long, John Sinton, Joe Dudley, Josh Garoon, Ian Manning, Ilse Mwanza, Robert Cancel, Nancy Andrew, Harry Wels, Moses Ngulube, Elvis Kampamba, Parker Shipton, Chief Nabwalya, James Peacock, Bram Buscher, Alan Savory, Adhemar Byl, Art Hoole, Thomas Lekan, Mwape Schilongo, Conrad Steenkamp, Dean Birkenkamp, and John McMurray. Their comments have encouraged and challenged me during the long gestation and revisions of this book.

I especially appreciate the encouragement of Professors Elizabeth Colson and Thayer Scudder who have contributed in many ways throughout these studies. As eminent scholars of the Zambian scene, they have set high standards for long-term anthropological research here and elsewhere. They have read and commented insightfully upon most if not all chapters in this book and suggested ways to sharpen my analysis. I hope this final form lives up to their standards of excellent, persistent scholarship. Professor James Peacock has been a fellow companion on many excursions of the mind and treks across the land of our common southern US landscape. I appreciate his warm and encouraging friendship over the years as we have walked to explore or talked and listened during many a hunting or fishing expedition. He and his wife Florence and their daughter Natalie accompanied my wife Martha and our two sons on a visit to Nabwalya during my stay there in 1989. Andrew Long, Bob Cancel, and Robert Serpell have also appeared on site.

My language facility in the ciBisa dialect was built upon my earlier absorption of the cognate Luba language during my youth in the Belgian Congo. It was honed on academic Bemba training under Professor Ivine Richardson and Mubanga Kashoki at Michigan State University, followed by that of Michael Mann and Grenson Lukwesa at the School of Oriental and African Studies, as well as by a term of Nyanja under Dr. Guy Atkins in London.

During my periods of research in Zambia, I was affiliated with the Institute for Social Research or Graduate Studies of the University of Zambia. During the compiling and writing of this book I was associated as a Research Associate in the Department of Anthropology, University of the Free State, Bloemfontein, South Africa.

Appreciably, I acknowledge those foundations that have made possible and facilitated my studies at different times. I remain grateful to the Ford Foundation, which funded through its Foreign Area Fellowship Program my basic grounding and networks in the social sciences and the humanities within the School of Oriental and African Studies and the London School of Economics (1965–66) and made possible my initial interdisciplinary studies and research. The Social Science Research Council and American Council of Learned Societies (1973) funded additional studies on local hunters and hunting and its symbols at Nabwalya and within Zambia's Northern and Eastern Provinces. A year's faculty appointment with USAID's Development Studies Program introduced me to development through its studies of these processes in North Carolina, Sri Lanka, and a brief revisit to Nabwalya (1978).

Consultancies with the UN's Food and Agriculture Organization (Forestry Division) in West Africa and Zambia were on natural resource management and economics (1984, 1987). Grants from the Council on International Exchange of Scholars (Senior Fulbright Research 1988–89) and the National Geographic Society (1989) enabled an additional year of residency at Nabwalya to set up baselines for continuing studies on village activities and uses of common resources including wildlife. Consultancies under NORAD (Norwegian Development Assistance 1989) and the World Conservation Union (1989) provided opportunities to interview Kunda hunters in Eastern Province and to evaluate the Luangwa Integrated Natural Resources Management Project, while those with World Wide Fund for Nature US (2001) and Mano Consulting (2002) enabled visits to proposed conservation projects on other Bisa landscapes on the Zambian plateau. Research grants from the National Endowment for the Humanities (1990, 1993, 1997) and the H. G. Guggenheim Foundation (1994) assisted with archival research, additional field work at Nabwalya, and analysis at home.

I acknowledge the personal attention and assistance from the staffs of the following archives: Public Record Office (London), Zambia National Archives (Lusaka), National Archives of Zimbabwe (Harare), and the National Archives of Botswana (Gaborone). In addition, the librarians and libraries at the following institutions assisted in making materials available for readings: Library Department of the British Museum (National History), Rhodes House (Oxford University), the Zoological Society of London Library, University of Zambia (Special Collections), and the Sterling Library at Yale University. The Wilson and Davis Libraries at the University of North Carolina at Chapel Hill and the University of Virginia Library solicited manuscripts from other collections for my reading while I was in residence on their campuses. The National Endowment for the Humanities stipends for Summer Seminars for College Teachers supported me for two months on campuses at the following universities: University of Southern California at Los Angeles (archaeology 1976), University of North Carolina at Chapel Hill (southern history 1980), Yale University (African American folklore 1984), University of Texas at Austin (British imperial history 1988), and the University of Virginia

(African history 2003). Each of these occasions involved participating in seminars as well as presenting papers, reading, and working on the documentation assembled in this book. I wish to thank each of these seminar directors, particularly William Roger Louis and Joseph Miller, for support and timely commentaries on this work.

The current book has taken shape through fits and starts over many decades as I have assembled its facts at intervals through the encouragement, participation, and kindness of many others. Its themes are processes, cultural as well as environmental, eventually revealing their changes and dimensions within the lifetimes of individuals and groups as well as within habitats and wildlife populations through events over time. My intentions began with hunches about circumstances and relationships, built upon trust and consent to establish guideposts and baselines, and persisted through conviction and collaboration as information accumulated meaning through directionality and interpretations. In reflection, writing this book has been a prolonged HUNT *tout court*, not only a search for the right words but also an exploration in assembling means and materials, documenting, interpreting, questioning sources all along the way, negotiating the thickets and savannas of its pursuit, appropriately differentiating between a charge and a compliment, all the while aspiring to ground the quest in beneficial ways for both hosts and readers. In recognizing that gaps and conclusions remain open or unattended in the multiplicity of events, histories, and lives within this significant Zambian place, I acknowledge my own inadequacies in observing and communicating. My hope is that this record of history and of local lives will enable residents to perceive their ways to more promising tomorrows. Moreover, I hope that this book's northern and other readers might reflect upon their own unexamined ways of how their lives and welfare are connected with and impinge upon distant persons as yet unknown.

Martha Damaris Singletary Marks has been my moral compass often, my collaborator from the beginning, and has remained so as the mother of our two sons, as well as my wife, my best friend, and the love of my life. Since Martha and I completed our first trips together and experienced the births of our two children, she has created a home for us all, a place to which we could return from experiences, both in our neighborhoods and from more distant

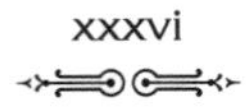

locales, to make the necessary readjustments in our lives within a rapidly changing world. In these intervals, she has created a successful career and supportive network for herself as a teacher, administrator, and counselor, and she has helped me discover the limits of recent insights. She knows that the time devoted alone to transcribe, read, write, and edit in order to make sense out of experiences and abstract thoughts combined in distant work away has its ups and downs, and it takes its tolls by depriving other planned or worthwhile ventures and relationships. She knows about the peaks of inspiration as well as the dales of despair, and she has accommodated ours bravely and graciously. I originally intended to dedicate this book to her, but with the passage of time, there are others, for our sons have matured into men and begun life as adults. Therefore, it remains dedicated to Martha, as well as to Stephan Singletary, Jon Stuart, Cara Deckelman, Anna Clare, Meagan Damaris, Erin Keiffer, and Benjamin Stuart Marks for everything they have clarified and mean to us, as well as what they may experience and become.

Stuart A. Marks
Henrico, Virginia
2 March 2015

Abbreviations and Glossary

ADMADE: Administrative Management Design for Game Management Areas, the Zambian "community-based" wildlife program (ADministrative MAnagement DEsign).

biological diversity: variations in biological organisms at the ecosystem, species ad gene levels.

BSAC: British South Africa Company.

CBNRM: "community-based" natural resource management.

CRB: Community Resource Board, an organization representing residents within GMAs or chiefdoms with an alleged common interest in wildlife conservation and recognized by the Zambia Wildlife Authority. Replaced the Wildlife Sub-Authority under the Zambia Wildlife Law of 1998. Membership included two "elected" members from each Village Area Group (VAG) within a GMA and representative(s) of the chief. Members elected a chairman and those serving on various board committees.

decentralization: the allocation of authority and responsibilities for specific decision-making and operations to lower echelons of government, such as a district or local community organization.

DG&F: Department of Game and Fisheries, Northern Rhodesia; Republic of Zambia.

DG&TC: Department of Game and Tsetse Control, Northern Rhodesia.

EU: European Union.

game: commonly hunted wild animals; also a designated legal category.

GMA(s): Game Management Area(s), usually bordering National Parks and allegedly holding significant populations of wildlife. Various statutory instruments feature the GMA as a focus institution for integrating conservation and community developments on customary or common land.

HIV/AIDS: human immunodeficiency virus/acquired immune deificiency syndrome.

hunting concession: the right to legally hunt within an area given to a commercial firm or individual by NPWS or ZAWA for a specific period of time; requires payment and assumption of some management responsibilities, typical in GMAs on communal lands.

MCC: Member of Central Committee, an administrative governing body established by the UNIP during Zambia's First Republic.

Mipashi Associates: an informal core of associated individuals, within and outside the Luangwa Valley and Nabwalya, who participated in, assisted with, or were instrumental in gathering the materials presented in this book. The name is based on the ciBisa vernacular meaning of *mipashi* (or sometimes spelled or pronounced as mipazhi) as "good spirits," whose activities are readily known and visible (in contrast to its antonym *cibanda,* "bad spirit" [omen] whose meaning remains hidden or uncertain).

MMD: Movement for Multiparty Democracy, a Zambian political party, which under Frederick Chiluba in 1991 ousted the initial Zambian government of Dr. Kenneth Kaunda (UNIP) that had governed Zambia since independence in 1964.

MT: Ministry of Tourism.

MTENR: Ministry of Tourism, Environment, and Natural Resources.

Nabwalya: includes those villages and settlements in the immediate vicinity of the chief's palace and local primary school along the Munyamadzi River, within the Munyamadzi GMA; the space for many of the studies documented in this book.

NCRF: Natural Resources Consultative Forum (Zambia).

neoliberal economics: current mantra of global economics that promotes the link between the "efficiencies" of the private sector and "free markets" over public controls.

NGO: nongovernmental organization.
NORAD: Norwegian Development Assistance.
NPWS: Zambian National Parks and Wildlife Service, successor to the DG&F and immediate predecessor of ZAWA.
Patriotic Front: Zambian political party under Michael Sata, who became Zambia's president in 2011 by defeating the MMD party, which had been in power since 1991. He died in office in 2014.
SPFE: Society for the Preservation of the Fauna of the Empire.
sustainable use: exploitation of a biological organism and its ecosystem at a rate within its capacity for continual renewal.
TFCA: Trans-frontier Conservation Area, an international scale of conservation space encompassing land of two or more adjacent countries.
UL: unit leader, usually a wildlife police officer and government civil servant, responsible for wildlife management (mainly wildlife protection and anti-poaching) operations; supervises local wildlife scout activities within a GMA.
UNDP: United Nations Development Program.
UNIP: United National Independence Party, the political party under Dr. Kenneth Kaunda; the governing party as Zambia became an independent state on October 24, 1964.
UNZA: University of Zambia.
USAID: US Agency for International Development.
VAG: Village Area Group, a political alignment of villages for representation on the Community Resource Board, the new local management authority under the 1998 Zambian Wildlife Act. Within the Munyamadzi GMA, there were five VAGs. Each VAG elected two representatives for officers on the CRB, who then elected each representative to serve on its committees. These elections were incentives for encouraging a fairer distribution of revenues throughout a GMA and to overcome the concentration of projects around a chief's village.
WCRF: Wildlife Conservation Revolving Fund.
WMA: Wildlife Management Authority, a political authority at the district level that was to supervise wildlife management within the GMAs of its borders; chaired by the district governor, a political appointee. Became inactive when UNIP lost the 1991 national election.

wildlife scouts: locally employed by the Sub-Authority and later the CRB from the percentage (40 percent) of the revenues received from safari hunting, allocated through the WCRF. They were trained by NPWS or ZAWA to enforce the local game laws and protect wildlife within their GMAs.

Wildlife Sub-Authority: an organization of local leaders within a GMA, chaired by chief and ADMADE unit leader as executive officers; subordinate to the short-lived district Wildlife Management Authority chaired by the district governor with the wildlife warden as executive officer (1988–91); predecessor of the CRB, which replaced the Sub-Authority under ZAWA in 1998.

WPO: wildlife police officer, a field staff officer under ZAWA.

ZAWA: Zambia Wildlife Authority.

ZK: Zambian Kwacha, the Zambian currency; rates of exchange variable.

Introduction

On Poaching an Elephant

Calling the Shots and Following the Ricochets

During our tour, we found that an elephant was poached on the 24th July, 1998 between Chibale and Poison [sic] village. The act is believed to have been done by local people. All flesh was removed from the carcass leaving behind the ivory intact. Suspected persons were taken to Mpika Police. There was a buffalo in a snare the same day these people were skinning an elephant. An ambush was made but nobody came to check the snare. There is evidence of poacher of small species like impalas which is high. There are a lot of guinea fowl traps in Munyamadzi River about 100 meters from the unit headquarters. We removed some during our morning wash-up and we also witnessed one fowl in a snare.[1]

This book is about how the issues involving and surrounding wild animals can separate people who value them for different reasons. For most northerners[2] and those living in the world's cities and farmlands, the realm of large wild animals, commonly referred to as "nature" or "environment," exists largely at a distance and external to their daily lives. An unlucky few may incur inconvenience if a deer runs into their vehicle on a highway or eats a valued shrub, or may suffer a more devastating loss if a child or relative is stalked and killed by a bear or cougar in a suburban backyard or along a running trail. Many northerners may visit foreign places briefly as tourists on vacation or to study, yet they do not live entirely within or depend materially for their livelihoods upon what is cultivated and gleaned from their immediate domiciles. Overlooked, perhaps even dismissed, during their brief interludes to remote places are the complex cultural, political-economic, and social realities of people living within the lower latitudes and rural environments. Some of these residents may possess rich webs of local knowledge, practices, and ideas about neighboring biological resources, which have supported them and helped to maintain their environments for decades, their ancestors for centuries. Today their plights and livelihoods must become significant parts of any resolution to sustain these resource flows and habitats. For this reason, their recent histories and management practices are worth learning as they provide different perspectives on environments and wildlife and reveal the cultural limitations of northern management models and the current strategies to sustain them (Comaroff and Comaroff 2012). This book is about the life histories and wildlife management activities of a small group of Valley Bisa men who reside in Zambia's Luangwa Valley. Their households, villages, and fields are often visited by wildlife, as their homeland is surrounded currently by three national parks. In recent years, both Valley Bisa welfare and resident wildlife have declined through government inattention and mismanagement.

As a depiction of how an African people have coped with abundant wild animals in the past half century, this book searches for a different narrative in global wildlife conservation. The events in the epigraph occurred a decade after the Zambian National Parks and Wildlife Services (NPWS) initiated what it promotionally labeled as a "community-based" wildlife program in 1988. Backed by generous American and European subsidies, this program was a global response to the extensive killings of elephants, rhinos, and other wildlife after the Zambian economy imploded during the late 1970s and 1980s. Under this plan, cadres of enforcement scouts, drawn from local communities, were recruited and given brief military training. These scouts were paid to enforce new restrictive entitlements to wildlife based on regulations that criminalized customary uses and practices to protect the "game" from residents within their homelands. Beyond employing local scouts as enforcers through its program, the NPWS promised to promote village development as well as protect local properties and lives against wildlife depredations. Among its pledges to residents for protecting wildlife was that they would accrue wealth by letting these wild animals be shot by safari hunters or be observed by others. The exchange was to work this way: "wildlife killed by local residents had only short-term value as meat for consumption (i.e., of no formal economic benefits); wildlife shot by safari hunters or observed by tourists was worth cash, its proceeds would be used for development, so everyone would benefit." Beyond the initial elusive donor funding, revenues for community developments never materialized in the amounts promised, everything took more time than expected, and community funding was not dependable and diminished through time. In certain years, all funds were absorbed by the government regulatory agency while the rural communities received none. Yet materials and supports for "anti-poaching," the visible essence and deterrent of centralized wildlife management and focus of donor attention, remained consistent and insensitive. Despite the initial pledges, twenty years later the program had failed miserably in its promise to enhance village welfare as well as achieve sustainable conservation.[3]

Rural people were not consulted about these procedures that perversely affected the very core of some identities and livelihoods within this chancy valley environment. The imposed rules made normative sense only within the limited frame of a wildlife narrative backed by centralized state power and supported by resources and experts from an unknown distance. Local residents learned the new rules painfully over time through harassment, imprisonment, and intimidation. Inhabitants were encouraged to comply; offers included meager incentives and promises, intermittent revenues for proscribed community "developments," and, for a very few, engagements as casual laborers. The cultural and environmental worlds, which generations of residents actively created and from which some derived their livelihoods, became transformed into an alien landscape, a playground of fantasy and commerce for strangers who appeared periodically as hunting tourists. Some local people benefitted and helped to build por-

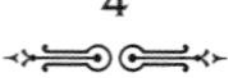

tions of this new world, a world that they could never master, but one that they might join as dependent subordinates. Their homeland, embedded with significant histories and identities, the wildlife and other resources that gave their lives meaning and sustenance as well as their flexible social institutions through which individuals mediated their conflicts and cooperation, were no longer theirs to husband or to extend. This book is about a world that was tragically lost, about a fork in the road of development, and sensibilities not taken by some perhaps well-meaning but ultimately insensitive distant others.

On Official Accounts of Poaching Elephants

The wildlife officer's brief in the epigraph tells about his passage through this landscape, yet he and his entourage remain silent on many contextual issues about this place—its history, politics, and local culture. In his brief sojourn, he jumps to conventional, convenient conclusions, and he reinforces professional stereotypes that implicate local people in several criminal offenses. Yet who are these residents and why does he suspect them? What are their recent histories and backgrounds that put them at odds with and incur threats from this official and his wildlife agenda? Do these injunctions, represented in his authoritative voice, indicate any boundaries or limitations on his conceptions of wildlife, about life, about other people and their relations to resources? What are the relationships of these communities to the abundant wildlife surrounding them? How did residents sustain themselves in this place and what are their options now? This book is about more than just "poaching" an elephant, yet this large beast, imaginary, dead or alive, stands metaphorically at the heart of sustainable natural resource issues throughout Southern Africa as well as elsewhere.

The wildlife officer's report is bland reading, the kind of succinct and superficial script that we might expect from an itinerant official or journeyman. As a transient, he appears on a scene for a brief moment, inscribes a cursory account, and declares the infractions he witnesses resolved. He notes other activities out of place and passes these enduring problems along to attending subordinates. Through such blips of pre-scripted observations recorded during brief mandatory expeditions from his office, everything seems explained and back in order within a single paragraph. After all, his audience includes his superiors and even those more distant who might be impressed with his verve and control. Such a narrow focus and lineal flattening is typical of outsiders who pass through landscapes created and inhabited by others, whose presence and memories these foreigners eventually seek to silence, if not to erase (Scott 1998). Yet local memories and identities persist and may be more sustainable in the long run than the impressions and visions of momentary strangers.

Within these supposedly officially silenced, vibrant spaces, life remains vigorously interconnected, difficult to keep within bounds or flattened on a page or two. Such life is multidimensional and persistent, with highs and lows, with inconsistencies and differences, with victims and victors, maybe even paradoxical with inconclusive evidence leading to more hesitant resolutions. Understanding these facets takes time and exposures, as people expend their lives in practices that itinerants cavalierly miss, dismiss, or judge by their own norms. This book has taken its own time, my lifetime and over half a century of intermittent observations and conversations with and by others, to connect the dots and meaningfully interpret this environment and its inhabitants. Its writing has required this time to find the appropriate words to express what has been learned and then unlearned even when inscribed. It has been edited and often rewritten in the long procedures of translating and communicating its connecting stories.

These real lives and the cultural differences between northerners and southerners are the "elephant within our room," the boardrooms wherein executives make decisions, and in the living rooms where citizens make contributions affecting the lives of others, people they don't know or even care to know existed. "Elephant in the room" is an idiom for something that so threatens privilege and presumption that it becomes impossible to ignore, except for a persistent conspiracy to discount, silence, or change its presence. This expression entails an assessment that the looming subject is significant and inevitably will imperil everything else around it.[4] Within these pages, this proverbial elephant becomes the cultural soul and cultural differences of some who still live on the land with elephants and other wild animals.

As a youth coming of age in the Belgian Congo and nurtured by another group of rural Africans about wildlife, I began to suspect the existence of such an elephant. My experience as a youth immersed in three cultures with three separate languages (and worldviews) and as many biodiverse environments grounded me as I pursued my childhood intuitions later through formal education in animal ecology and anthropology.[5] Later, while a graduate student, I began tracking this metaphorical beast in the recently independent state of Zambia slightly southeast of where I had become aware of this creature's plausible existence.

Animal ecology taught me the conventional northern wisdom about wild beasts, about how they supposedly behaved, about how they should be managed and by whom. Anthropology sensitized me to the human side of that endeavor—to the meanings about and uses of animals as well as the silences inherent in any group's cultural consensus about them. Understanding one's own culture takes most of a lifetime. Insights into another society and its ways require a host of enthusiastic interpreters as teachers, attentive listening and familiarity with different activities, a receptive heart, and good fortune. Anthropology's gift is in its ability to distill some cultural knowledge from one group of people and make it available and interpret it reasonably for others. Preferably this analytical act takes place without losing essential information during the attempts to make it relevant for a new audience differently oriented. As one perceptively listens to indigenous plights and follows their leads, local constraints are found frequently in the channels of distant policies and in the power and profits of earlier intruders on their landscapes.

Myths, subliminal stories infused throughout a culture, form the architecture upon which empires are structured, demolished, and resurrected or reaffirmed. International wildlife conservation is one of the world's great myths. It is a compelling narrative, especially for those living in the Northern Hemisphere whose funding and writing promote its proactive scripts. They assume its messages and means have universal applications, bespeak a "global common good," offer "win-win" resolutions for everyone, and at least save the "game" for its promoters while its losers are silenced or demoted from their "fields," deposited on the rubbish heaps of archaic livelihoods and identities. Yet such precepts are beneficial mainly for a small minority who think they can afford to separate themselves from the plights of others throughout the rest of the world. The composition of these global conservation initiatives and wildlife narratives is really about some of us (northerners)—about our heroes, our needs and deeds, our careers, our histories, and how we spend our time. We immerse all these elements in our expectations that others aspire to become like us in all our different ways. We base our visions in northern imperial experiences with wildlife and in professional lives with its imaginations, "spirits" (intuitions), "demons" (capitalism), "rituals" (peer reviews, best methods), "sacred texts" (scientific publications, vocabularies), and membership (professional) behaviors. This vision has yet to incorporate the dreams and experiences of others, although we might wonder about their stories and practices.[6] We seldom position ourselves to hear alternative voices or place ourselves in situations to learn from or about them. They exist nonetheless, never as privileged as professional conventions and, some might say, overwhelmed by the cacophony and discursive imperatives of our ongoing environmental and extinction crises fueled by climate change. Like a fish that becomes aware of its limits only as it struggles out of its watery medium, most humans seem to muddle along within their cultural liabilities and routines. Expansions in understanding may come, if at all, from duress or after devastating failures. Thus our "elephant" has yet to metastasize or transform the discourse within the international chambers of conservation and "sustainable" development (Marnham 1987; Garland 2008).

Wild elephants in Africa are one of the great symbols of international conservation where they figure prominently in organizational efforts to protect them, particularly on someone else's turf. Under the "umbrella species" promotion that supposedly protects elephant range habitats for all other wildlife, the survival saga of elephants is spun as a struggle between northerners' expert knowledge [read science and technology] and the criminal greed [read degenerative] of insensitive (unspecified) men. The "unspecified" reference here is to unrestrained actors, mainly a generic African or Asian, given the characteristic pejorative vocabularies through which most northerners describe "poachers." These men occur in "gangs," engage in "brutal" and "unlawful activities" (poaching), and sell products in "black" markets. The main middlemen and consumers of these products are crafty Asians in Africa and elsewhere. Forgotten in this context is that not too long ago, northerners and colonials were the main consumers of elephant products as status symbols (piano keys and ivory billiards) for an expanding middle class (Parker and Amin 1983; Spinage 1994). The main strategy for countering this world scourge has been a militaristic anti-poaching surge involving massive infusions of new technologies, funds, and expertise for surveillance and prosecution. During these military expeditions on the ground, local residents in wildlife areas suffer the worst effects in the short and longer term.

Despite the promotional and emotional appeals of such depictions, elephant slaughters continue to occur throughout most of Africa. There are other ways to comprehend these tragic narratives and other connections to make for us all. In some ways, these wildlife slaughters are about our demands as northerners [including some Asians] and about our failures to read between the lines of our press releases or even to learn from our histories. This book is about different relations with wildlife in a particular place, relations that have been silenced, if not covered over, by monolithic stories about wildlife wars of "rights" and "wrongs." It is only one story among many that should be told.

An Ethnographic Synthesis of Some Local Experiences

Missing from mass conservation appeals are the histories of people who have coexisted with large, dangerous beasts for ages, and how these residents have cultivated the environments where they and elephants currently live. African conservation texts remain mainly crisis and discursive narratives, constructed more to generate revenues for expanding imperial economic designs (incorporating technologies and various tourisms) than for crafting sustainable conservation practices in conjunction with those currently living with these animals. Rather than simple struggles between criminals and civilized men and women, these frontier accounts reflect deeper political and economic tensions between the Northern and Southern Hemispheres as well as those within both hemispheres. For centuries, those in the North have sought and acquired the natural resources of Africa and elsewhere. When northerners entered as colonists, they arrived as strangers with designs to change the "nature" of Africa—to exploit and even to export its "natural resources" for material and financial benefits. In the process, the colonists sought to "(re)create" a world in which they were more familiar and comfortable. Those processes continue today, often in collusion with smaller enclaves of African beneficiaries, who themselves live behind walls separating them from the majority of their differently connected compatriots. Much of the power behind conservation's incursions into this continental hinterland still originates in the passions, myths, technologies, strategies, and finances bearing northern imprints. These examples of landscapes, animals, and peoples bear the residues of a colonial world perpetuated now under the covers of newer vocabularies and priorities. Both the history of northern intrusion and the "looming elephants" of its effects remain linked in the industrial, commercial, and conservation boardrooms in New York, London, Paris, Beijing, Tokyo and in other northern cities where financial and strategic decisions are plotted.

I begin by unraveling "cultural processes" in which something that was not supposed to happen did happen, during which something hidden unexpectedly surfaced and then was forced to appear as if the disruption was resolved. Yet the repercussions from this particular real-life elephant's demise rippled on for years, repeated itself later in a similar event, and affected people in ways that outsiders rarely imagine or hear. The event is the same elephant that the itinerant official in the epigraph cavalierly catalogued as "poached" before unilaterally indicting neighboring residents as the "criminals."

This story originates with those who took their time to inform me of its particulars and is (re)assembled here from the diaries and accounts of five individuals, all of whom I knew.[7] This chronicle is about more than an illegal kill of an elephant, a protected species in Zambia, something that legally never should have happened. The account begins with some residents awakening abruptly into an indeterminate and unanticipated space. It continues with how these individuals sought to position themselves in an uncertain world that constantly shifted as decisions made by distant outsiders drifted in along with rumors that impinged upon the broader dynamics of their daily lives. What do you think were the conservation messages delivered and received locally?

Some unknown person shot an elephant in the neighborhood of a cluster of villages during the evening of 24 July 1998. The fatally wounded beast bled profusely as it wandered through the villages, crossed the Munyamadzi River, and expired near Chifukula stream opposite the villages of Paison and Chibale.[8] Early the following morning, women collecting water and washing at the river noticed blood along the riverbank and vultures circling nearby. They returned to their homes and reported the scene to their headman. In accordance with his responsibilities, Headman Paison proceeded immediately to Kanele Wildlife Camp, some five kilometers away, to inform the scouts of the incident. In the meantime, other villagers hastened to the dead elephant "to keep the vultures away" and proceeded to flay the carcass.

On his way to the wildlife camp, Paison passed the school where he found the Community Conservation Project supervisor and senior counselor, Mr. Njovu, who informed him that the chief, accompanied by the wildlife unit leader, was in Lusaka. The deputy unit leader was also off post and on patrol with the warden and others from Mpika (the district's administrative center). With the chief away, Mr. Njovu, as "self-ascribed acting chief," called a special meeting of the Munyamadzi Sub-Authority Wildlife Committee to deal with this quandary. He sought a consensus to explain the circumstances to outside authorities, who he suspected would eventually hear about this elephant's demise.

On the same evening that the elephant expired, Mumbwa, the pastor of several Pentecostal Holiness churches as well as Paison's grandson and acting headman in an adjacent settlement, officiated over the wake and funeral of a nephew. Because most surrounding villagers had heard of the elephant and the windfall of meat it provided, only six men and sixteen women attended the burial the following morning. Before his departure for the wildlife camp, Paison instructed Mumbwa to inform those assembled at the elephant carcass not to butcher or take away any meat until the wildlife officers arrived to examine the site.

After concluding the funeral about noon, Mumbwa crossed the river and encountered people carrying meat to their homes. They shunted past as Mumbwa informed them of his grandfather's message. At the carcass site Mumbwa faced an unreceptive audience, even those who were his kith and kin. The butchers remained inattentive as they flayed and distributed the meat. Perplexed, Mumbwa recrossed the river where, not finding his grandfather at home, he jumped on a bicycle and rode in the direction of the wildlife camp. He found his grandfather, Mr. Njovu, and other local authorities at the school debating the slaying of the elephant and its likely consequences. In the meantime, Paison learned that all the wildlife scouts at the camp were out on patrol.

Mr. Njovu commanded Paison and Mumbwa as headmen to collect elephant meat as tribute from each household in their respective villages. Upon his return from Lusaka, the chief, as traditional "custodian of the land," would find tangible evidence that his subjects had respected him and, in his absence, his office. Mr. Njovu assumed that this traditional tribute might provide some cover for the scrambled butchery prior to the official inspection to determine the cause of the elephant's death. As instructed, Mumbwa returned, solicited meat, and noted the names of donors on a sheet of paper. A few days later while taking a fifty-kilogram bag of smoked elephant flesh as tribute to the chief's palace, Mumbwa encountered Kanele Wildlife Camp's deputy unit leader in the company of another scout. They were returning from several days in the bush allegedly searching for poachers. Mumbwa informed them what he was doing and why, who had instructed him, and ended by mentioning his recent appointment to the local Wildlife Sub-Authority committee. The scouts asked him to accompany them to Kanele Wildlife Camp so they could write a report on the incident. Upon their arrival at the camp, the scouts arrested Mumbwa and charged him with possession of elephant meat, a government entity illegally taken from a mammal classified within Zambia as a protected and endangered species. He was further charged with flagrant disrespect for the office of the president, as all wild animals were vested in this sovereignty.

The scouts demanded that Mumbwa take them to the kill site. While there, they recovered the ivory still attached to the skull. After prolonged and sometimes violent discussions between these two parties, the scout allowed Mumbwa to return to his village on the condition that he give them a list of the households that had butchered the carcass and had provided tribute.

At midnight, the wildlife scouts knocked on Mumbwa's door and ordered him outside to reveal his list. Mumbwa pleaded with the scouts that they return during the daytime when he could assist them. They refused and began beating and abusing him. Other households in the village heard the commotion, and their men quickly ran out and disappeared into the dark night. In the ensuing melee, the scouts captured a young man, one of Paison's sons, who confessed to donating meat to the chief. Scouts escorted both Mumbwa and the young man to Kanele Wildlife Camp. The captives remained handcuffed for days, were repeatedly beaten and interrogated as the scouts waited for transport to carry the accused to the magistrate's district court in Mpika, some 140 kilometers away and on the plateau.

Mumbwa's mother went to see Mr. Njovu, a close relative. She assailed him for allowing these arrests as Mumbwa was following Mr. Njovu's instructions. To console her, Mr. Njovu proceeded to the wildlife camp where he intended to obtain the immediate release of his nephew by bellowing accusations. Among other things, Mr. Njovu allegedly shouted that he, as "acting chief," had the authority to send Mumbwa on his mission. Further, he threatened to dismiss all village scouts, including all civil servants, as they only brought trouble. Among his alleged quotes were, "You people from the plateau come here very poor, like water monitors [large lizards] with tails, and, when you become rich after getting our money, you start doing what you want. If you don't release Mumbwa, I will do something to you [a veiled curse of intending witchcraft]." Although Mr. Njovu's outburst resulted in Mumbwa's release, the deputy unit leader was compelled to report these happenings and the "poached" elephant to his warden of the Bangweulu Command.

By coincidence, the warden was then on tour in the valley along with a member of Parliament and some national administrators from the NPWS. While surrounded by this company, the warden directed the deputy unit leader to proceed with his investigations. Therefore, upon his return to camp, the deputy rearrested Mumbwa and Paison's son, trucked them up the escarpment, and placed them in prison at Mpika. At the police compound, Mumbwa retold his version of how Mr. Njovu involved him with scavenging tribute and delivering messages. The police kept Mumbwa and Paison's son in prison to await their arraignment in court. In the meantime, three Mpika wildlife police officers (WPO) with three local wildlife scouts commandeered a vehicle from the Wildlife Unit, left Mpika in the late afternoon, and made plans for arresting Mr. Njovu. The vehicle arrived outside of Mr. Njovu's door in the valley shortly after midnight.

The men surrounded Mr. Njovu's house and ordered him to come outside. Mr. Njovu responded by asking if the order implied "war." When the scouts' rejoinder was negative, he opened his door and appeared on the front step. The scouts ordered him to clothe himself for a trip to Mpika, as he was being investigated for killing an elephant. Mr. Njovu gave money to his young granddaughter to support her in his absence; she reminded the officials to respect and not beat her grandfather. The scouts ordered Mr. Njovu into the open back of the vehicle rather than to his normal space within the enclosed canopy next to the driver. They refused Njovu's request to pass at the chief's palace, as they suspected the chief would order his immediate release. On their way up the escarpment, the officials derided Mr. Njovu about his presumption of chiefly authority, as they knew he was not of the royal clan, and about his monopolizing the important positions of all major community committees.

When the party reached Mpika in mid-morning, they encountered a party of ten villagers awaiting transport to the valley. Among them were two local civil servants, both of whom Njovu vociferously accused of tattling on him to the wildlife authorities. He accused them of spreading rumors about his incessant demands and assumed prerogatives. They admonished him for his assumed authority and for threatening their dismissals from government service. After these exchanges, the WPOs remained with their ward for the rest of the day until they had delivered him into the custody of the Zambian police. Intimidated by Mr. Njovu, the Mpika wildlife staff granted the privileges he demanded, but once incarcerated, the police treated him like everyone else, as a prisoner for eleven days.

The wildlife vehicle carrying the ten villagers returned to the valley from Mpika and made the customary initial stop at the palace to greet the chief. The driver presented a letter to the chief from a former provincial officer containing information on Mr. Njovu's plight and on the seriousness of the pending case. The chief angrily told the driver to inform the unit leader to return immediately to Mpika and return with Mr. Njovu. A few days later, the chief was present when Mr. Njovu appeared in court. The chief prevailed upon the warden, who intervened before the magistrate on behalf of the chief's senior counselor. Since the scouts had little evidence to prosecute, the court released both Mr. Njovu and Paison's son without compensation and without an apology for their hardships. In contrast, Mumbwa spent twenty-one days in prison, faced the magistrate alone, and had his case dismissed "innocent and up to date." Before returning to his pastoral duties, his village, and his household, he proceeded "into the bush to fast and to pray" [his words].

A special meeting took place at the palace on the afternoon of 10 November 1998. Its attendees included the warden, the unit leader, the chief, Mr. Njovu, and some members of the Wildlife Sub-Authority, including Mumbwa. Among the issues discussed was why Mr. Njovu had been "under the hands of the wildlife scouts and taken to prison." The warden formally apologized to both Mr. Njovu and Mumbwa. They said nothing about the other innocent victim, Paison's son. Instead, all the blame fell on the unit leader who had failed to follow protocol, as the local committee should have sorted out the case initially before taking it to the district.

The chief agreed that the unit leader had not followed his advice, but remained angry over a national radio broadcast that held him responsible for beating members of the unit leader's family as a consequence of Njovu's arrest and imprisonment. When both the chief and Mr. Njovu stated that the local wildlife scouts accused the unit leader [an outsider] of torturing them and that the unit leader could no longer work "to the satisfaction of the community," the unit leader responded that his wife had been beaten and her clothes torn off. As a distant chief's relative allegedly had assaulted the unit leader's family, he desperately sought transfer elsewhere.

A month later, wildlife scouts arrested a "notorious poacher," Kazembe, whom the wildlife officials had employed as a village scout in order to reform him. Although several others were accused as well of illegally killing another elephant, scouts detained Kazembe and sent him to prison.[9] After his release from prison, Kazembe, because of his reputation for fierceness and bravery, continued his employment as a wildlife scout. The new unit leader depended on him "to control" (kill) specific elephants, buffalos, lions, and crocodiles that had damaged human lives and livelihoods.

In March 2000, both the chief and Mr. Njovu fell sick simultaneously from malaria and suffered other complications. Residents suspected something ominous, as both men were inseparable within the local political sphere. Although not fully recovered, Mr. Njovu "felt compelled" to travel to Mpika as the chairman of the chief's Malaila (a recently resurrected "traditional" ceremony) to consult with committee members there. After arriving at Mpika on the plateau, Mr. Njovu succumbed and died within the day. After securing a coffin and making arrangements with the warden for transport, the committee brought Mr. Njovu's body back to the valley for burial. Hundreds of mourners, including a member of Parliament, district officials, and police officers, were in attendance at the funeral.

The night before Mr. Njovu's burial, elephants trumpeted in the bush near the chief's palace. Everyone who heard these noises became apprehensive. In the morning, they found an elephant dead within the shadow of the Kanele Wildlife Camp. A large bullet wound proved its unnatural death. Members of the funeral procession consumed this windfall of elephant flesh, but the beast's assailant remained unknown. Some mourners associated this incident with the elephant killed two years earlier and Mr. Njovu's imprisonment and demise.

The day after Mr. Njovu's funeral, wildlife scouts arrested a local hunter and Mr. Cottoni, a retired soldier who had arrived the evening before the elephant's death. They accused Mr. Cottoni of providing the bullets to the local resident as part in a "business venture" [an informal contract for the resident to secure ivory].

Wildlife scouts handcuffed both suspects and handed them to the authorities among the official mourners returning to Mpika. After spending some time in prison before their court appearance, the police released both men as no evidence linked them to the alleged criminal acts. Having endured repeated beatings by scouts and police, Mr. Cottoni allegedly extorted a large sum of money from the arresting scouts as compensation for his suffering.

"I was arrested by the scouts allegedly for shooting an elephant, which died near my house," the local hunter told me in 2006 as he reflected upon this ordeal. "The elephant was shot by some unknown person—somewhere! Somehow! The elephant only got tired and died near my place. I was taken to Mpika where I remained in detention for two weeks. I was released on free bail, tried, and found innocent by the court. During my arrest, I was really annoyed, angry for them [the scouts] taking me as an accused person, arresting me only on rumor."

A Cultural Introduction to Some Enduring Conservation Issues

This synthesis of conversations, notes, observations, letters, and storytelling was, for me at least, something of a symbolic Rubicon passage, a tangible crossing of a cultural watershed into a different world. Although I had traversed this ethnic threshold in some ways before, some residents were now revealing more intimate details of how their daily lives intermingled with mine and how they were profoundly influenced by the murky decisions and policies from beyond local horizons. Something new had surfaced in our conversations as their voices and feelings became more audible, reflective, and personal.

There was no compelling evidence that any specific resident was connected criminally with the killing of either elephant, yet many of them bore the brunt of the state's prosecutorial fist. Some paid a very heavy price in time and labor lost by arrest and imprisonment, in suffering, and in beatings. In his annual report that same year, the wildlife official quoted in the epigraph further elaborated on his initial report about what he had detected and inferred. Unlike what he had observed elsewhere in other valley

GMAs, snaring and poaching at Nabwalya were infrequent events as "only one buffalo was snared and one elephant was killed using a gun." His indicator that local people were involved was that they had removed only the meat and left the tusks behind. This behavior suggested to him that hunger was the motivation behind their action; this elephant was, in his report, "poached for nutrition purpose [*sic*]." The white professional hunter had reported that trophy animals were easy to find within this hunting tract and that he was happy with this local community's commitment to wildlife protection.[10]

To my knowledge, neither case was resolved. Both cases were dropped as soon as it was safe to do so. The real perpetrators were never pursued or arraigned; outsiders implicated insiders, yet insiders knew differently. Most probably, the first elephant was fatally wounded by an outside gang of commercial poachers, who, together with dozens of carriers, weekly descended the Muchinga escarpment from the adjacent plateau searching for bushmeat and ivory within the Luangwa Valley's expansive national parks and GMAs. These groups minimize their time within this vastness and typically seek large mammals, quickly flay the carcasses of those killed, punctually load their carriers with trophies and meat, and retreat back to the plateau. Once there, they offload their wares to other businesspeople, who carry the meat to markets within Zambia's cities. These gangs still persist [as of 2015] as few are captured and successfully prosecuted. Elephants fatally wounded during these incursions wander and persist for days before succumbing near a river or village. The first elephant apparently fell victim to this circumstance before being butchered by the villagers shortly after its demise.

The second elephant kill seems to have been contrived quickly as a customary tribute, an insider's scheme with outside supporters (or perhaps the reverse), during the unexpected passing of a significant dignitary. In this sense the elephant's demise became a "respectful" [if only a "traditional"] means for hosting and feeding a large party of residential mourners and regional celebrities, the latter swarming to celebrate Nabwalya's image as a "wild and different place" with plenty of bushmeat. Marginal culprits became the temporary scapegoats to protect this "expanded community" against the possibility of an official judicial proceeding should

powerful and distant persons inquire and require more about this elephant's death.

Understanding how members of a local community encountered, interacted with, suffered from, and endured events that are unexpectedly thrust upon them is just the beginning in discussions of conservation dilemmas in Africa. These events, generated externally, rupture local routines, cause a bustle of activities and reciprocal accusations, and provoke ground swells of questions, many of which remain unanswered and unresolved, especially at the local level. Such provincial conservation difficulties are not resolvable exclusively by applying additional force ("anti-poaching"), although this force may be necessary if deployed as a protective envelope for GMA biodiversity and residents against the destructive exploitation both by foreign and commercial gangs of bushmeat traffickers. Governmental intransience, the presumed privileges of its officials, and the untold instances of corruption by its agents on "wildlife frontiers" have produced deep distrust of government initiatives and promises. Within these distant if only mythical Southern sites, additional speculative links may fade beyond the local horizons into the consuming markets of the Northern Hemisphere, where the appearance of valuable commodities and promoted tourist fantasies appear without acknowledging their origins or burdens. Both local environments and their residents are intertwined and influenced by outside national and global groups, who in turn respond, or are driven, by demands for minerals, protein, energy, or fabled journeys and adventures. Consequently, a plausible resolution necessitates a broader vision and more extensive consensus before these dilemmas and their complex interconnections are appropriately described or witnessed. Such resolutions require commitments across a wide range of disciplines, worldviews, and states: "all at once, a matter of culture/power/history, nature" (Biersack (2006: 27). Hence this dilemma's elephantine dimensions and the requirement of a new arena of sobering thoughts and plausible activities.

How actors and practices play through time are respectively socially significant aspects of their successes or failures. An extended time frame assists a researcher to capture and link several episodes of an ongoing cultural process and its theatrics. Attributing meaning to such events also requires an interpretive and investigatory lens. When George Orwell wrote on shooting an Asian elephant, the narrator in his narrative responded within the expectations of his colonial class in Burma and of the supposed stance of his spectators. Orwell's narrator was conscious of his official role as a representative of an oppressive imperial power and its local, visible authority, and as such he felt compelled to act in prescribed ways. The Burmese hated the narrator as an imperial police officer and baited him, as Orwell explored this cultural baggage that the officer carried on his lapels. The story of how this subject came to shoot a magnificent animal "to avoid looking like a fool" was perhaps an emblematic straw on the "camel's back" of Orwell's colonial duties and compelling consciousness. Such a perceptive story may have been instructive later as Orwell became a critical observer and sensitive writer of his own society and culture (Orwell 1936).

In his role as the European game ranger stationed at Mpika in Northern Rhodesia (now Zambia) during the late 1940s until the mid-1950s, Maj. Eustice Poles shot many elephants in the Luangwa Valley. Yet to my knowledge he never crossed such an intercultural or introspective boundary as I suspect Orwell and I did. During his posting, Poles had responsibilities that included overseeing the "colonial wildlife estate" in the Central Luangwa Valley (including the Valley Bisa homeland) and supervising the African Elephant Guards stationed there. Killing two large-tusked elephants each year was an official privilege for supplementing his official salary. In March 1956, while serving as host to a distinguished visiting scientist, Poles shot a large bull elephant across the Munyamadzi River near Paison Village, close to the site where the initial elephant in our narrative expired in 1998.[11] Having successfully pursued "his" elephant for personal remuneration, Major Poles summoned Chief Nabwalya to his camp and explained to him the changing colonial policy regarding elephants raiding "gardens."

The colonial state, as represented by Major Poles, pursued its monetary interests in elephants in the name of "conservation". Henceforth, Major Poles instructed this chief [Kabuswe Mbuluma] that the shooting of "garden-raiding elephants" was the responsibility of "the government's elephant control guards" and no longer a right by which local residents could protect their proper-

ties or themselves.[12] Poles sanctioned this new policy with a severe penalty: should a local person kill an elephant subsequently while defending his crops ("flagrant shooting of an elephant on the pretext that they were raiding gardens" is how Poles expressed the offense in his official report), the chief must oversee that carriers ferried all the cured meat and ivories to district headquarters, a five to eight days on foot and up a steep escarpment. The ivories and protein were colonial properties and were not to remain in the valley.

In reports to his director, Major Poles confessed reservations about the legitimate claims of his elephant control guards and that they misconstrued his orders when killing these large, supposedly marauding beasts. Many kills were never confirmed as "crop raiders." In particular, the activities of Sandford Njovu [no relation to the Mr. Njovu previously mentioned; see below] concerned Major Poles, for this guard expended a lot of ammunition and killed many elephants. In his reports, Poles expressed his official displeasure with Njovu's "indiscriminant shooting," suggesting that the loss of so many immature elephants would cut "into our capital."[13] Yet in describing the "resource" (elephants) in this way, Poles missed the point of Njovu's strategic and cultural objectives (or cultural game).

Sandford Njovu was a contender for the Bisa chieftaincy in the Luangwa Valley. Years earlier in claiming an economic stringency, the British South Africa Company government subordinated his descent group of the chief's lineage (Kazembe), as well as those of several other small chieftaincies west of the Luangwa River, to that of the reigning and appointed Valley Bisa chief at Nabwalya. The contending currency in the ensuing political contest between two chiefly lineages for official recognition and dependents, culturally coded "respect" (*mucinshi*), was animal protein and protection for clients, not ivory. Sandford Njovu was well aware of his responsibilities in reference to his employment and with respect to ivory. He knew that any compromising behavior, if it appeared in "official light," might cost him his existing role together with his anticipated goal of one day becoming the important valley chief, rather than a descendant of his rivals. Sandford's strategy to secure the chief's title was a long shot, for, although he was younger than the incumbent, he had to wait for the older chief's death before he could advance his candidacy. The incumbent chief reigned for fifty-one years.

Unlike George Orwell, Major Poles never saw beyond his own interests and role, beyond his employment, nor glimpsed the motivations within a different cultural world, even one that he presumed to manage.[14] Neither did Sandford Njovu when he sought the Nabwalya chieftainship title upon the death of its incumbent in 1984. Sandford Njovu lost his competitive edge by limiting his spatial alliances to valley clients and district-level patrons where, as some stories have it, in addition to using the meat from the large number of elephants which he officially killed as a protector of the properties and as a provider of bushmeat for the many matrilineages throughout the central Luangwa Valley, he greased the palms of local politicians and district officers with ivories and other wildlife products. The eventual successor to the valley (Nabwalya) chieftaincy had lived for years on the Zambian Copperbelt and possessed a wider range of significant contacts as well as resources that enabled him to outmaneuver Njovu politically on the national level. When the dispute between the two chiefly contenders was brought to the Zambian High Court, the judges confirmed the precedence of the earlier colonial state.[15]

Like it has also done to Asian cultures through the phenomenon of "Orientalism," the western world of scholarship, development, and popular media has also characterized the societies and cultures of Africa as alien and different through the use of specific vocabularies that isolate and widen the gaps between "us" and "them." This discourse depicts African culture and Africans as devoid of complexity and agency (Said 1978).[16] In such texts, Africa and its people become a political foil where nothing seems to work, at least in a deterministic western, northern, and rational way (Chabal and Daloz 1999; Ellis and Haar 2004). Africa becomes the "dark continent" yet again, full of mysteries, despair, and chaos as the media and some experts continue to describe its people in pejorative terms such as "magic," "superstition," "tribes," "chaos," and "corruption." Through such depictions, all Africans clearly lack all the civilized attributes that make northerners the superior world citizens we assume. To ditch such facile descriptors and deceptive representations involves developing another vocabulary of ideas and identities, as well as the use of more nuanced ways to converse about human differences and values (Klein 1996; Kohn, 2013; Das, Jackson, Kleinman, and Singh 2014).

Northern discourse further follows the Cartesian tradition of separating humans from other life forms, cartelizing the webs of life and thereby mystifying our intricate relations with the rest of the world, which is sanctioned further by the silences in our religious spheres (Connolly 2013; Uhl 2013). This dichotomous order is expressed in the geography of our lives, in the distances between where we live and work, in our language, and in our play. Our (in)sensitivities expose other forms of life to exploitation by some, as they are enabled to extend their reach as their access enhances differences in health and welfare between neighbors and those further afield. Prodigious northern contributions to protect charismatic megafauna, such as elephants and rhinos, in sanctuaries and national parks along with the promotion of egregious wildlife policies such as blatant anti-poaching campaigns and shoot-to-kill suspects on African homelands further separate human communities and minds. By immediately accepting superficial and promotional discourses on "poaching" and "poverty" at face value as the local cause of wildlife decreases, rather than searching for how these local activities and adverse conditions might connect to more distant demands, we erode capacities in other life and in others' lives as well as in our own. Our worldviews and livelihoods further distance northerners from other peoples. Labeled benignly as "resources," elephants and other forms of life become tourists' cherished momentary vistas and wealthy sport hunters' trophies. Other sojourners crave environmental products from beyond their own limits. Life forms are turned into revenue-generating objects by weak postcolonial states and rent-seeking agents dependent upon foreign financial venality. Silenced in the newer vocabularies, or falling between northern bounds of perception, are the histories, wisdom, plights, and values of rural residents; the "others" who cope daily with these creatures as well as with their detractors on very uneven political and economic playing fields.

Deep fissures and contradictory divides abound in these "African gardens of Eden," not just within national parks or adjacent geographies (game management areas) but also significantly between residents living within them and all others. I venture no romantic return to the past or any exclusive vision for the future; neither do I suggest a future sequenced exclusively of "indigenous" origin, but rather one in which these two ideas play a creative and enabling part. I imply that we have a lot to learn from and about others in order to escape our own narcissisms. Conservation as an ideology and as practice must begin with what's on the ground and in residents' minds. Blending and negotiating "the local with the outsiders' ideal" is largely the option of the residents. I imagine a future grounded in culturally meaningful ways as rural residents are incorporated within conservation efforts and other global links as respected and respectful actors, as privileged players sustaining their identities while creating their own futures in association with the rest of us.

For such a world to materialize, outsiders seeking to conserve significant areas must extend their cultural boundaries, experiences, and vocabularies. These expansions demand that they assume roles of listening and learning instead of imposition, becoming, in William Easterly's terms (2007) "searchers" rather than "planners." Planners advocate chauvinistic cultural approaches with "good" self-directed intentions, work to develop global plans centered within the hegemony of those who employ them, think they know the questions and the answers, and flourish in situations where they remain largely unaccountable for their actions. Consequently, their worlds become increasingly one-sided, uncertain, and unstable. Searchers seek alternative and adaptive approaches in variable conditions. As agents of change, these mediators accept responsibility for their interventions; they never privilege *a priori* knowledge but acknowledge the complexities of living within local and global environments. Searchers willingly negotiate, experiment, listen attentively, and iteratively incorporate what they learn. They recognize the durability of homespun commitments and livelihoods rather than those leveraged from afar. Such individuals may seem idealistic, yet the label assigned to them is intended to project the progressive, incorporative attitude of an active verb rather than a static assumed noun of passive delivery. At the moment, I can imagine no sustainable world without striving for greater equality and social justice.

This book is a contribution to a reasoned discussion linking conservation in the social and life sciences as well as in the humanities in Africa. It elaborates upon life-long processes of learning about human interests in wildlife, about different cultural histories, about divergent livelihoods and the environmental processes of a people residing in a small, yet significant place. Therefore, I write for the reflective, general reader and questioning professional hopeful of an interdisciplinary renewal within and outside of academia as well as practical and real-world resolutions incorporating a future vision for us all. I also write for rural residents, for they were active parties in my own transformative processes, and their livelihoods, identities, and prospects remain crucial concerns at risk within a harsh world bereft of soul.

The Present Book

I begin chapter 1 by describing the *identities* of those who constitute themselves today as the Valley Bisa. Their myths do not translate readily, leaving much still unknown and perhaps unknowable to outsiders; yet shadows of their past history and heroes are etched within their landscapes and inscribed in the shadowy texts of foreigners. In their dynamic and often tragic past, itinerant strangers often became their patrons, captors, or fates—symbolized as "mother" in the chapter's epigraph.[17] For colonials, the central Luangwa Valley was a "wilderness" of unknowns, which they sought to reconstruct in an image of expansive wildlife sanctuaries that obliterated earlier histories and entitlements. For residents, this landscape remains full of memories and meanings—a thicket of diverse spiritual, social, political-economic and ecological insights that each generation of residents porously bounds as they remake and blend portions into their daily livelihoods.

Chapter 2 dismisses some of the *images* and *ideas* employed by outsiders to convey aspects of Valley Bisa social organization and life. Static terms convey neither an understanding of residents' interconnections nor of their embedded values. I describe the basic architecture of clans, lineages, and leadership to show how these categories appear in the daily lives and activities of residents. Persons, both gendered and generational, have used their environments as places to reproduce and sustain themselves, as places from which they acquire food and collect materials while working through their differing aims and claims with others. Based upon protracted observations during one agricultural season, I followed the daily routines of several women and men as they reveal "respect" within their work, with the land, and with each other.[18] Extended and briefer snapshots of these same participants in later chapters describe how some of these actors have fared in subsequent years. Individual activities affect how they perceive, ponder, and exploit the "goods" that surround them in the village, within their fields, and within the nearby bush. Valley Bisa society is not without its endemic conflicts and struggles; therefore, I discuss some age and gender inequalities, together with witchcraft.

That Valley Bisa society is (or has been) isolated and removed from the rest of the world's activities is an *illusion*, often cited by outsiders to convey their difficulties of getting there and to depict the "dismal" human lives and conditions they encounter upon arrival. Yet the people living within Zambia's central Luangwa Valley always were connected to and intertwined with their neighbors as transients in good as well as traumatic times. As in the past, wildlife is the major entity attracting outsiders today. It is also an important "good" upon which residents have depended for survival. In the recent past, local people largely managed wildlife in their interests; now their entitlements and endowments have been rescinded by the state in its own interests. Today, state wildlife management remains linked to the technologies of colonial domination expressed in the perception of the central Luangwa as "wilderness," its wildlife as a "resource, "and its residents as "cheap labor." Each of these labels as "properties" is assessed and expressed primarily in "economic" units and in calculations of value. Outsiders never recognize this landscape as intricately bound to the identities of its inhabitants, as embodied in their beliefs about ancestors, or as the repository of their cultural memories. Chapter 3 describes some hazards of the Valley Bisa as primarily hoe cultivators living side by side with large wild mammals, of outsiders' ventures with wildlife representing foreign interests, and of state plans to dispossess residents of their land and of practices that enabled them to prosper in the past.

A society's environmental knowledge and practices reflect how people understand the world and their places within it. These appraisals change as people cope with novel experiences and develop new technologies. As the main passion of some Valley Bisa men, hunting is a significant way in which they connect to life around them, both in the village and in the bush. Chapter 4 introduces some of the ways in which residents (notably a local cohort of hunters) perceived, ordered, and ritualized their world in terms of prey and space. Hunters' prey metaphorically shares many attributes with their pursuers. In addition, prey may act either on their own or as surrogates of maligned humans. Yet no cultural catalogue remains the same for long, as the depository changes with the times and with the experiences of its carriers. As the understandings and exposures of residents have broadened, so have their contentions and uncertainties.

Local knowledge constitutes a wealth of proficiency and strategies that seems to work in this unpredictable African environment. Its shifting contents become momentarily tangible as local actors think and perform in ways that often overwhelm northern categories and academic divides. Such know-how is constantly evolving, developing through personal practice, shaped through contacts and events, altered in responses to local happenings and affiliations, as well as reactive to outside political and economic interventions. As an open-ended, unpredictable encounter, hunting accumulates fluency and practice with fellow humans and other sentient creatures; the latter are aware of being pursued and may respond with evasion, deception, and even dangerous confrontations. Hunting is about learning and honing skills at every stage, beginning with its preparations and ending with sharing as well as exchanging its products for other essentials of well-being. As actors in environmental and social processes, hunters are aware of their dependencies and relations with others and concerned with their standing in their respective lineages as well as further afield. Information on three generations of hunters, their lives and accomplishments for over half a century, comprises the materials in the next five chapters.

Chapter 5 explores the cultural history of the twentieth century through the life histories, contributions, and memories of three generations of local hunters. These individuals were more than just informants; they were also friends, whom I accompanied in the bush and came to trust on occasion, even with my own life. For these individuals the quest began with a dream, developed into a vision, blossomed into social wealth through performance and patronage (or not), leaving memories and a legacy. Their histories show how individuals achieved their local identities, how they absorbed, tested, improvised, and expanded their repertoire of skills in the bush and in the villages, or were overwhelmed by the assets or social maneuvers of their competitors and kin. When these local roles became criminalized as "poaching," state prosecutions challenged the culture of their craft, their traditions, their husbandry, and their identities. Although the identities and practices of those depicted have now become largely history, their spirits remain in local memories, resurrected, remembered, and valued as needed.

Chapter 6 captures some of the excitement of learning and sharing as I follow an articulate hunter on two hunts during different seasons. This tutor anchors history and stories spatially while traversing his accustomed terrain in the nearby bush. In addition, I track local knowledge about a major prey species, the buffalo, as its structures and roles have changed its practices and goals over several decades. Gathered from informal conversations, local records, and participant observation, this practical knowledge and its deployment manifests in a small group of hunters in a particular place (the Nabwalya Study Area) and time (1966–93). Such information is porous, constantly changing with the experiences, ages, and circumstances of its practitioners. It is no panacea, but rather a litmus test of identity, action, and shifting goals, which are later clarified quantitatively and chronologically for this assemblage of hunters and their associates in chapter 9.

The customary products of local hunting by a few men were protection against dangerous beasts and the anticipated delivery of a steady supply of meat to kin and clients. Their yields complemented the agricultural activities of related women and the wages of other lineage men, who participated in labor markets elsewhere. Chapter 7 examines dimensions of local hunters' decisions while in the bush, facets of others' powers to influence their successes and failures while afield, and what these hunters brought back to share and to distribute as "goods." A number of factors acting in synergy changed earlier orientations in the targeting of species and the disposal of its proceeds. These causes included a deteriorating national economy combined with outside pressures on local practices and products. As bushmeat acquired increasing monetary value beyond local "lineage" worth, some residents and many outsiders capitalized upon the cash values of bush products in the wider national market. As law enforcement increased, residents largely shifted from firearms to previous more (re)sil(i)ent technologies [snares] and from taking the larger species (buffalo) to smaller game (impala, warthogs), whose carcasses were easier to hide. Subsequently, buffalo became the privileged focus of licensed safari hunters and local elites and the extralegal targets of many others.

Chapter 8 chronicles how the technologies of destruction, particularly muzzle-loading firearms, became intricately woven into the customary fabric of Valley Bisa society as metonyms of lineage affiliation and identity. Guns turned into a potent symbol of a gendered stronghold of power and authority, as well as a representation of its vulnerabilities. These weapons connected groups of related people to the productivity of their land and symbolically served as mediators in the social links between the living and the dead (spirits). As special properties vested in relationships among people, ancestors, and wildlife, some guns were beyond sale.

Chapter 9 describes the efflorescence of the "traditional" hunting system through tracking iteratively the accomplishments, qualitatively and quantitatively, of all local residential hunters between 1950 and 2000. My narrative follows the political fortunes of young men with their patrons as they engaged in the politics of provisioning protection and animal protein. Changes brought about by national economic declines, through education, and by shifting demographics, together with an imposed "community-based" wildlife program, turned this world on its head beginning in the 1990s. This latter intervention shifted the local currency of "respect" (*mucinzhi)* for lineage seniority and meat to money and goods (*goot milile*), a moral economy where "wealth in people" still mattered. Local employment and the unparalleled accumulation of things and cash by a few prominently placed men, who challenged the earlier matrilineal boundaries by accumulating fungible wealth, allowed different options. This gathering of flexible wealth permitted Big Men to diversify their business undertakings, to purchase urban supplies, and to sell or distribute these products on credit while accumulating clients, dependents, and wives. As this new social development gathered momentum, many younger entrepreneurs found meaning in the powerful symbols and religious practices of Pentecostalism rather than in the power of the ancestors and the astuteness of elders. Serving as the upholders of a universal faith, these successful entrepreneurs also parlayed their wealth and influence into status within the ranks of these rapidly expanding local churches. Tensions between the new and older forms of wealth brought a plethora of witchcraft allegations to the surface as leaders mobilized their clients during repetitive and enduring generational conflicts. The local saying that serves as this chapter's epigraph conveys some of the "community-based" wildlife program's tragic effects on local welfare, identities, and livelihoods. This adage, "In killing a buffalo, the game guard likens it to his mother," cautions wildlife scouts to avoid sacrificing the welfare of their local heritage in their enthusiasm for employment and their deployment in outside interests.

Chapter 10 reveals that much more was at stake in the inception of the "community-based" wildlife program within Zambia than improving rural community welfare or promises of restoring rural management and entitlements to wildlife. A review of national documents and political decisions to initiate this program in the late 1980s and to renew it in the late 1990s shows that the program's priorities included expansive national agendas and the control of the lucrative wildlife traffic through an alliance between public institutions and some in the private commercial sectors. As a consequence and despite its promotions, sustainable conservation of wildlife and improvements in village welfare never were primary concerns. The reoccurrence and interpretation of two events witnessed in 2006, an altercation over land boundaries and the killing of a prominent village elder by an elephant, indicate the continuing depth of the dilemmas and distrust between the state and local authorities as well as the jeopardy of living with large, dangerous mammals. They also reveal the despondency, dependency, and poverty felt by many residents, especially among those who remember the earlier times. All these topics and concerns remain interrelated and must be recognized and reconciled with an unfortunate past before any sustainable future becomes plausible. Their effects impinge upon the futures of us all even if we remain unaware of their claims and imprints.

Notes

1. Unit Inspection Report for the Munyamadzi Game Management Unit, 23 August 1998; prepared by the Nyamaluma Institute for Community-Based Resource Management.
2. "Northerners" is a term I use to differentiate people currently residing in the Northern Hemisphere, mainly within Western Europe and North America, as well as those typically urban dwellers. "Southerners" refers to the inhabitants of the Southern Hemisphere, many of whom still live in rural villages and small towns in Sub-Saharan Africa. That the South also has its very large cities as well as large concentrations of refugees is not denied.
3. The inception of this "community-based" wildlife program at Nabwalya in 1988–89 and its imprint on the people involved there in 2006 is described in Marks (2014: 238–74).
4. http://en.wikipedia.org/wiki/elephant in the room; referenced August 15, 2007.
5. My parents were southern Presbyterian missionaries in the former Belgian Congo from 1948 to 1961. My father, a dentist, and my mother, a nurse, together with others, established a medical and dental institution to teach some Congolese (Lulua) how to become health workers, and aspire to live a "Christian" (if not Presbyterian) lifestyle. The boundaries of their work and engagements were defined two generations earlier in a court trial between a despotic monarch, Leopold II, and the pioneering efforts of an earlier generation of Presbyterian missionaries, particularly Dr. William H. Sheppard, an African American, and William M. Morrison. See also Benedetto and Vass (1996); Hochschild (1999); Kingsolver (1998); Kennedy (2002).
6. Examples of these stories, tragedies, and ideals are presented in the Ken Burns film, *The National Parks: America's Best Idea*, shown on Public Broadcasting Stations beginning in 2009. Rosaleen Duffy (2010) describes some tragic and draconian practices of international wildlife conservation and NGOs in the developing world.
7. Conversations with other residents at various times helped to clarify these events and the connections within the written accounts. I use

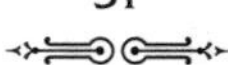

a writer's license to shorten the story and resolve some less tangible elements, and I elaborate upon some nuances within endnotes.

8. The protocol by which villagers could obtain meat from such found carcasses demanded that they first report the carcass position to the nearest wildlife camp and wait while the scouts investigated the scene. Once they completed their investigation, scouts often required villagers to butcher the meat and carry it to their camp. For these villagers, the unexpected disposition of such a large carcass in their midst was a treasure trove from which a distant bureaucracy might seek to disenfranchise them. Their hurry to flay the carcass indicated their delight in this find; their subsequent behavior showed their anxiety over its consequences. The reference to where this elephant carcass lay, between Chibale and Poison [*sic*] village, was probably not only a misspelling of Paison but a Freudian slip given the normative vocabularies used by game officials to denigrate GMA residents' behaviors.
9. Although the sentencing of Kazembe seemed to "officially" settle the earlier poaching incident, it remained unclear if Kazembe was the actual killer. Kazembe had a reputation as a local "elephant slayer," for he had several prior convictions as a "poacher." According to the people who knew him best, Kazembe's deployment never showed any signs of transformation toward becoming a "conservationist" (see his reappearance in chapter 10,endnotes 19 and 23).
10. Community Development Officer, Evaluation and Monitoring Reports for 1998, ADMADE Units for Mwanya, Nabwalya, Chitungulu, Chifunda, and Chikwa (period 14/08/1998–05/09/1998), typed report in author's possession.
11. Both Fraser Darling [an imminent British ecologist that the Northern Rhodesian Game and Tsetse Control Department engaged as a consultant to help them strengthen their case for more funds and new policies during 1956–57] and Eustice Poles describe these events in their field notes. See John Morton Boyd (1992: 40–41). In describing the local Africans, including chief Nabwalya, encountered on this safari, Fraser Darling absorbed many of the prejudiced attitudes of his host, Eustice Poles. Upon learning of the death of this large elephant shot by his host, Darling (1960) revealed that his more cosmopolitan sentiments about wild animals were "poles" apart from that of the provincial ranger.
12. The colonial government recognized the right of local owners to kill wildlife (*damage feasant*) in the defense of life and properties (see Faunal Conservation Ordinance CAP 241 [1964 edition], section 26). Once killed, the carcass and trophy theoretically belonged to the

state for disposal. Given the distance from Mpika and difficulties of travel, the latter provisions were rarely enforced in the central Luangwa during earlier periods [as explained later in chapter 1 and 3]. In his few recorded interactions with Chief Nabwalya, Poles sought to strengthen his department's status as the agency for killing raiding elephants, thereby assuring possession of the ivories as state properties. Under indirect rule, the native authorities retained considerable power as to which game violations they prosecuted within their territories.

13. Eustice Poles, "Report of a Tour of the Munyamadzi Controlled Area, Report 1/1947, under Annexure 1—Elephant Control Guards, Duties and Responsibilities 1 March 1947" (copy of report, typescript). In this report, Poles records that Sandford Njovu killed twenty-one elephants in 1946 and five more in January 1947. Poles had no idea about other game that Sandford Njovu might have shot, nor did he understand what "game" he was playing. Poles had "no complete records" of the ammunition Njovu possessed in 1946, but that the latter was known to possess "a balance 110 rounds of ammunition " on 1 June 1946 and had withdrawn subsequently another 72 rounds. More information in endnote 15.
14. While I was studying in London in 1965–66 preparatory to fieldwork, some retiring Zambian officials with whom I corresponded suggested that I write to Major Poles, then retired and living in England, to seek his advice for living in the Luangwa Valley and review his extensive and personal records. I wrote him an introductory letter telling him of my intended studies and inquired if it would be convenient for me to visit him. As he preferred fishing in Ireland, he declined to meet me. Instead he offered the following advice: "News out of Africa percolates to me from time to time and I am led to believe that the inhabitants of this part of the Luangwa Valley eschewed cannibalism since their final supper party with the white P. A. [provincial administrator]. You should therefore be reasonably safe. Also I understand they have acquired the dubious blessings of Christian superstition and education. Whether or not these changes in their environment make them more or less interesting for study must remain a matter of personal taste; doubtless it has enhanced their natural cunning and added to their notorious rascality. In my time communication with this area consisted only of a native path but I am told than [*sic*] an all weather motor road has since been constructed which makes your mission easier, safe and more agreeable." (Personal letter to author dated 30 March 1966).

15. The eventual ascendancy of the Nabwalya lineage of the Ng'ona clan as the sole recognized Valley Bisa chieftaincy between the Luangwa River and the Muchinga escarpment was undoubtedly a product of its central location, adjacent to the site of an early colonial post (see chapter 1). In this region, the colonial state had recognized previously at least eight Bisa chiefs, but the main rival to Nabwalya was always Kazembe—and the district initially bore the name of Kazembe. As the main contender in the Kazembe line, a younger Sandford Njovu, the elephant control guard, contended with the chief by attracting clients and by solidifying this patronage with protection and with wild meat. The chieftainship was sanctioned by the state, and its occupant, also a renowned hunter, possessed material wealth that made him attractive to women (and their lineages) in all sections of his "enlarged" chiefdom. That Njovu outlived the chief (who died in 1984) and the four-year interregnum before the state appointed the new chief in the Nabwalya line is testimony to the strength of the Njovu's strategy.
16. In his influential book, Edward Said (1978) portrayed and critiqued "Orientalism," which he depicted as an assemblage of false assumptions and romanticized images informing western attitudes and writings about Asia and the Middle East. He argued that the western experiences of colonialism and political domination distorted these images and attitudes in unflattering ways, reducing their complexity and agency while continuing to serve as implicit justification for expansive and military ambitions. Westerners also wrote these histories and constructed their identities in which the West was the norm from which the foreign and exotic them deviated. Said was likewise critical of Arab and Oriental elites, many of whom had internalized these "Orientalist ideas". While engaging and profound, Said's ideas are not without their detractors and critics. I initially took his perspective to reexamine my engagements in Southern Africa and those of colleagues.
17. Translation: "An important traveler may become your mother." Meaning: You can ask anything of important strangers, as it is through showing appropriate "respect" that one might find new connections, knowledge, and challenges to enlarge one's world of opportunities.
18. For details and analysis of these studies see Marks (2014: 72–114).

Section I

On Becoming, Being, and Staying Bisa

Chapter 1

History and Circumstance

On Becoming and Being Bisa

An important traveler may become your mother.
Bisa proverb (recorded 1993)[1]

The Valley Bisa homeland has been an extensive space on the floor of Zambia's central Luangwa Valley, a parsimonious environment that was known and inhabited centuries ago by their ancestors, who settled at key sites in their search to survive. For most of the nineteenth century, predatory raids by more powerful neighbors, the Bemba and Ngoni, and others linked with the East African coast slave and ivory trades under Swahili, Portuguese, and Arab influences increased residents' uncertainty. Later and during the early twentieth century, the destinies of residents shifted and fell under South African and British colonials who oversaw the labor and commercial markets of their southern trade routes. Colonial influences progressively marginalized Valley Bisa fortunes as they constricted the locals' homeland within a narrow strip of land between two large faunal reserves established to preserve a foreign hunting legacy. As the small valley population increased and Northern Rhodesia became independent Zambia in 1964, colonial resource bequests became entrenched and selectively enforced to support privileged foreign and national access to this valley's wildlife. Within this broad historical context, distant, unexpected influences and individuals have impinged upon Valley Bisa lives and livelihoods with ambiguous risks as well as uncertain opportunities.

As indicated in the epigraph, openness and "respect" shown to strangers might bring beneficial connections, new knowledge, and allies for chiefs and residents alike. Despite their initial promises, such arrangements often turn into subordination and, if prolonged, dependency. Inside this indeterminate valley crucible, the epigraph appears cogent as residents' persistence continues to perplex agents of change just as outside contacts have disordered and made most residents vulnerable in unprecedented ways.

The Valley Bisa humanized this landscape by investing their identities within it before their valley spaces became restricted and their livelihood opportunities defined by reformist colonial and national states. Such a landscape stands in sharp contrast to the conceptual "primal Edens" promoted in earlier narratives of colonials or in the current (con)quests of safari identities sought within the proclaimed "wilderness" concessions. This lived-in environment and its cultural spaces come alive as residents share their stories of past events and explain their meanings and organizing principles. One may gain this sense by listening attentively to the conventional chatter of village lives or by accompanying forays within the adjacent bush. Within the bush, the names of thickets, water sources, baobabs, patches of land, and scattered resources evoke memories of the living and the dead, as well as recent and past events, and they can also conjure secular and spiritual dimensions. Excursions in villages and over this terrain are introductions to important names, to oral histories and meanings in this place.

My studies with the Valley Bisa commenced in 1966, less than two years after Zambia gained its independence. Nabwalya, the chief's village, and its environs became the space within which many of my intermittent village and bush observations, as well as conversations, have taken place over the subsequent six decades. The terrain itself is a dense depository of Valley Bisa history, language, and practices, expanded and examined each generation by residents through their daily imprints of ideas, activities, and performances. Today the Munyamadzi Game Management Area

(GMA) contains the villages and settlements of the Nabwalya chiefdom within some 2,650 square kilometers. Toward the center of this GMA is a smaller quadrant of some seventy-five square kilometers along the Munyamadzi River that includes the villages and habitats immediately surrounding the chief's palace. This smaller space is a diverse environmental patchwork similar to most found within the larger GMA. This place is the ground stage for many of the personal dramas and faunal interactions portrayed in later chapters. I label this latter space the Nabwalya Study Area, yet the frames for some happenings here emanate offstage or in reaction to more distant agendas.

This exploration of recent Valley Bisa history and local management of wildlife begins with descriptions of the location, environmental, and historical components upon which this story unfolds. Whereas small Bisa and other ethnic groups' settlements were scattered at favorable sites throughout the western Luangwa Valley, the territory of the current reigning Valley Bisa chief Nabwalya is today confined geographically to a corridor of land, the Munyamadzi GMA, dividing the South Luangwa and North Luangwa National Parks. The Nabwalya chiefdom stretches from the Luangwa River and the Luambe National Park on its eastern flank, its boundary with Zambia's Eastern Province, westward some eighty kilometers to the steep Muchinga Escarpment within Mpika District, the largest region within Zambia's new Muchinga Province.[2]

Domesticating a Landscape through Language and History

I use the term "landscape" intentionally because it reflects the broader intertwined nature of cultural and ecological processes. "Landscape" extends beyond the natural and social debates of transient outsiders to include surroundings that local people recognize, experience, extract resources from, and understand more broadly in terms of their own welfare, livelihoods, and identities. These relationships are local (*in situ*) processes, absorbed into one's being by living within a particular place over time, learned through affiliation and practice, rather than as a mere physical

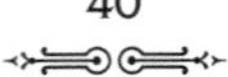

environment of "something out there" (Wolmer 2007).[3] People craft landscapes through their experiences within and engagements with the world around them. Some local residents (more so than others) know the plants, the animals and their movements, and the sources of water, building materials, and various foods as well as the "powers of place" contingent upon their roles and experiences. As they continually interact with these elements, relationships become embedded in historical promptings, local lore, and meanings. Compared with the multidimensional sensory values of residents, outsiders' and tourists' images of this landscape remain largely visual and superficial (Bender 1999).

During the nineteenth century, the Valley Bisa homeland was aligned within the main slave and ivory trade routes to the East African coast that passed through the Luangwa Valley; the initial establishment of a British South Africa Company (BSAC) post on Ngala Hill in 1899 was a colonial strategy to counter these connections. In 1908, the BSAC moved its district headquarters to Mpika, a site on the adjacent plateau that was linked to southern supplies and considered healthier for Europeans than the depressive heat of the valley. Later, sleeping sickness and other dreaded tropical maladies within the valley caused its closure to outside traffic for years after 1912, and a railroad built further south connected to a network of roads on the plateau resulted in new developing commercial links and massive influxes of men and materials on the plateau. With the British Colonial Office under international demands in the 1930s to preserve the fauna within all its African protectorates, Northern Rhodesian officials allocated much of the sparsely populated valley terrain west of the Luangwa River as two extensive game reserves. The scattered Valley Bisa villages were allowed to remain within a narrow corridor of territory including the Munyamadzi and Mupamadzi Rivers designated as Native Trust Land. Some villagers displaced from the game reserves were absorbed as residents within Nabwalya's chiefdom while others resettled near the base of the escarpment in the west or across the Luangwa River in Eastern Province. Subsequent generations have found the state's demarcations on the ground and restrictive access to the patchy assets of timber, wildlife, fisheries and fertile alluvial soils within this parsimonious valley inimical to their general welfare and livelihoods (Marks 2014).

The Valley Bisa Homeland, Yesterday (1966) and Today (2012)

When my studies began in 1966, the encompassing boundaries of the two large Luangwa Valley Game Reserves (defined in 1938) and the newly constructed bush track (completed 1960) connecting Nabwalya (the chief's village) during the dry season to Mpika and to a ferry crossing on the Luangwa River were prominent features of the state interests on charts and maps (figure 1.1). On the ground, the state's physical presence was minimal, consisting of a small dispensary at Muchinka, a primary school, and a local court at Nabwalya. Besides the resident chief, two teachers, a tax collector, and two itinerant game guards (all outsiders) represented state concerns. Few officials and safari hunters visited between that dry season and the next (1966–67). Official duties focused on conflicting ministerial agendas consisting of planning for a cooperative farm along the Mupamadzi River and a proposal to resettle residents based on the recommendation of an international wildlife team centered within the Department of Game and Fisheries. Because of its abundant wildlife, the Munyamadzi was among the initial Controlled Areas declared in 1945 in order to regulate hunting access by outsiders. In 1954, it became more restrictive as a First Class Controlled Hunting Area giving the Game Department oversight and control over game licenses issued to residents and outsiders.

Figure 1.2 shows later inscriptions and interventions on this terrain. When the earlier Luangwa Valley Game Reserves were promulgated as national parks in 1972, the South Luangwa National Park included an additional large chunk of Bisa homeland, the Chifungwe Plain, unilaterally inserted in the legislation as essential terrain for elephant survival and breeding grounds. The former Controlled Areas became GMAs. Under the new Zambian Wildlife Law (1998), villages were grouped into five Village Area Groups (VAGs) to promote "democratic elections" by selecting two representatives from each group of villages as members of the new Munyamadzi Community Resource Board (CRB). As the earlier mandated Wildlife Sub-Authority had been captured by local elites, donors pushed this political strategy to "assure" more effective representative distribution of funds and structures throughout the GMAs. Despite these

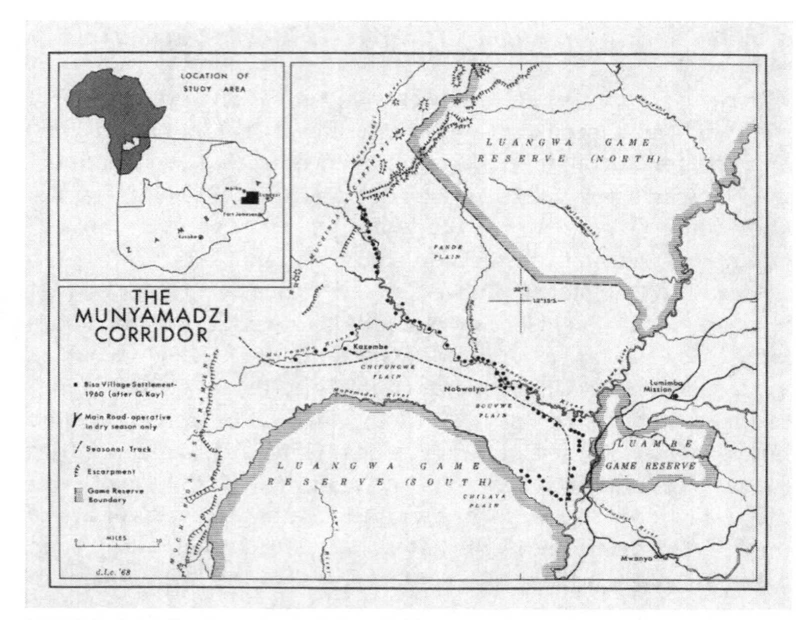

FIGURE 1.1. The location of the Valley Bisa homeland and Nabwalya, the chief's palace, as they appeared on maps within the geographic contexts of Africa and Zambia in 1966. This homeland was then known as the Munyamadzi Corridor, a "First Class Controlled (Hunting) Area," bounded by two massive Luangwa Valley Game Reserves. Note the siting of villages along the rivers, flood plains, and the vehicle track. (Adopted from Marks 2005 with permission)

intents, this new board was quickly mastered by elite intentions as well.[4]

During wetter decades, some residents constructed their settlements and villages in the hinterland along deep lagoons, which retained water during the dry season (Marks 1976: Figure 1.2, p. 22), and along the Luangwa flood plain. Later, access to the rivers' water and the fertile pockets of alluvial soils along riverbanks became crucial to meet villagers' needs and for cultivating subsistence crops. Conflicts between humans and wildlife occur almost on a daily basis in villages and fields along the rivers, particularly during the late dry season, as water depressions in the hinterland recede, and at the end of the wet season, as the cultivated grains ripen. In the 1990s, the drilling of water bore-

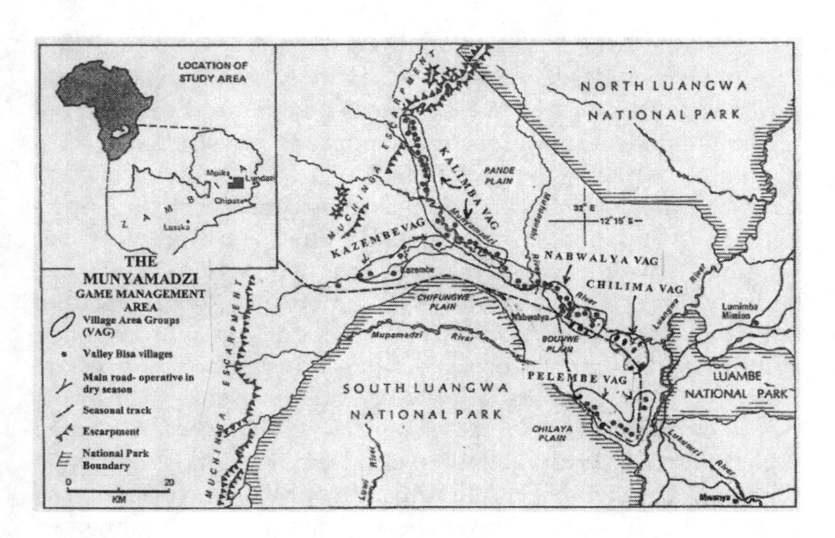

FIGURE 1.2. The Valley Bisa homeland as the Munyamadzi Game Management Area (GMA) in 1972 when the Chifungwe Plain was excised to become an extension of the South Luangwa National Park. In 1999, GMA villages were organized into five Village Area Groups (VAGs) under the Zambia Wildlife Authority (ZAWA). (Map revisions drawn by author published in Marks 1984, 2014)

holes permitted some villagers to resettle away from rivers and the dangers of crocodile attacks, where they could bathe, wash grains, and collect water more safely.

Figures 1.1 and 1.2 indicate some features of progressive state and outside inscriptions; yet other processes also become noticeable over time. Safari firms have bushwacked numerous tracks leading away from the main road to Nabwalya toward significant wildlife habitats where they search for suitable trophies for their clients each dry season. Beyond these confusing road connections and the foot tracks to the plateau, two aircraft landing strips exist to fly in the supplies and hunting clients for the GMA's two safari blocks. These hunting concessions generate foreign exchange for their operators, revenues for the Zambian Wildlife Authority (ZAWA), and some funds for the local Community Resource Board (CRB) to support law enforcement and state-approved developments.

The modernizing currents of improved access and outside funding have pulsed across this valley in waves of new constructions, material possessions, and ideas. Although the objectives of access and external funds may appear positive initially, their procedures, urgency to resolve priorities, and distant dilemmas tend to both weaken local capacities and thereby increase dependencies. Given external interest toward quick-starting something tangible as well as generating positive descriptions and statistics, promotional documents proclaim only affirmative results and delusionary market "win-win" stories designed to stimulate funding for short-term projects. Money and cultural expertise push these results, which allow outsiders to advance their priorities at the expense of residents, some of whom may be attracted by the initial promises and benefits. Local recipients must later cope with their decisions, with the culpability of their participation as well as the delusions of promises and dreams, which they can never own. By establishing structures of oversight and procedures, government officials contribute momentum and frequency to the repetitive "booms and busts" on frontiers, these indeterminate cycles further undermining local confidence, objectives, as they offer to a few the challenges of new initiatives. While proclaiming to accomplish community purposes through tangible things (buildings, gross revenues, numbers of arrests), foreign-funded projects deflect attention from the relational and social processes (violence, economic inequalities, scarcities) occurring simultaneously on the ground.

Funneled initially through an imposed Wildlife Sub-Authority chaired by the chief, some safari revenues infused quarterly to local communities resulted in the construction of a few permanent buildings (school, housing, wildlife camps), material improvements (wells, hammer mills), famine relief, and increased employment of local casual laborers during the 1990s. Many of the promised constructions materialized near the chief's palace; as the authority captured and dispersed some funds and employments on site, other benefits allegedly changed hands discretely elsewhere. Using "development speak," residents renamed the chief's palace "Nabwalya Central" focusing attention on the lack of structures and opportunities elsewhere within the GMA.[5] During this decade, successive cresting waves promoted wildlife protection, chiefly authority, religious proselytization, and temporary employment. Community and safari funds contributed to the operations of seven permanent wildlife camps commanded by national wildlife police officers and locally employed wildlife scouts. Despite occasional official meetings, most residents learned the new rules through the arrests of relatives and neighbors, violently captured during anti-poaching patrols or through collaboration with local informers. Under the Zambia Wildlife Act of 1998, all staff of the former wildlife department were retired early when it became the statutory Zambia Wildlife Authority. A few biologists and wildlife scouts were re-employed, but political appointees and businessmen headed the new authority mandated to find external revenues to make the new agency self-supporting. Reflecting the transient wealth of those forced into early retirement, floods of material possessions—new bicycles, boom boxes, furniture, TVs, cell phones, and an occasional vehicle—appeared in villages and were soon discarded as junk on refuse heaps.

As chair of the Wildlife Sub-Authority, the chief oversaw the distribution of community funds, the recruitment of wildlife scouts, and the building of his palace, and he sought to solidify his patronage through appointments and employment. As the earlier cultural and elder authority collapsed under the stringency of this new wildlife program, alternating droughts and floods, and sporadic disease outbreaks, a few residents dispersed elsewhere to find food or employment temporarily. Those remaining searched for answers to their dilemmas and uncertainties. Weeks of religious crusades by fundamentalist and Pentecostal groups sought new converts and enacted massive conversions during the dry seasons. Catholic priests and Protestant pastors increased their visits to serve and revive their congregations.

During the initial decade of the twenty-first century, more buildings and churches were constructed. Pentecostal churches became prominent in or adjacent to many villages. Nabwalya became a Catholic parish with a newly constructed mission compound near the new secondary school, with resident seminarians dispensing outreach services. A Baptist missionary couple built a residence at Nabwalya and flew in medical supplies for the health clinic. The CRB built an office complex for its officers. Some of these new compounds featured Internet service, satellite phones, television, and refrigerators. Traffic to the plateau increased from vehicles owned locally by the missions, chief, the CRB chairman, safari firms, and wildlife officers. A generation of "non-runners" on blocks provided spare parts for the next stock of vehicles. To facilitate valley communications among local cell-phone owners, an outside company discussed building a communication tower, and in 2014 a Chinese firm signed a contract to construct an all-weather road, connecting the plateau with Nabwalya, and a bridge across the Luangwa River. This new road and tower will increase further the tempo of differential exchanges across this landscape.

As in many marginal areas, census figures for residents within the Nabwalya chiefdom remain problematic, yet they indicate trends within an otherwise flexible and mobile population.[6] The first BSAC census in 1910 estimated 2,124 adults and 1,420 children (total 3,544), figures that showed slight variations throughout the 1920s and 1930s. On the eve of independence in 1963, the population was listed as 2,224 adults and 3,559 children (total 5,783) in 76 villages. Beginning in the 1940s and lasting until the 1980s, most men worked as regional migrant laborers throughout Southern Africa. During our survey exercises in 2006, we estimated the numbers of residents throughout the GMA between 8,800 and 10,000 individuals of all ages in 96 villages (Marks 2014: 34–35; appendix B).

The Nabwalya Study Area—Place and Overview, 1966–2006

As people have settled the area around Chief Nabwalya's residence for more than a century, I know more about their dynamics and impacts on this landscape than for the rest of the GMA. Beginning in 1966, each personal visit entailed inquiries of households on the Nabwalya Study Area, including their distribution in settlements, the extent of their fields, and their impacts on the nearby habitats and wildlife. We begin with estimates of residents on the seventy-five square kilometers surrounding the chief's palace and the main causes for their changes through time. An overview of the area's natural features and habitats highlights some significant imprints of Valley Bisa lives and memories on this landscape. The variegated landforms and habitats have contributed to the quality of human life and diversity of wildlife found here in the past. Yet the recent expansion in human numbers and activities, contributions by residents as well as outsiders, has impacted local welfare and the diversity and numbers of wildlife.[7]

Human Dynamics on Nabwalya Study Area, 1966–2006

My initial mapping of settlements and conversations with people at Nabwalya took place in October 1966 as residents were preparing their fields for that agricultural season. I recorded thirteen villages and seven scattered smaller settlements inhabited by 82 men, 162 women, and some 222 children (Marks 1976: table 2, p. 19; figure 1.3, p. 153). Matrilineal-related women formed the core of these villages and, along with their daughters, were the main cultivators of sorghums and other local agricultural products. One of their elder brothers served as headman, who along with a few elders and younger men, assisted in subsistence and construction chores, provided protection and animal protein (bushmeat), oversaw the disposition of wealth, and attended to the health of the village as they mediated conflicts within and between lineages. At any one time, 50–70 percent of men were away as migrant laborers; some remained away for years. They sent occasional remittances to relatives in the homeland and upon their return brought back cash and material goods. During 1966–67, nine residents (mostly outsiders) worked on salary at the palace, school, local court, and as game guards.

Until the mid-1980s, Nabwalya remained a sparsely populated and remote outpost (table 1.1). Most economic activities were subsistence with bartering or reciprocal work; a few local stores stocked with outside products, purchased through migrant laborers' remittances, absorbed most exchanges in local currency. Those on salary purchased prestigious capital goods elsewhere (i.e., guns, cloth, salt, cooking oil), while a few trafficked local products (chickens, tobacco, fish, bushmeat) in distant markets.

Local economic transactions changed noticeably after the death of the chief in 1984 and with his replacement in 1988. Revenues poured in from the new "community-based" wildlife program (ADMADE) administered through the Zambian state. As the main hub to (re)distribute these funds and materials throughout the GMA, Nabwalya began to grow in size and population. As

TABLE 1.1. Estimates of residents on the Nabwalya Study Area from intermittent household and settlement surveys, 1966–2006. (Adopted from Marks 2014: table 5.1, p. 74)

Month	Adult Men	Adult Women	Children <15 years	Description/Reference
1966 October	82	162	222+	Villages small, defined, few scattered settlements, households enumerated in all villages and most settlements. (Marks 1976)
1973 August	97 (150)a	165 (197)a	249 (311)a	68 households sampled in major villages, state considering combining villages into larger townships (numbers in village headmen registries). (Marks 1979a)
1978 August	29	93	343	93 households sampled in the larger villages and settlements around primary school. Economic survey. (Marks 1979a).
1988 October	65	170	433	135 households in villages, 35 households in settlements; not included are settlements and fields north of Munyamadzi River.
1993 July	126	169	479	Few defined villages, mostly scattered settlements. All households south of Munyamadzi River enumerated. Major droughts in 1990–2; cholera outbreak 1991.
1997 July	162	274	516	Few defined villages, settlements scattered, some households shifting from north to south of Munyamadzi River. Major droughts in 1996, 1997.
2001 August	267	359	703	Scattered settlements along river and roads, many from margins settling closer in, those from elsewhere expecting employment, heavy rainfall and floods in previous years. Government moratorium on safari hunting (2000–2) resulted in no revenues for communities, affected drop in numbers subsequently.
2006 May	173	228	536	Scattered settlements and clusters of households, all residents counted from Kanele to Paison and along the Nabwalya-Luangwa road. Refurbished school, permanent buildings at Kanele, community building, police post. Dirt airstrip actively engaged in bringing in safari clients. Heavy rains, floods 2002, 2003, 2004; drought 2005, 62 days of rain in 2006.

Sources: population and household studies during research stays at Nabwalya, 1966–2006
a = number of residents according to village headman's records. (Modified from Marks 2014: 74)

migrant labor became noncompetitive elsewhere, men remained or returned to compete for the few local jobs under the wildlife program; parents within the GMA sent their children to the expanded basic school at Nabwalya (table 1.1). Much of the windfall revenues from the new program that were filtered through Nabwalya or the safari camps never got far afield. Comparatively few men controlled and circulated this wealth, a circumstance that increased women's dependency for cash and essential household items, clothing, school and clinic fees, and general welfare. As in the rest of the valley, repetitive droughts and floods affected agricultural production, thereby acerbating competitive economic struggles among all residents and furthering their transition into a market economy. Many local products (labor, firewood, bushmeat) became commoditized for sale locally and in distant markets; those who could not produce these items increasingly depended upon others or outside assistance.

All segments of the Nabwalya population showed increases through 2001. That year, the Zambian president's two-year moratorium on safari hunting ended that source of community funding for development. Fearing a massive assault on "unprotected wildlife," the European Union (EU) provided salaries for wildlife scouts and supports for anti-poaching operations during this moratorium. The EU had previously pushed for the creation of a new wildlife institution to control the rampant corruption within the former wildlife department (National Parks and Wildlife Service—NPWS). Upon the assumption of the new agency, the Zambia Wildlife Authority (ZAWA) reduced the amount of revenues from safari hunting for community development and made the restructured local institution, the CRB, responsible for wildlife protection, support, and funding. The reduction in numbers of residents in the Nabwalya Study Area in 2006 reflects these changes (Marks 2014).[8]

These developments at Nabwalya and residents' demands for land and local resources have impacted the biodiversity within nearby environments. The diversity of habitats here, its natural salt licks, year-round water sources, and other resources attract and hold an array of wild vertebrates throughout the year. Several waves of Bisa ancestors chose to settle here because they could depend upon its varied natural assets, particularly its abundance of large and small mammals, which sustained them in normal as well as uncertain times.

An Overview of the Nabwalya Study Area

The best vantage point to view the central Luangwa Valley at Nabwalya is from the colonial *boma* (administrative) ruins on top of Ngala Hill (figure 1.3). Discarded as a BSAC post in 1908, this hill site served as an early medical research center specializing in trypanosomiasis and as a quarantine post briefly for "sleeping sickness" patients until it was abandoned in 1912 (Hall 1910). The government designated the valley a sleeping sickness zone and closed it to outside traffic until 1924 (Hall 1950). Once the valley was opened for visits again, this elevated place, with its evening breezes, proximity to the river, and above the hum of village life, became the choice encampment for occasional colonial

FIGURE 1.3. Aerial view of northwestern section of the Nabwalya Study Area in January 1967. Note the fullness of the Munyamadzi River flanked by Ngala (middle left to river) and Chongo foothills (beyond river), the positions of some villages and cleared fields on the flat riverine savanna, and alluvial soils near the river. The seasonal vehicular track (bottom center) leads from the government school and chief's palace (not shown) toward Kanele game guard site (Nawalia *boma* ruins) atop Ngala ridge. (Aerial photograph taken by author in 1967)

and Zambian itinerant administrators during brief visits. During the mid-1960s, this hill was used as a surveillance site for wildlife enforcement units and is currently the headquarters of ZAWA operations within the Munyamadzi GMA. Residents' named this site Kanele Camp, institutionalizing the persona of an earlier colonial administrator who forcibly sought to alter and discipline their lives with beatings. William P. Kennelly was the Native Commissioner here (1902–4) and, by his own record, is officially credited with constructing 130 miles of roads during his tenure (Lundazi District Notebook). Kennelly's brother allegedly was a member of Major Wilson's Shangani Patrol pursing the Ndebele King Lobengula during the 1893 insurrection in Rhodesia. This patrol was surrounded by a Ndebele regiment and annihilated.

Ngala is the last ridge in a series of eroded escarpment hills as one travels on foot along the Munyamadzi River from its tumble down the Muchinga Escarpment to its convergence with the Luangwa River, still some twenty kilometers further east. The hot salt spring that seeps from this hill's eastern base to fill shallow and salty puddles gives this ridge its name. As in the past, local residents occasionally gather and burn its reeds to distill its precious salt, as well as visit the spring to bathe in its warm waters. At night, wild animals pause briefly to sip its salty water and taste its soil.

Ngala is just one in a series of foothills that extend at right angles to the Munyamadzi River in both directions. Immediately north of this river is Chongo ridge, also the name of another hot spring at its base that bears the name of an initial settler. Rather than face servitude under a challenging stranger, Chongo and some of his followers allegedly committed suicide here. Nearby residents enliven his name and the event by invoking Chongo's spirit in rituals and stories during stressful times. "We respect the spirit of Chongo even as we have our own chief of the Ng'ona clan," wrote one young resident in 1989. "When the rains are late or there is little rainfall, we send the great granddaughter of Chongo to make a shrine near this spring. To honor his name, she pours beer and places sorghum flour on the soil while humbly asking his spirit to bless the land so our crops will grow and we can survive in this land."[9] Still hot and gushing from the hillside in 2011, Chongo appeared a mere shadow of its past. A BSAC officer stationed at Nawalia in 1905 noted that this spring "sent a steady stream of hot water down to a considerable lagoon of warm water in which numbers of hippo lived and also numerous crocodiles" (Hall 1950).

This "considerable lagoon" is a diminished oxbow a short distance east of the ridge where its crescent shape now mirrors the upper bend in the current Munyamadzi River as it twists back on itself on the flat plains after its long run through the foothills. Locally, this lagoon is *Chinama* ("the place of animals"), named because its dense reeds, thickets, and water gave sanctuary to hippos and other beasts during the daytime (1966), although it was reputedly an important refuge for residents hiding from the Ngoni in the late nineteenth century. As of 2006, this lagoon had become a shallow depression, and much of its water had evaporated by early dry season. The schools of hippos that earlier had harbored here during the daytime, had served as an accessible source of bushmeat for trade for generations of residents. Rather than consume these butchered carcasses themselves, they used the smoked flesh during past famines to exchange with people on the plateau for grains.

Residents recognize that the lagoons and depressions now adjacent to the Munyamadzi River were formerly connected as riverbeds of this waterway (*Munyamadzi mufwa,* literally "dead river"). They are also mindful of the generalized drying and erratic trends in rainfall, which have affected their livelihoods over time. As he reiterated their names (Katobo, Mukwinda, Chifuvia, and Chipulu-pulu), the elderly headman Chibale informed me in 1967 that in his youth he fished on these slightly depressed sites where his fellow villagers then planted crops. The drying trend has affected the fates of other larger lagoons, Mupete and Kapola, which in the 1960s and 1970s retained sufficient run-off from the rains to last throughout the dry seasons. By the 1990s, farmers had progressively encroached on their margins as Nabwalya had become a major service and employment center for the GMA.

The Munyamadzi River (literally "water place") and the somewhat parallel river further south, the Mupamadzi ("yields its water"), today bear these Bisa names from their respective origins on the western plateau until their merger with the Luangwa River in the valley. These continuous cartographic descriptions reflect the products of colonial mapping and enterprise to open (to exploit) the connectedness of this landscape rather than the provinciality of indigenous names. Given the smaller perspectives of earlier valley residents, each section of the Munyamadzi in the valley was known by previous sectional names of Kapamba, Kasompola and Kapani much like those residual lagoons names given me earlier by Chibale.

Bemba Stream originates in the hot spring at the base of Ngala Hill and meanders in a torturous eastern course to eventually join the Munyamadzi downstream. During the 1950s when the valley was wetter, several villages and fields straddled its course. Now the stream holds water only during the rains. In the late 1990s, the chief encouraged villagers to relocate from the river and settle along the main road east of his palace. These villagers now get their water from wells dug by Irish Aid.

On the southern side of Ngala and across the Mpika road is a conical peak, *Kapili Ndozi* ("hill of witches"), reputedly the site where earlier generations decapitated those individuals convicted of innate witchcraft. Closer to the river and west of Ngala is a small stream near a baobab tree. This site, Chambolo, is allegedly where adulterers received corporal punishment for their offenses. Although neither these nor earlier sanctions are endorsed by the current national court system, local belief still remains in the power of witches and in the pervasive attractiveness of extramarital relations. Both crimes remain prime presumptions in social strife.

Another hot spring, Malanda, occurs along the junction of the foothills with the sandy plains some four kilometers southwest of Ngala. This boiling, sulfurous stream arises in two reedy hillocks and flows southeast, creating open grassland bordered by mopane woodlands. Streams like Malanda within the grassland retain their mineral waters year round, attracting wildlife during all seasons and at most hours of the day or night. Herds of buffalo, eland, elephant, and impala, as well as smaller aggregations of wildebeests, warthogs, and zebras, may be found here if one waits quietly and if these mammals are not recently disturbed. Malanda stream connects east with an eroded streambed that channels sheets of water from the elevated lands west and south of the springs during the rains. During heavy downpours, the water floods out across the sandy woodlands reaching the larger Mbouvwe Plains further east. East of Mbouvwe Plain, the water outflow structures the shallow channels of Mwebe stream (a small village formerly stood along its banks) and the deeper channels and holes of Muvuzye. The shallow links connecting these parched watercourses are barely visible during the dry season.

The expansive Mbouvwe Plain occurs some three kilometers south of the Munyamadzi River. Its heavy clays expand and become slippery during the rains, then contract and harden during the intense heat of the dry season. The inevitable hot fires of the dry seasons move through this terrain, keeping the plains open as grasslands. Wild grazers and browsers are attracted to its openness and its stunted shrubs and grasses. This massive plain has three tapered extensions or "offshoots". One arm, mwana waMbouvwe ("smaller ('child of') Mbouvwe"), stretches narrowly westward toward Malanda. Chela, another semi-detached arm, stretches north, almost reaching Bemba Stream before bending westward toward Kawele thicket, and then further eastward. Many species favor Kawele thicket, an extensive tangle of various trees on the sandy soils overlain by the outwash sediments from the foothills. Mammals come here to escape the heat of the dry season and, at all times, to find cover after grazing or browsing in adjacent habitats or the nearby fields. Somewhat south and east of Mbouve is another clearing, Citema Leza (literally, "God's opening"). The many damaged, broken trunks of mopane trees suggest that this species was once part of the surrounding woodland canopy before the soils dried and their saplings became selectively battered by elephants.

Depending on the season, the vast landscape stretching eastward below Ngala Hill is a crazy quilt of habitats combining assemblages of recognized plants that are each recognized locally with a vernacular name. The name of this habitat composition is similar to the general vegetative and topographic designations employed by plant ecologists in the same region. These vegetative types listed in terms of decreasing area around Nabwalya are: *busenga* (*Brachystegia* upland), *ilambo* (mopane woodland), *cisanze* (riverine savanna), *nyika* (grassland, plain), and *lusaka* (thicket).[10] A plant association at a particular site may also have local name bearing the name of a person, a nearby event or activity, or a prominent physical feature or tragedy with which it is associated. Traveling across this landscape with some residents, and by listening carefully one becomes aware of a host of local heroes, histories, and tragedies, kept alive in the conversations among one's sojourners. These landscape lexicons and sites are not just shadowy hallmarks of a past; they orally express meanings, precautions, signposts and virtues that are continually updated with new stories and contemporary associations (chapter 6). Learning the names and stories associated with places such as such as *Maida* [where an elephant killed a prominent returning migrant in 1966], *Jela,* [final field cultivated by a favored chief's wife in *cisanze*] or *Kusompola* [where several smaller streams join as they enter the Munyamadazi River allegedly renown as site where young girls were seized by (*-sompola*) crocodiles prior to their *chisungu* (puberty ritual)] inform about the significance of these sites for listeners.

Less than a kilometer east from Ngala Hill, a large fig tree (*citunduma)* once stood out among the clearings of fields and settlements in the vicinity of the primary school and the Catholic church. In the shade of this big tree, some Ng'ona elders erected a small, thatched ancestral shrine (*mvuba*). Before planting each year's crops, some elder women poured ceremonial beer into and spread sorghum onto holes in the ground under the shrine's roof. Each hole bore the name of their lineage ancestor. When I lived in a house facing this tree during 1988–89, some residents allegedly saw lion forms (*nkalamo ya ciaalo*—representatives of past "owners of this landscape") in the dark shadows of its low-lying limbs; others, including myself, heard only plaintive lion roars reminding listeners supposedly of their dispositions.[11] In October 2010, a bolt of lightning demolished this towering tree the same evening of a lineage wake nearby. Witnesses to this powerful explosive force told me that witch potions spewed from the trunk including animal familiars such as owls, cats, a large lizard ("*mbulu*"), and snakes. Disturbed bees from a hive within the tree killed two cocks in an adjacent settlement where local rumor declared that the elderly, lonely man there was a residential wizard.

The wide, distinctive limbs and towering trunks of mature baobabs stand above neighboring trees. These long-lived, awkward-looking trees provide fruit, flowers, nesting sites, and

fibers for many species. Current baobabs have survived dense populations of elephants who hammered them for the scarce elements within the bark. Residents use baobab bark and fruit and exploit the beehives. Many local baobabs have names—Chakanga, Katankila, Musokota, Chipi, Somo, and Chile. Some names evince historic stories such as Banakazabwe, named for an elderly woman, who reputedly found shelter within this baobab's hollow trunk to escape Ngoni raiders. Another hollow tree near Mupete became the crypt of a famous hunter, who requested burial within its open trunk. Kabuswe Yombwe is a baobab near Paison Village. "Today, our custom is to hold a ceremony in Kabuswe Yombwe's honor when soliciting rain," wrote a young Bisa student in 1989 for a class project requiring an interview with his parents. "People brew a special beer that ripens the day of the ceremony. We bring this beer to its base, dance, shoot arrows, throw spears, and fire bullets into the baobab's trunk and branches. As the tree bleeds, we sing special songs. We expect lots of rainfall and a good harvest by remembering Kabuswe Yombwe, the late "priest" of this land whose spirit is still present."[12]

The land in the shadows of Ngala and Chongo Hills is known as Chitaba, allegedly named after a once plentiful evergreen tree. The Ng'ona chief remains the sole state-sanctioned "owner of this land" (*mwine ciaalo*), yet important dimensions of this ownership depend on the disposition of its reigning regent. In 1967, as I was contemplating an overnight encampment in the bush with two local hunters, they suggested that I consult with the chief. These men told me that recently a Zambian game ranger had tried to secure a buffalo and failed.[13] So I asked the chief for his permission and protection, and I inquired about the ranger's situation. His response reminded me that there were protocols for living off this land, as well as spiritual preparations required in seeking protection from harm.

> [The ranger] failed because he did not get my permission. I wondered why he didn't come and ask me as *mwine ciaalo*. I would have asked the spirits [*mipashi*] to bless his efforts saying, "Ancient spirits [respectfully pluralized], your children want meat to consume. Give them safety while they hunt in your land." When you go into the bush, take some sorghum flour [*ubunga*] and put it

> under a tree. Address the spirits and say "We have come here on a mission. Cleanse our bodies and as your children, let us hunt successfully in your land."[14]

The fertile pockets, habitats, and floodplains of the Munyamadzi, Mupamadzi, and Luangwa Rivers, as well as the shadow lands of the Muchingas and its foothills, have served as home for generations of Valley Bisa. There were strategic and provisional reasons why Chongo and his Ng'ona contender selected this location, just as there were similar calculations in the minds of the BSAC agents and later the wildlife officials for selecting the top of Ngala for their *boma*. These hills provide a vantage point for scrutinizing human and other activities on the broad level land below them in a westward direction. Initial Bisa arrivals used this overlook to spot raiding warlords from a distance and to give them enough time to fade into the thickets and other redoubts to escape. For the colonials, the strategic and elevated site provided a command post for supervising those whom they considered lower in their new political-economic order. For wildlife officials, the elevation enabled them to hear gunshots, detect fires, or locate sites of suspicious activities (e.g. vortexes of vultures).

The diverse habitats in Ngala's vicinity, with its resources and covers, offer a cornucopia of opportunities to sustain human welfare in an otherwise frugal environment. The river delivers a constant source of water, the hot springs generate precious salts, a scattering of dense thickets provides cover for escaping enemies and for stalking prey, alluvial soils sustain a rich substrate for producing food and medical plants, and abundant forests supply materials for construction, tools, firewood, weapons, and additional wild products. Yet for this landscape to promote human wealth and welfare there must be a plan, a system of ideas, organization, and behaviors, as well as time and experience to pull it all together into a working synthesis—what anthropologists label as the domains of culture and social structure. This book describes such a strategic plan, a locally constructed arrangement of people and natural resources that endured for some five decades before its demise under more compelling inside and outside circumstances.

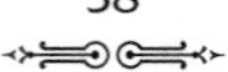

Migrations and Stories to Sustain a Social Order

The emergence of Bisa-ness in oral history endures, and the transmissions are important foundation events in social distinctions, such as those of ethnicity and between rulers and subjects. Other records by outsiders provide some approximations for when Bisa chiefs settled on the plateau or in the Luangwa Valley. During the late 1890s, agents of BSAC established an administrative post at Nabwalya (initially spelled "Nawalia"), adjacent to a Bisa chieftainess. A review of the archival records, recorded rationales, and chronologies provides a colonial perspective on events and their placement of additional demands on valley residents.

Orally transmitted traditions and histories occur in most societies. Those in Central Africa explain who the group's founders were, where they came from, their relationships to others along the way, and how the group came to settle within their present environment. Such traditions often include dramatic events or occurrences, some of which are mythical, others plausible (Vansina 1965, 1966; Roberts 1973; Miller 1988). Migration stories dramatize a lengthy process in which telling incidents shaped and continue to influence relationships within the group (Cunnison 1959).

Some Bisa Founding Parables

Along with other groups inhabiting Northern Zambia, the Bisa place their origins centuries ago in the Lunda-Luba Empire, then located in what is now the south-central Democratic Republic of the Congo (Zaire). This empire, called Kola, was ruled by Mwata Yamfwa, who was described as the emperor of the western Lunda, a dynasty begun with the marriage of a Lunda princess to a Luba hunter, Chibinda Ilunga.[15] Disputes over land and the dispersion of people following the death of a significant patron fueled an emigration that colonial sources place around 1650 CE. Colonial agents, who recorded earlier materials, interpreted these movements politically as of rather recent origin as that chronology fitted their own mission and myths (Brelsford 1956; Thomas 1958). Later studies have suggested that earlier settlements on the plateau originated differently beginning in the eleventh and twelfth centuries CE, merging when another group who eventually became their chiefs arrived. This research implies that the immigrants from Kola may have subverted and supplanted the knowledge of other clans among whom they took up residence (Roberts 1976). How this subvention took place remains speculative, yet the story of Chongo and the establishment of a Bisa chiefly lineage (clan) in the Luangwa Valley (recorded on pages 66–7) insinuate how some earlier settlers may have responded to these strangers.

Historical stories about primal groups splitting into separate identities often find texture in terms of geographical features where inappropriate behavior differentiates the lineages of leaders from those destined as commoners. Oral traditions state that the Bisa and Bemba royal clans (Crocodile) traveled as one group until they reached the Luapula River. Here according to a Bisa tradition, the Bemba built a fish trap, caught a crocodile, and stayed to consume it; hence their royal clan name became that of the *Bena Ng'andu*. The Bisa discovered a ford, crossed the river, and separated from the Bemba while moving south and east, becoming known as the Bena Ng'ona.

Other legends tell how this Bisa royal clan further divided. One such account begins when two sisters paused to collect mushrooms after a heavy rain. With her supply soon exhausted, the younger sister petitioned her elder sibling for mushrooms to feed her starving child. The older sister refused, saying that she had none, an act interpreted as causing the child's death. Later, the elder sister stumbled and broke her pot, revealing that she did indeed have mushrooms to spare. By violating the norms of propriety and kinship, the elder sister's descendants became known as Bena Ng'ona Samfwe,[16] and her descendants were denied the possibility of becoming chiefs.

The initial Bisa chief to settle on the plateau at Chinama was Mwansabamba. His followers included his sisters' sons. Mwansabamba dictated that the meat of all wild game be cooked in a large central pot and then redistributed to the wives of his sisters' sons. This custom angered his nephews and their dependents for it questioned their maturity and abilities to manage households. When these complaints surfaced, Mwansabamba responded that if his sisters' sons demanded their own cooking places, then they should find land and become chiefs on their own. This story became the founding charter for the initial scattering of Bisa chiefs on the plateau and eventually into the Luangwa Valley. It is recited today during the cooking of meat in a common pot for the annual Chinama-nongo ceremony, which commemorates Bisa unity and promotes Chief Kopa as their senior chief.[17]

Some Bisa chiefs and associated clans settled on the plateau near Mayense Hill, a series of granite outcroppings near the headwaters of the Mutinondo and Mupamadzi Rivers (Brelsford 1956; Thomas 1958; Roberts 1973). From here, they made gradual and seasonal forays into the adjacent Luangwa Valley to exploit wildlife and other resources, and they also participated in trade expeditions to the East African coast. Among the first settlers along the Munyamadzi River near the Ngala foothills in the valley was Chongo, whose exact origin remains vague, as he may have been a former slave. The group around Chongo prospered until a Ng'ona clansman, searching for territory and subjects, unseated him.[18]

> Long ago, the ruler of this place and land was Chongo; he belonged to the Nsoka [snake] clan and his wife came from the Nsofu [elephant] clan. They lived a good life and multiplied. Later another man came looking for new land to rule. Rather than enter Chongo's settlement immediately, this stranger hid himself at Malanda, an open glade in the bush. There he collected feathers from all the different birds found in that place. He concocted a mug of clay and affixed these feathers with beeswax. Then he approached the village.
>
> He asked its residents to tell him who was the "owner of this land" [*mwine ciaalo*]? Everyone answered, "Chongo is the one who has found this place." When the stranger found himself in the presence of Chongo, he asked the same question of him.
>
> "I came here alone a long time ago and found this place," replied Chongo. "It's a good place with salt, fertile land, and plenty of animals. Others joined me here and we have prospered. Why are you questioning me?"
>
> "If you are the owner, where are your symbols so I can believe you?"
>
> Upon hearing this challenge, Chongo was ready to fight, yet he was hesitant because of the sudden appearance of this stranger. So when the stranger suggested that they both go to the counselors for judgment, Chongo agreed. The wise counselors heard both claims. The counselors asked, "Chongo what symbols do you have that you are the real owner of this land?"

Chongo had nothing to show as he thought these others would recognize the rightness of his oral claims. When the counselors posed the same question to the stranger, he produced a feathered mug. The fertile clay and its symbols of many feathers astonished everyone. Certainly, they thought, this stranger was clever, must have important connections elsewhere and accepted his claims. The stranger then introduced himself as a Ng'ona clansman, the traditional lineage of Bisa chiefs.

The stranger had upstaged Chongo and that hurt. He returned and told his wife that there were two choices—to become a slave to this Ng'ona or die. His wife replied, "It's better to die than to be ruled by a stranger."

So they rapidly gathered up all their close relatives and household things—including mortars, chickens, pots—and threw themselves into the nearby sulfurous hot spring. Then some strange and surprising things happened. Upon his death, Chongo turned into a large crocodile recognized by a white stripe on its upper jaw. When some hunters found the lagoon, they left their marks by cutting down the trees along its edge. Returning the next morning, they found no cut trees. This restoration made the hunters believe that Chongo had a powerful, vengeful spirit that protected this site.

Later people found beads along the spring's shore, picked them up, and became lost until they restored the beads to their original places. Some say Chongo appears in dreams telling of new places and resources. Others say that flocks of white egrets at the spring may be Chongo's chickens. Even today near the hot springs, people hear the pounding of mortars and the crowing of chickens.

Such stories of chiefs and the founding of a polity are central in the practices and choices "of becoming and staying Bisa." Further, the role of retelling these stories at the resurrected *Malaila* [victory] celebrations demonstrates their importance within the negotiated conditions under which individuals retain this ethnicity. They legitimize the political and ritual authority by which chiefs, with their headmen and counselors, claim to "own the land," as well as their abilities to effect the land's wellbeing and that of its residents.

Different "ways of knowing," "cleverness," and "respect (*mucinshi*)" are ascribed to Ng'ona chiefs. Their abilities to outwit contenders and remain current on information are hallmarks of their status, and this demands deference. In the story of Chongo, an important ancestral Ng'ona took something (land and power) from an initial settler and used it to "gather people (clients)" and to establish a new polity centered on the consent of those around him.

This Ng'ona was Shilambwe. Kazembe Mukulu, together with his sister Pelembe, became the dominant force in the central Luangwa Valley.[19] As in the wisdom embodied in the epigraph of this chapter, one's destiny may reside in the potential powers of strangers and depend upon one's cunning cultivation of such opportunities. You can ask a stranger for anything, as you would your mother, and thereby become surprised or challenged in what materializes.

The myth of an original settler, such as Chongo, implies ritual power over the land, rain, and fertility just as the shedding of blood by Ng'ona combatants. Serious dissention among Ng'ona contenders for power often occurs upon the death of an incumbent. Rebellious factions are serious threats, must be dealt with forcibly, and, like Chongo, are inscribed on the memory landscape in physical form and renewed periodically by rituals of respect. The following story is about Kabuswe Yombwe, another powerful Ng'ona rival whose spirit resides in a baobab tree, currently a rain shrine.

> Kabuswe Yombwe was a brave man. He wanted to be chief when his uncle died. He put forward his claims and persisted even when not chosen. Kabuswe Yombwe began looking like a ghost [powerful presence], which caused people to take him seriously. They thought that maybe Kabuswe Yombwe was the *nsimapepo* [priest of prayers] who controlled the rain.
>
> The new chief realized that he must kill this pretender or suffer the consequences. He knew that Kabuswe Yombwe possessed the powers of a shape-shifter and could turn himself into other things when facing danger. When the chief's counselors came looking for him, they found only his wife at home. She told them that Kabuswe Yombwe had just turned himself into a rubbish heap and that's where they might find him.
>
> The loyalists gathered to discuss new strategies for catching this evasive contender for, if they failed, his cleverness might mean their loss of leadership. With a plan in mind, they proceeded to his house where again they found only his wife. Kabuswe Yombwe had sensed their arrival and turned himself into a large baobab tree. The loyalists mocked and beat his wife, threatening to kill her unless she told them exactly where her husband was. She pointed to the large baobab tree. They went there, fired a gun into its broad trunk and filled its limbs with arrows and spears. Thick, red "blood" [sap] gushed from its trunk and branches. Kabuswe Yombwe did not reveal himself but remained inside the tree. When the assaulting group exhausted its ammunitions, they returned to their leader and told what had happened.

We do not know for certain when the various Bisa chiefs established themselves on the plateau and in the Luangwa Valley, yet recent records suggest some threads and dates for weaving together an interpretation of when they did and for how they lived.

Being Bisa during the Precolonial Period

Sometime around 1760, Bisa traders began participating in the export of ivory and copper from the interior, exchanging them with the Portuguese and Arabs on the East African coast for cloth and beads (Alpers 1975; Roberts 1970). Two Portuguese expeditions in 1798 and in 1831–32 sought to establish direct relations with King Kazembe, a Lunda potentate on the Luapula River, to connect their colonies on both the Atlantic and Indian oceans, and to dominate these continental trades. Both expeditions found King Kazembe unsupportive of their missions and returned to Mozambique.

Portuguese Accounts

The first Portuguese emissary, Dr. Lacerda, encountered Bisa hippo trappers and settlements in the Luangwa Valley, as well as a chief on the plateau, in 1798 (Burton 1973; Roberts 1973: 87). Upon his walk back from Kazembe, a desperate Lacerda noted exchanges of slaves for bushmeat, ivory, and grains (Astle 1999: 3). Some Bisa served as intermediaries for King Kazembe.

In 1825, the Portuguese governor of the Sena Rivers acquired land on the Luangwa River near the valley's center, which he used as a trading post with residents and caravans. A small Portuguese military force occupied this site, called Marambo, for a few years before abandoning it (Gamitto 1960: 136–37). In 1831, the Monteiro-Gamitto expedition found a few survivors of Marambo subsisting on wild fruits, roots, and buffalo meat scavenged from lions. Gamitto noted here "two strange things" about the relationships between humans and wildlife. First, he observed that groups of unarmed men easily intimidated nonaggressive lions. Gamitto attributed this trait to superstitions that these beasts were reincarnated (chiefs') spirits; he then suggested the abundance of prey and the lions' infrequent encounters with people as the plausible cause. Second, he noticed that large crocodiles fled as swimmers disturbed them in river pools (Gamitto 1960: 138–40).

This expedition paused on the east bank of the Luangwa River to consider their next move. A powerful Bisa chief "Kasembe"(sic) and his subjects living near the western escarpment had a reputation as "notorious highwaymen" and subsisted only on wild fruits and game meat. Rather than visit this village to replenish his supplies, Gamitto decided to send tribute instead. Before crossing the river, the Portuguese went to see a Bisa, Chewa, and Senga encampment of hippo trappers to explore their mode of hunting and production. These trappers resided in temporary camps away from the hippo pools where their prey spent the day. They used weighted poisoned spears suspended by ropes from trees along well-defined hippo trails to take their quarries. He noted these trappers believed that hippo brains contained a poison and made no use of this animal's ivory. They sun-dried the meat to use for their provisions as well as to trade for slaves, elephant ivory, and cloth. The trappers dismissed Gamitto's offer to kill hippos, claiming that the noise of guns would scatter their prey and cause the trappers to spend additional time relocating the dispersed hippos, to build another camp, and to (re)deploy their downfalls (Gamitto 1960: 151–53).

Crossing the Luangwa, the expedition followed the Mupamadzi River westward, encountering only small, uncooperative Bisa encampments. After haggling with the locals, the Portuguese secured a guide from a hunting camp and made their way through the Muchinga Escarpment. Within this escarpment and on the adjacent plateau, the expedition saw reminders of recent skirmishes with the Bemba, who engaged in dominating trades of ivory, cloth, and slaves. Many Bisa hid in the bush or lived in large stockades for protection against marauding bands and warlords. Food was unavailable for the expedition from some plateau villages. These Bemba and other raids accelerated Bisa settlements throughout the Luangwa Valley, as Gamitto noted Bisa settlements even east of the Luangwa River.

As the Lacerda expedition had, this mission also failed to secure trade and passage rights from the Lunda king Kazembe, and so they retreated to Mozambique. From Gamitto's accounts, we glimpse Bisa life and organization. Several significant chiefs residing in territorial palaces (*musumba, singular*) ruled over people referred to as Bisa. He notes that a supreme chief may have existed at one time, for in each chief's territory subordinate chiefs (*fumo*) paid tribute.

It is clear from these accounts that the Bisa were either traders or cultivators. Those who were merchants were always on the move in caravans, yet "did not like trading in slaves." The cultivators were mainly women, older men, and children. They produced plentiful crops and cultivated widely (Gamitto 1960: 155–208). Bisa possessions were limited to traded necessities, and the people were "hard working." They possessed only offensive weapons (bows, arrows, and spears) and wore distinctive bark garments and hair styling.

Gamitto (1960: 145–46) noted that famines frequently caused people to migrate to other places for food and to offer themselves as "slaves for hunger." A descent group (lineage) survived hard times by exchanging its labor or kin for food with more prosperous groups. They bartered labor for food (*kucinjizya*: "to exchange"; *kubombela filyo*: "to work for food") or pawned (*muzya* [singular], *bazya* [plural])[20] junior kinsmen for food or goods. Persons given in compensation lost some of their rights and privileges.[21]

Beginning in the eighteenth century, men from a broad range of interior ethnic groups (including some Bisa) became slaves attached to the Portuguese crown estates (*prazos*) along the Zambezi River. These diverse groups of men initially served as militia to protect the estates and later as armed bands known as Chikunda, principal elephant hunters throughout South-Central Africa. When the Portuguese estates collapsed toward the middle of the nineteenth century, these bands, under the leadership of warlords, became semiautonomous and powerful military and political forces throughout the region. Their hunting expeditions, status, and warrior-hunter cult, celebrated in verse and dance, readily set them apart from the more traditional traders and cultivators in the Luangwa Valley. Some of these specialized hunting groups depended on the sanction of chiefs as they produced wildlife goods traded to slave and ivory caravans (A. Isaacman 1972; A. Isaacman and B. Isaacman 2004; Wiese 1983). Under the direction of itinerant warlords, the Chikunda forged alliances with some residential groups and chiefs, and it was from these migrant hunters that the Valley Bisa absorbed new rituals, relational (hierarchical) ideas, and techniques for hunting elephants and other wildlife and for defending themselves. When British colonial expansion caused this system to collapse at the end of the nineteenth century, stranded Chikunda hunters receded into the general population (Marks 1976).[22]

Recent Scholarship and Histories

Judith Kingsley's studies within a Plateau Bisa chiefdom found that local histories were not as extensive as those of their chiefs. Her findings imply that clans with different histories became "Bisa" by acknowledging Ng'ona ownership of the land and by adopting deferential attitudes (Kingsley 1980: 77). Kingsley's studies indicate that individuals and kin-based groups shifted their residences and alignments among chiefs through marriages, intra- and intergroup conflicts, early or suspicious deaths among relatives, and for security. Some longer-term alliances between commoner and specific Ng'ona lineages resulted in institutionalized arrangements, such as those for marriage, and for joking, burial, and ritual functions. Since chiefs did not possess consummate control and depended upon other clans for advice and support, these arrangements produced a political system in which local autonomy and loyalty were strong while chiefly control remained comparatively weak. The practices of giving goods (*mitulo*) and services or labor (*milaza*) to a chief, as well as the chief's complementary gift-giving, reflect such economic relationships.

Bisa society benefited from its participation in long-distance trade, as many current staples came from South America. The Portuguese brought maize, cassava, groundnuts, sorghum, and beans to the shores of Angola, and these crops spread throughout Central Africa during the sixteenth and seventeenth centuries (Jones 1957; Kingsley 1980: chapter 5). Yet early Bisa involvement in long-distance trade led to the depredations by the Bemba and later by the Ngoni, Chikunda, Arabs, and to eventual domination by the British. Weakened by the Bemba wars beginning in the

1820s, Bisa chiefs lost extensive territory and sought refuge in the Bangweulu Swamps, in the Muchinga Escarpment, and within the Luangwa Valley. The effects of the Bemba, Ngoni, Chikunda, and other attacks are dramatically reflected in oral accounts, as the disintegration of community relations, security, and general viability of place.[23] The effects of these armed itinerant groups reorganized regional household and village economies, constantly shuffled political allegiances, and produced ecological disasters in their wakes. Yet these calamities did not affect all Bisa chiefs to the same extent. Kambwili, a Ng'ona clansman and trader from the plateau, saw the political turmoil as an opportunity. He intervened on behalf of a contender in a Chewa dispute east of the Luangwa River in the late 1880s and gained possession of territory there (Lane-Poole 1938).

Beginning about 1830 and lasting for the rest of the nineteenth century, the Bisa both within the Luangwa Valley and on the plateau witnessed a "rising tide of violence" (A. Isaacman and Rosenthal 1984). The Bemba sought trading goods as well as territory and resources from the Bisa heartland, and the migrating Ngoni within the valley sought recruits for their expanding society. Beginning about 1840, roving parties of hunters, traders, and slavers, such as the Yao, Swahili, Arab, and Chikunda, ravished the smaller societies in the Luangwa Valley and the adjacent plateau in their search for bound labor and ivory for the East African trade. Slaves became a major export by 1860, reaching record numbers by about 1880 (Langworthy 1972; Roberts 1973: 197–98) when firearms and innovative military strategies began to play an important role (Fraser 1923).

In 1866, David Livingstone passed through the central Luangwa on his last trek into the interior and visited the Bisa chief Kazembe, whose village he described as a "miserable hamlet of a few huts," its inhabitants "very suspicious" and inhospitable. Nearer the escarpment, Livingstone encountered Chief Kavimba, who had recently repelled an Ngoni attack and was unfriendly (Livingstone 1874). Later British agents in the 1890s were impressed with Kambwili's stockade east of the Luangwa River and his alliances with various Arab traders who protected him. From Kambwili, most of these agents traveled to the chief's associates on the plateau rather than through Kazembe (Marks 1973a).

Brief Summary of Precolonial Bisa Life and Society

While successful long-distance trading may have led to economic and social stratification among some powerful Lunda and Bemba, the Bisa and smaller groups living in the shadows of these dominant tribes never developed similarly. Instead they and their traders were frequently targeted for raids by more powerful warlords (Roberts 1973: 109). Smaller chiefs had little control over most ivory hunters and traders. Hunters' payments in tusks depended on a chief's reputation and ability to demand such tribute as well as the hunters' need for patronage. Yet some chiefs' powers may have increased over vulnerable villagers during the turbulent times of the nineteenth century when many people assembled in fortified locations for protection against raiders. Others, such as Chongo, eked out a living by hiding in thickets or caves in the Muchinga Escarpment near varied resources.

"Being Bisa" in precolonial times meant aligning one's identity and group prospects with an existing Ng'ona clansperson, who allegedly ritually owned a landscape. "Ownership of the land" reflected the chief's responsibility for the welfare of the land, the fertility of its resources, and the general well-being of its residents—a metaphorical interdependence assumed during prosperous times and sought during troubling ones. Group identities were flexible as individuals and groups shifted in residence and loyalty, contingent upon life cycles and circumstances.

Tumultuous times during the late nineteenth century expanded habitats and wildlife throughout the region. As wildlife increased in the Luangwa Valley, large caravans of traders and slavers sought these animals for food, as agricultural products were difficult to obtain during their passages. Wildlife also spread the range of the tsetse fly (*Glossina sp.),* the insect host responsible for spreading a blood parasite, a trypanosome, that is lethal to domestic stock but wildlife is largely immune. The existence of this fly and endemic parasite was the reason that the valley inhabitants do not keep livestock and occasionally suffer from outbreaks of sleeping sickness from tsetse fly bites. The African Rinderpest Pan-zootic (cattle and wildlife plague) scourged wild and domesticated animals throughout southern Africa during the 1890s (Spinage 2003). Although these rinderpest plagues reduced susceptible mammalian species to low levels, the trends of most Luangwa Valley wildlife populations increased progressively throughout most of the twentieth century, reaching their zenith in the mid-1970s (Freehling and Marks 1998). Tsetse flies remain a legacy with which the Valley Bisa must still contend. Ever since their settlement within Luangwa Valley, the Bisa have used an arsenal of weapons and developed local strategies to protect themselves and their properties against marauding wildlife.

At the time they were first encountered by agents of BSAC, Bisa chieftaincies extended eastward from the swamps and islands of the Lake Bangweulu depression, across the Great Northern Plateau, and into the grasslands, extensive woodlands, and flood plains of the Luangwa Valley. These extensive and differing ecosystems and resources together with their recent histories have created divergent economic and political destinies for various Bisa chieftaincies.[24]

The Valley Bisa under a British Company (BSAC) and Colonialism

During the 1890s and early 1900s, a few widely spread agents of the BSAC sought to control what they perceived as a fractured and disturbed landscape in Central Africa. After years of hesitating on the sidelines, the colonial, commercial, and missionary agents gained ascendancy over leadership elements within the militant Bemba and Ngoni groups (Gann 1969; Wills 1967). Impostors and charlatans abounded in this transitory world and the colonials needed allies to structure and manage their difficult political and economic positions. BSAC officials contributed to the confusion through their haphazard and often contradictory as well as subjective interventions (Roberts 1973; Wiese 1983).

Generally and within the Central Luangwa Valley

The colonials placed their organizational hierarchies and lexicons on this seemingly contingent and supple landscape. They recognized and subsidized leaders (chiefs and headmen); delineated boundaries (chiefdoms, tribes, districts, provinces); created new demands and conditions for labor (taxes, discipline, migrants); established ranks (rulers and subjects), new rules, and enforcements (courts, messengers, police); introduced formal international currency; and formalized language. English became the official language as missionaries and linguists helped foreign administrators come to grips with the subtleties and formalized peculiarities of Bisa speech. Given time, the "*b*" sound gained hegemony over what the uninitiated ear perceived as a "*w*" in English; the earlier "Wisa" became "Bisa" and the colonial outpost, "Nawalia" became "Nabwalya" (Madan 1906). The creation and placement of district and provincial boundaries made sense mainly for the colonial engines of administration and commerce. The building of the railroad further south linking the colony with Rhodesia and South Africa and the movement of the district administrative center from the valley (Nabwalya) to the plateau (Mpika) indicated these shifts in power and commerce from an east-west to a south-north axis.

The BSAC sought to end slavery so that freedmen and women could provide labor to expand commercial markets and to build new state and commercial institutions. Yet terminating the slave trade was easier than ending informal slavery, a common and widespread institution. An administrator noted in 1910 that "probably quite a large number" were still in slavery, many with the status of "almost that of relatives."[25]

Colonials also sought to capture the trades in ivory and other valuable resources for their own coffers. They made subjective decisions about and frequently acted cavalierly toward indigenous institutions and customs. Colonial administrators needed the chiefs as intermediaries to manage labor, gather taxes, and garner supplies; consequently, they appointed those with whom they felt comfortable. BSAC officials recognized a number of small chiefs, often appointed favored candidates, applied seemingly abstract rules, established ranks, and often arbitrarily delineated chieftainships and their subjects (Chanock 1998; Marks 1973a, 2004b).

As commoners rapidly dispersed from their protective stockades into smaller scattered settlements, officials needed local knowledge of numbers of people and of places to extract labor and taxes. Initial administrative interventions centered on transforming traditional agriculture, building roads, recruiting labor, and stabilizing the minimum numbers of adults required to establish and sustain a village. The imposition of taxes compelled most adult men to seek regional employment. Officials disarmed Africans and restricted their customary livelihoods through new rules about taking game and prohibiting the destruction of what outsiders considered natural vegetation. Famine and new diseases became rampant (Vail 1977). Locals resisted many of these imposed regulations, forcing the administration to modify some of its policies particularly concerned with their relationships with Northern Rhodesians and with their tropical environment (Chipungu 1992; Chanock 1998).

A few other political changes tangentially affected people's lives in the valley. In 1911, the territories of Northeastern and Northwestern Rhodesia were unified as Northern Rhodesia and linked administratively to Southern Rhodesia, until the latter became self-governing in 1923. With apparently few resources, BSAC considered administering Northern Rhodesia as a financial liability and a burden with few viable exploitable resources. Negotiations between the company and the British government placed the territory as a protectorate under the Colonial Office in London, represented by an appointed resident governor. When Northern Rhodesia became an imperial protectorate on April 1, 1924, the BSAC retained its interests in mineral rights and extensive tracts of land (Gann 1969; Wills 1967).

In 1953, the Colonial Office united Southern Rhodesia, dominated by white settlers, administratively with the territories of Northern Rhodesia and Nyasaland into the Federation of Rhodesia and Nyasaland. This federation was bitterly opposed by Africans who feared domination by Southern Rhodesians. African discontent and widespread violence convinced the British to break up the Federation in December 1963 and to grant political independence to Malawi (6 July 1964) and to Zambia (24 October 1964). Southern Rhodesian settlers announced their own unilateral declaration of independence (UDI) in November 1965. African rule was finally established when Zimbabwe gained its independence in 1980. The earlier colonial decision to connect Northern Rhodesia to southern routes became one of the major constraints on Zambia's development during its first sixteen years of independence.

The Valley Bisa Specifically

About 1900, the BSAC established their *boma* atop Ngala Hill close to the village occupied by a Ng'ona named Chisenga Kalonda.[26] In colonial accounts, Chisenga Kalonda was described as "a feeble old woman of considerable character . . . the type that commands respect." When asked her name, the chieftainess replied respectfully "NaBwalya [Na—the mother of—Bwalya, her eldest daughter] Chiombo [her father's name]." Her name (initially heard and recorded by BSAC officers as "Nawalia") became that for both the chiefdom and their constructed *boma*. Initially, BSAC administrators also recognized other small valley Ng'ona chiefs, Saidi, Kazembe, Mutupa, and Kambwili, and provided them with subsidies. Later the BSAC reduced the number or recognized chiefs until only Nabwalya remained west of the Luangwa River to the Muchinga Escarpment, within what later became the Munyamadzi GMA.[27]

The Nawalia *boma* operated for some eight years as the main administrative center in the central Luangwa and served as a residence for European officials (Marks 1973a). Hector Croad, with the title of collector, was its first resident. Locals remember Philip. E. Hall fondly while William P. Kennelly, who allegedly despised Africans and caned local residents, is not. Hall, a BSAC commissioner at Nawalia between 1904 and 1907, described that "the population was so small and the land so large, that generally there seems . . . no idea among the natives of tribal property or sovereignty over definite areas." He depicted "certain centers of tribal influence" with rapidly decreasing effects and authority with increasing distance from these prominent cores.[28]

At an official gathering at Nawalia in August 1908, His Honour BSAC Administrator Wallace announced that the western half of "Nawalia District" was to be transferred immediately to Mpika District (a part of Northern Province) while some Bisa territory east of the Luangwa River (Chief Kambwili) was to become part of Eastern Province. Situated atop the Muchinga Escarpment, Mpika was considered a healthier location for European administrators and better situated to facilitate the developments and commerce expected to follow from the construction of the railroad further south. Wallace mandated that smaller villages must regroup into larger ones to encourage the recruitment of laborers for Southern Rhodesia and urged valley men to find jobs as carriers so they could pay taxes. At this meeting, residents complained that Bemba employees at Mpika were "overbearing" in their duties and that adultery was rampant within villages. Administrators promised to employ only Bisa messengers and police "as far as possible . . . so long as the people behaved themselves and that the *boma* should hear adultery cases only when chiefs failed to settle them."[29]

An African Lakes Store that had supplied Nawalia closed there in 1909 and in 1910. When two Europeans died of sleeping sickness, the administration closed the Luangwa Valley to outsiders for several years. The abandoned station at Nawalia served as a tropical site for early research in trypanosomiasis, but failed as a detention center for sleeping sickness patients. This was the reason given for its abandonment in 1911 (Hall 1910; 1950). The closures of valley routes to outsiders because of repeated sleeping sickness outbreaks and later in the massive movements of troops and materials during the First World War through Mpika confirmed the administrator's "higher-ground" commercial possibilities for the plateau and its connections with the south. This switch in administrative attention insulated the Valley Bisa homeland from the intensive scrutiny ("gaze") of the few Mpika stationed officials. Unlike most administrative tours on the plateau accomplished in short order by motorized transport over a network of roads, each valley journey required weeks of foot travel, descent into a hot and uncertain valley environment, employing and managing dozens of "cranky" recruited carriers with heavy burdens of camping equipment, food, cutlery, chairs and other essential executive materials. These long "foot safaris" reminiscent of a recent past appealed to a few officials, who as sportsmen appreciated its hunting opportunities, but were disdained by most for its associated risks and the time such expeditions required for them away from more compelling office duties.

When Northern Rhodesia became a British Protectorate in 1924, its colonial administration extended and increased the powers of chiefs under Indirect Rule. Chiefs and their headmen became the local eyes, ears, and voices for the district administrators. Chiefs received subsidies, emblems of office, and paid staff (*kapaso*) to monitor behavior and enforce regulations within their demarcated territories. As head of a designated Native Authority, chiefs supervised the collection of taxes and fees, determined nearby hunting rights, reported on activities and health, and sat in judgment of their subjects in local courts. Many chiefs continued to expect the traditional tributes of goods and labor required ritual respect as "owner of the land."

Throughout the 1930s, the global economic recession meant severe regional administrative entrenchments with few officers to administer Indirect Rule as official policy over vast rural areas in which chiefs and Native Authorities mainly governed. In expansive rural districts covering diverse terrains such as in Mpika District, these circumstances left the small, marginal Valley Bisa population within the Luangwa Valley largely disconnected and largely on their own. The comparative isolation of the Valley Bisa homeland of these years was the political-economic envelope under which a newly appointed chief and his allies reconfigured their cultural practices to cope with increasing wildlife populations and their subsistence needs. Their location on the eastern edge of the colony's largest district and along the boundaries of its Northern, Eastern, and Central Provinces encouraged this marginality and insularity. Surrounded on three sides by game reserves (1938) and separated by a steep escarpment and at least six days of walking from their supervisory district officials in Mpika encouraged parochial initiatives. This separation and infrequent oversight created an experimental space for residents to innovate and employ existing social structures and local incentives to overcome problems with neighboring wild animals, which brought diseases, destruction to human lives and properties yet provided potential animal proteins to their diet and a safety net during droughts. Most men participated as laborers within the colonial economy to pay the state's annual levies and to obtain other materials (salt, iron pots, cloth, weapons, hoes) required in villages. The indeterminate absence of most adult men (with up to seventy percent each year) shifted daily village tasks mostly to the elderly, women, and children and this had longterm social and environmental consequences (Vial 1977: 155).

In 1933 the colonial administration, disappointed by the rapid turnover of elderly chiefs at Nabwalya, summoned a young Ng'ona clansman from his work in Southern Rhodesia and, after checking his local lineage affiliation, appointed him as Chief Nabwalya. This individual, Kabuswe Mbuluma, remained as chief until his death on 28 April 1984. In conjunction with local and distant allies, he integrated local management of wildlife into the valley's problematic subsistence agriculture in ways that I identify as "lineage husbandry". I use this informal marker with its grounding in oral, experiential as well as small scale, face to face social and cultural constructions to distinguish it from the more formal written, national regulations that colonial and bureaucratic states attempt to enforce against the cultural grains of more subsistence and provincial groups. Valley Bisa men competed for status and wealth by gathering matrikin and clients within their spheres of influence (villages or settlements). These masculine transactions within its rural system of prestige was a lifetime career and objective, measured in terms of numbers of dependents and an individual's judicious disposition of his accumulated materials, cash and provision of security. While migrant labor remained a main source of new lineage wealth, a few individuals minimized their time as labor migrants to become local hunters or clients under the patronage of a significant elder. This elder controlled the lineage's weapons (primarily muzzle-loading gun(s)) while the younger client(s) provided time and expertise to acquire local bushmeat for lineage consumption and to protect properties and lives. How several generations of these hunters managed their aspirations within village life and on the gaming fields of wild animals is the gist of our developing story. Most but not all local hunters worked within this husbandry of lineage incentives and social rewards that lasted for most of this chief's fifty-one-year tenure. Yet the pervasive "winds of change"—including modernization, individualization, technologies of communication and transport, work experiences elsewhere, resource scarcities, demographic changes, "parochial dreams about a recent past" and centralized management began to blur earlier boundaries during this chief's declining years, culminating in his death in 1984. The undermining of his local regime's "assumed autonomy" accelerated with the completion of the rugged vehicular track down the Muchinga Escarpment in the early 1960s and the granting of political independence to Zambia in 1964.[30]

Before beginning this story, I examine some organizing structures and ideas pertinent to how group lives were shaped on the ground and a few glimpses into the daily lives of representative villagers (chapter two). Then I consider how Valley Bisa marginality may have worked initially to their advantage, yet eventually became disadvantageous for most residents. As wildlife developed into a significant global and tourist resource, supported by national and international interests and expertise, promoting claims of "sustainable conservation" and improvements for local welfare, national policies became structured in subtle ways that were detrimental both for wildlife conservation and against the "cultural grains" and benefits of the majority of these GMA residents (chapter three) at least in the short run.

Notes

1. *Mulumendo mukulu, ninoko.*
2. Muchinga Province was created in 2011 by Zambia's president Michael Sata and includes several districts formerly in Northern Province that straddle the Muchinga Escarpment rim.
3. I am aware of the complicated etymology and academic uses for "landscape." Some of these differences are explained and referenced by William Wolmer (2007). I use this term in his nuanced sense.
4. Marks (2014: 115–44) describes how this occurred.
5. Similar dispositions of structures and welfare from allocations of safari hunting revenues were noticeable throughout many of Zambia's GMAs.
6. What these census figures represent from the standpoint of residents' movements and their census takers' methods are discussed in Marks (1976: 14–18; 2014: 34–40)
7. For more detailed information on numbers of residents, settlements, impacts on habitats, and wildlife at various times see Marks (1976, 2014).
8. Marks (2014) provides additional details accounting for these population shifts within the Munyamadzi GMA (pp. 34–47), on the Nabwalya Study Area (pp. 72–102), and transitions in state management regulations for wildlife (pp. 18–33).
9. These remarks were submitted by a seventh-grade student at Nabwalya Primary School in response to my sponsorship of a writing assignment on local history in 1989.

10. See Marks (1976: 149–66) for earlier descriptions of these plant communities including their associated species and local human uses. Figure 13 (Marks 1978: 153) shows the disposition of habitat types with reference to villages and cultivated sites on the Nabwalya Study Area in 1966.
11. The processes by which a chief's spirit turns into a lion medium (*kupakishya mfumu*) occurs upon his burial. A Bisa counselor informed me that the chief's body was wrapped in a lion or zebra skin with ten male lion claws placed on his fingers and ten claws from a lioness on his toes. These claws are the activating particles (*cisimba, visimba* [plural]) that transformed the chief into a lion form. Marks (1984: 1–2) describes the three types of lions recognized by the Valley Bisa.
12. This composition was from another paper submitted through the local primary school as an assignment I sponsored in 1989.
13. *Ciaalo ciakanga*: literally "the land has refused, withdrawn" is the local expression. This incidence of rejection of the game ranger's "disrespect" in not seeking permission to hunt from the chief, as traditional "owner of the land," was related to the government's discussions then about eventually removing the Valley Bisa from their valley homeland.
14. These blessings were "*Tupapatile mipasi yakale. Baana benu munani bakofwaya, balunge mutende baana benu mumpanga yenu*"and "*Twaiza mukubomba, mwitusya pambilibili, fwe bana benu, tulunge mutende mu ciaalo cienu*". The chief indicated the site for our bush encampment, gave the prescription, and said the words to keep any malicious animals and bad spirits away as we slept. Interivew with Chief Nabwalya CN, SAM Field Note 17 May 1967 (2).
15. The name Chibinda, a term of respect for hunters as well as for skilled craftsmen and traders, has a long and noble history in Central Africa. Leaders of itinerant traders and the hippo and elephants guilds in the nineteenth century also used this term.
16. *Samfwe*: a small, edible mushroom that germinates and rots typically within one night. It is difficult to remove the dirt in which these mushrooms grow.
17. These recently (re-constructed) and state supported ceremonial events and their stories may justify current arrangements rather than necessarily providing knowledge of past movements and eventual settlements. Territory, rank, and seniority were issues when the British colonial authorities sought to establish a hierarchy of chiefs among the Bisa similar to that for the Bemba.

18. Conquerors and conquered defended their respective roles through the telling of such stories. Each embedded his/her identities within these oral narratives. I collected several accounts of Chongo and Kabuswe Yombwe. These two stories contain the most details of Chongo's overthrow and of the reason why Kabuswe Yombwe was killed. Both were written by the sons of Ng'ona parents in response to a competition that I sponsored through the local primary school in 1989. This competition was open to all students and challenged them to engage their parents in questions about Bisa landscapes and its history.
19. Interviews with Chief Nabwalya, 29 October, 17 December 1966, and 17 May 1967 deal with genealogy of the Nabwalya chiefdom beginning with Kazembe Mukulu. Interview with an elderly Ng'ona resident (C.C.), 27 April 1967 confirmed that Silambwe was the first Ng'ona to confront Chongo; as did an elderly Ng'ona woman (C.), 5 May 1967. Additional information on Valley Bisa valley settlements and relationships among the Ng'ona were sought through informal interviews with members of Kazembe's Ng'ona lineage in August 2006.
20. The Bisa term *muzya* covers a wide range of non-kin dependents whose position varied in terms of their status deprivations, making the translation of this term into English as "slave," together with its related term *buzya* ("slavery") misleading. The term applies to individuals given in local compensation, those treated as quasi-kin, and those sold to traders.
21. If a pawn or slave married another of equal standing, their children retained the same status. Yet if woman slave married a member of her owner's descent group, then her children belonged to the owner's group. Unions between female *bazya* and male clan members contradicted the normative rules of matrilineages by permitting men to claim their biological offspring as belonging to their own descent groups. Often young women were more valuable as pawns in slave transactions.
22. Carl Wiese (1983) discusses the activities of elephant hunters, including the ivory trades of the Chikunda and Bisa in Eastern Province (Zambia) during the tumultuous years 1888–91. I found that not much of their specialized lore remained among the Valley Bisa, as the elder generation of local hunters I interviewed could recall little. (Marks interviews with WD, 17 August 1973, and with Chief Nabwalya, 18 August 1978). Marks (1976: chapter Five) discusses elephant guilds and the ritual treatment of an elephant killed in 1967 (Chapter Seven).

23. *Ciaalo cikokalipa,* translated as "the land is becoming hot," is a ciBisa idiom for expressing dangerous and uncertain times, such as political chaos or when the rains fail.
24. Residents in Bisa chieftaincies within and along the margins of Lake Bangweulu subsist on staples of cassava (*Manihot* sp.), maize, and smaller amounts of millet (*Eleusine coracana*). Fish and formerly abundant swamp-dwelling mammals (black lechwe, sitatunga, otters) were the mainstays of survival and commercial exploitation as the Copperbelt and provincial centers developed. Within the woodland on the plateau, slash-and-burn cultivation (*citemene*) of millet, maize, and cassava is the dominant practice. A major surfaced highway and railroad was constructed through the center of Mpika District making the town a major mechanical workshop for the plateau. Beyond agriculture and development projects, cash incomes are largely from urban employment. In the Luangwa Valley, varieties of sorghum (*Sorghum vulare*) and maize (*Zea mays*) are the main staples grown on alluvial soils adjacent to the streams and rivers. Millet and cassava are grown in small amounts, as are sunflowers for oil and eventually some foreign legumes. The presence of the tsetse fly prevents residents from keeping livestock in the Luangwa Valley. Yet large and small wild mammals are common and largely immune to these parasites.
25. Zambian National Archives, file KSM 2/2/1–4 slaves, dated 14 July 1910.
26. The establishment date for Nawalia (Nabwalya) *boma* is uncertain. The Lundazi District Notebook (Zambia National Archive, file KST 3/1) mentions that Hector Croad was the first collector and lists his residency as "1898 or 9 to 1900." The Mpika District Notebook (Zambia National Archives, KSD 4/1) lists official residence beginning with E. Methuen Savage in 1901. These notebooks were kept in each district and contain specific (taxes, ivory procurements, population within the district, dates of signifcant visitors and events) and general (comments about staff, places and status where elephants and other wildlife could be found, weather, trade, summaries of meetings) that replacement officers needed to know.
27. Kazembe's district originally contained the divisions of Kazembe, Ndombo, Mumpempa, Chitala, Mpelembe, and Nawalia, numbered forty-six villages, and covered about 2,550 square miles. Saidi's district, northeast of Kazembe's near the Luangwa River, was a smaller division of 520 square miles and seven villages. In 1940, with much of the land declared as the North Luangwa Game Reserve and the chieftaincy reduced to two or three villages, Saidi's villagers moved

across the Luangwa River into Lundazi District, although the Ng'ona clan retained an affiliation with Nabwalya. Mutupa's district of six villages was located northwest of Kazembe's in the hilly ground of the escarpment. Portions of this district may have been absorbed into that of Chief Mukungule (1950?). Zambia National Archives, Mpika District Notebook, vol. 2, 4/1.

28. Comments attributed to P. E. Hall in the Lundazi District Notebook. This note suggests that in the distances between chief's villages (centers of tribal authority), there was plenty of frontier space in which individuals and lineages could make decisions about their affinities. Hall was formerly stationed at Nawalia and transferred upon its closure across the Luangwa River to Lundazi *boma* in Eastern Province. If he took the original Nawalia Notebook with him to Lundazi rather than transfer it to Mpika, this action may account for the discrepancy between the accounts mentioned above (n26) about the dates on Nawalia.
29. Lundazi District Notebook, Zambian National Archives KST 3/1 under *Indabas* page 41.
30. Leroy Vial (1977) reviews archival materials and describes the deterioration in both social welfare and environments caused by colonial policies in the Luangwa Valley.

Chapter 2

Creating and Sustaining a Good Life within a Difficult Environment

Be proud of Nabwalya, you are Bisa people
You can not be proud of other people's land
Just look how Mr. Somo suffered and was in difficulties
Many people were arrested and some died
Now Mr. Somo, you will be the chief
Be slow to anger and always remember that
A child is like an axe
It can cut you or you can place it on your shoulder[1]
Bisa song (recorded 1988)

In important ways, the quest for understanding small-scale societies in Africa, such as the Valley Bisa, begins with dismissing much of what we think we know about them. The standardized categories and knowledge by which outsiders have typically described such rural societies, a legacy of a colonial past, assumes a sameness and definiteness that dissipates under closer scrutiny. The continued use of pejorative concepts, such as tribe, traditions, magic, and "culture," embedded in colonial assumptions of homogeneity and difference, detract from seeing the vitality in others' lives, from noticing more than just one kind of African or Africa, and from understanding the nature of changes and possibilities.[2] A tribal lens never captures the diversity and meanings of place and time even within a single Bisa village. Whereas most Africans may be in sync with the tempo of their place and time, our more distant vocabularies and concepts have not kept pace with local or regional developments and circumstances.

Colonials assumed that the basic unit of African society was the "tribe," which they self-servingly constructed, reinforced, and combined with race and culture to explain why Africans were different, acted in the ways they did, and needed guidance. The concept of tribe enabled the fiction of indirect rule for discrete and corresponding political units (albeit inferior), each with authenticated customs (culture) and binding membership codes (social organization). For some anthropologists working within this colonial frame, tribes were a distinct evolutionary type of social organization somewhat akin to feudal Europe with an egalitarian economy that employed simple tools to produce largely subsistence goods. The identities within this level of organization revealed their customs, whose mutually supportive elements of ideas and behaviors (expressed within the theoretical frame of structural functionalism) ensured that changes were difficult, maybe even traumatic. Moreover, the "ethnographic presence" in anthropological writings presumed a constancy that diverted attention from change and the real power, perks, and impacts of the colonial state. Yet the colonial codifications of "customary laws" and "traditions" were not solely external creations, for their applications in supervised courts created spaces in which some African collaborators responded knowingly against the intentions of these codes, if not the well-being, of their neighbors in settling personal grudges (Chanock 1998).[3]

These common-sense notions of tribe and customs, along with the current commonplace usage of "community" to describe rural lifestyles, endure in popular articles and discussions about Africa and development. This deployment of a term to depict the lives of others focuses our attention in certain ways; it may prevent more appropriate vocabularies for the poignant details of different conditions, and it may also silence other significant processes in others' lives. Vocabularies, like maps, take particular forms and are judged ultimately on their utilities. The scholarly dialogues between the representations (descriptive vocabularies) used and the social realities observed must be continuous, particularly if they are to become observers' charts depicting the inevitable eddies troubling the lives of others intermittently known.[4] This chapter begins a search for appropriate nouns and verbs, as well as a narrative, to show that while their lives may seem restrictive in some ways, the Valley Bisa have remained creatively responsive in other practices to the contingencies they face every day.

The opening epigraph summarizes the recent political scuffle over the Nabwalya chieftaincy and ends with a metaphorical warning about children, experiences, and time. These relationships and arenas are where the main personal struggles and transformations take place, the real compositions of life stories and legacies. In portraying the gender, status, and relational roles in this chapter, I describe ideal expectations, the ways they are usually discussed as representations in ordinary conversations, rather than depicting their fuzziness in real time or circumstances. Age and time change perceptions as well as goals. Normative depictions relate generally to the earlier periods of my studies until subsequent visits and rapport allowed me to focus on individual histories within the context of lineage ideals. As outside connections proliferated through government, foreign, and local initiatives, a few well-situated individuals focused on accumulating personal wealth as a means to accrue clients and influence, while incessant droughts undermined the livelihoods of many others. Over decades, prolonged observations and persistent forces slowly revealed alterations in roles, gender relations, and stresses between young and old. Personal strife surfaced in accusations of witchcraft, resentments became spatially apparent in the placement of scattered settlements and households, while wholesale conversions to global charismatic religions expressed different exposures and uncertainties within an expanding world. Some variations of norms and expectations appear in the sequential glimpses of individual daily profiles in this chapter, while the life histories of individuals in subsequent chapters reveal the longer processes of incremental changes.

The Importance of Extended Relationships: Power, Knowledge, Agency

The sequential migrations of clans from the north (now the Congo) centuries ago and their distinctive histories shaped the language and identities of those currently inhabiting the west-central Luangwa Valley. The largest political unit during this period was probably the clan, a group of kin and alliances. Although chiefs' clans emerged by claiming territory, special rituals, and perhaps trading and intermediary roles, their alliances with other social units remained in flux during these turbulent times. In this section, I discuss some of the means and processes by which chiefs, colonials, and clans (lineages) have managed their members and alliances.

The Bisa expression *mutundu* is sometimes translated as tribe, yet there is no English equivalent. More appropriately, the term specifies subdivisions such as species, group, kind, or clan. When used to describe people, *mutundu* is not associated with the same fixity (name) and place (territory) that is associated with the colonial use of "tribe." The dynamics and constituents of this indigenous political unit are conceptually open-ended and unstable rather than permanent.[5] In both oral and written accounts, the malleable complexities of shifting political and historical processes were twisted, lost, forgotten, or suppressed or selectively worked by different parties. Consequently, deciphering the "truth" in any accounting is always a matter of cautious interpretation and imagination.

The Chieftaincy

In chapter 1, the oral history of a contentious encounter between Chongo, as the earlier settler, and his Ng'ona contender, as a late arriver, suggests one way in which land rights might have been settled between a late presuming dominant representative and earlier settlers in earlier epochs. Given that Chongo and his wife committed suicide rather than opt to live under new rules, others in Chongo's group had similar choices: they could follow Chongo's example, move elsewhere, or accept the intruder's "cleverness" by negotiating "to stay" (*kwikala*) with him. Apparently, some clients negotiated to stay, as their descendants continue to pay "respects" to Chongo's spiritual powers and presence. At the core of these unstable relationships on frontiers is the ability of some individuals to assemble followers or achieve ascendancy under the umbrella of common (lineage) goals.

In the process of establishing an administrative center near this same hot spring around 1900, BSAC officers reported finding a Ng'ona chief, "a feeble old woman of considerable character . . . the type that commands respect." Administrators built their outpost (*boma*) near her settlement in a commanding position atop adjacent Ngala Hill. For a while, these commissioners also recognized four other neighboring Ng'ona chiefs, including one who bore the name of Kazembe, who claimed superiority over Nawalia. Whereas these officials initially named their created district for Kazembe, they discredited his claim, describing him "as an intelligent man of considerable ability . . . but not, I think, over straight" (Mpika District Notebooks; Marks 1973a).

The emergence of a singular chiefdom on the west bank of the Luangwa River was the result of BSAC administrators asserting their political power to consolidate their scattered "subjects" and to control their fiscal exigencies. Rationalized initially in terms of economic stringency, BSAC decisions to reduce the numbers of chiefs, to legitimize claimants, and to define territory were upheld later by the Zambian government, thereby assuring the ascendancy of the Nabwalya descendants within the confines of this territory. However, the demarcation of the Luangwa Valley Game Reserves (1938) and National Parks (1972), together with the state's "ultimate powers over people" and local resources, have subsequently and progressively circumscribed the extent and powers of this reigning Ng'ona chief. Recent reminders of the chief's subordination to the state include the long interregnum after this chief's death, the contentious battles between contending claimants (1984–88), and the loss of local entitlements in natural resources, particularly in wildlife and in choices of settlement sites. Yet in other ways, recent alliances with the state have broadened and standardized the chief's powers over other clans and individuals within his sphere of influence. A chief serves as the main representative for government services and subsidies in rural areas. Whereas chiefly control appears more formal than

FIGURE 2.1. "The cleverness of chiefs." Chief Nabwalya [Kabuswe Mbuluma] sitting on a mat outside of one of his wives' huts in Pelembe Village. Surrounded by four of his wives, some of his children, and two counselors, he entertains them with stories, gesturing with his hands as he speaks. The men are drinking sorghum beer. Note the beer pot and drinking straw within a white dish between the two men. (Photograph by author in 1973)

in the past, individual commoners and groups still retain their rights to move and to affiliate wherever chiefs and headmen welcome them.

"Becoming, being, and staying as Bisa" is a historically constituted political choice made consciously by individuals and groups. The powers by which chiefs and clans "gather people" are unstable, contingent, and relational. In practice these processes depend on the continuing recruitment of others to their respective *mitundu* [groups, singular is *mutundu*]. Whether new recruits are outsiders or are part of the clan by birth, they are welcome as long as they demonstrate their loyalty to the reigning chief. Similarly, chiefs make alliances and are thereby incorporated into larger political realms that extend their practical knowledge and strengthen their "stays" as "owners of the land." Both Ng'ona chiefs I have known have used their "powers to gather people" in different ways.

FIGURE 2.2. "The cleverness of chiefs." Chief Nabwalya [Blackson Somo] during a week's walking tour of his chiefdom, during which he lectured and answered questions about development and wildlife conservation. Here he talks to an assembly composed mostly of men in White's Village. He was appointed chief in 1990. (Photograph by author in 1989)

Confirmed in 1933, Chief Nabwalya [Kabuswe Mbuluma] was faced with consolidating his mandated chiefdom after the state withdrew its recognition from the small Ng'ona chieftaincies in the valley. As a practicing polygamist, he maintained households in each section of his territory (figure 2.1). His client hunters, together with the state's elephant control guards and later foreign safari hunters, kept his wives' households well supplied with bushmeat. He championed local rights to wildlife and defended some subjects' claims rather than those advanced by the state's wildlife department. In July 1967, perceptive valley orators meeting with Bisa senior chief Kopa unanimously refused to consider resettlement (under a Zambian state proposal) because their "ancestors had found this peaceful and productive land and buried their umbilical cords here," and that they were "not prepared to give up our elephants and buffalo . . . and the thick gravies (*muuto*) they provide" (Marks and Marks).[6]

Upon Chief Kabuswe Mbuluma's death in 1984, the surviving contenders confronted a determined attempt by another Ng'ona segment (heir to the Kazembe line) to secure the chief's title. This bitter confrontation lasted several years, and included an appeal to the Zambian Supreme Court before the state appointed (1988) and confirmed (1990) Blackson Somo, a descendent of the Nabwalya "womb" as Chief Nabwalya. As a consequence of conflicts with his predecessor, the chief spent most of his adult life as a businessman in Kitwe before he was appointed as a chief (figure 2,2). With his incumbency secure, he became the chairman of the Munyamadzi Wildlife Sub-Authority, an important and lucrative GMA that produced large government revenues from safari hunting concessions. Under a new donor-pushed "community-based" wildlife program, a portion of revenues from access and hunting licenses were allocated for GMA wildlife management and community improvements.

The chief's new roles, enforcing wildlife laws and pushing wildlife department-sanctioned developments, have caused tensions with many of his subjects. As a wildlife protector and conservationist, the chief, along with his advisors, oversees the disbursement of cash and employment from the "community funds" under the ADMADE program. Unlike the previous chief, whose patronage depended upon dispensations of protection and provenance (local meat), championing local causes and rights in wildlife, and his subjects obtaining employment elsewhere, the present chief's patronage relies mainly upon cash distributions, appointments, and local employments. His largesse depends upon the disposition of the state's wildlife agency and upon the ostensible restriction of his subjects' access to wild animals. The chief sanctions outside hunters legally pursuing game within the GMA, maintains links with influential friends elsewhere, and disciplines foreigners and his subjects suspected of disrespectful behavior. As a monogamist and a Roman Catholic, he is an active participant in the local parish. Yet neither his power nor affective reach appears as extensive among his subjects as that of his predecessor.

The recitation of myths and stories, particularly the founding rituals of the polity, are central to a chief's practice of "gathering people" and of his subjects remaining Bisa. In 1984, Zambia reversed its initial policies reducing chiefs' powers and created a place for annual "traditional" ceremonies wherein histories, storytelling, dancing, and feasting play critical public roles.[7] Organized by the reigning Valley Bisa chief and attended by national, provincial, and district dignitaries, as well as by other regional chiefs and residents, this local ceremony is known as *Malaila*. On 11 September 1999, the reigning chief addressed his *Malaila*:

> I take this opportunity to welcome you to our chiefdom and especially to the *Malaila* ceremony . . . *Malaila* means victory and is an important annual event for the Bisa people in the Luangwa Valley. This traditional ceremony plays an important role in reminding us of the hardships that we have gone through and of our triumphs over them. These hardships include droughts, attacks by wild animals, and sicknesses. During tragic periods, our people worshiped the ancestral spirits, like Kabuswe Yombwe, to assist them through their problems. It was after successfully overcoming the problem, of killing a menacing wild animal such as an attacking lion and showing off its head or upon receiving rains and a good harvest, that people gathered to celebrate their victories.
>
> These highlights of our *Malaila* do not in any way contradict with the *Chinama Nongo*[8] ceremony for all the Bisa living in Zambia. The primary objective of *Chinama Nongo* celebrates the time when the Bisa came from the Congo to Zambia. It is my strong belief that people gathered here will see the difference between the two and take *Malaila* as a unifying factor for the Bisa in the Luangwa Valley and *Chinama Nongo* as a complementary ceremony for all the Bisa in Zambia.
>
> You will agree with me that Nabwalya area is the most underdeveloped in Mpika District with a bad road network, no health facilities, and without agricultural inputs like seeds. Some schools could not open this year because we lacked certified teachers, who shun the area because of its underdevelopment.

The chief thanked the government for, among other things, sending "untrained" teachers, providing aid, placing two grinding machines (reminding "at present none are working"), purchasing two vehicles, and constructing classrooms, staff houses, and the palace. He pleaded for upgrading the valley's network of roads and cautioned that if the district did not "find a formula to our problems . . . we have no option but to request a sub-boma." He concluded by asking the government representatives, "Is it cheaper to give us relief food each year or to provide us with good roads, better health facilities, and agricultural inputs?" For the rest of the festivities, he welcomed, hosted, and feted official guests, championed local needs for development, and substantiated local history, identities, and the legitimacy of his authority.[9]

Bisa ways of building power and and acquiring knowledge are negotiated and relational, initiated locally and extended outward to regional and national (even international) political spheres. Developing power and *mitundu* are processes that last a lifetime and may be accumulative. In similar fashion, local knowledge (knowing) is active, searching, pragmatic, personal, and usually vested with interests. Furthermore, much of the authoritative arrangements between chiefs and government are also secret, guarded, and closed to public commentary or scrutiny. Since much of these agreements between powerful chiefs and other agents remains undisclosed, those active within the more encompassing worlds of power and knowledge demand deference from those of lesser status.

The founders of the Valley Bisa polity emerged from the complex political and trade networks of the nineteenth century. This period of upheaval and mixture of peoples, regionally and even internationally, is reflected in their language, which incorporates components of Bemba, Nyanja, and Chewa speech. Chiefs still welcome other people from dissimilar *mitundu* (plural of *mutundu*), speak different languages, and offer diverse skills, contacts, and even resources enhancing their leverage "to stay" as the "owner of the land." They solicit individuals for advice, turning some into advisors who might requisition resources from elsewhere.

The authority of chiefs ultimately depends on the continuous acknowledgment of groups who reside and work on Bisa land as well as the polities of the Zambian state. Individuals and villages exist as Bisa because of an alliance struck with a Ng'ona chief. The continuation of these alliances depends daily on the "respectful" behavior and voice by which residents affirm the practicalities of their daily lives and defer to the greater power and knowledge upon which their survival ultimately depends. Relatives and neighbors carefully monitor each other's behavior and attitude; they may report aberrations in the chief's court.

Commoners show their "respect for" (*mucinzi*) and loyalty to the reigning chief by using appropriate terms and by showing deference in his presence, properly conducting themselves and their conversations at all times, actively participating in rituals, and providing their labor when summoned to the palace. Comparable performances and attitudes of "staying well" (*kwikale bwino*) are expected through the daily demonstrations of "respectful" relationships among kin and neighbors, among young and old. These mundane performances can be examined through vignettes of representatives' activities.

Lineage and Clan

I consider the concepts of "clan" and "lineage" as the same level of social organization, with the main differentiator between them being dispersion over space and time. Lineage is more immediate, its adherents known by at least some of its current members, most of whom generally reside close by. Clans are less concrete and may be fictive as well as widespread and patchy in distribution.

A village headman is the "owner of a village" (*mwine muzi*) and is responsible to the chief regarding its management. Most headmen are selected from the available elders of the matrilineal group around which a village acquires its identity. Each successive headman assumes the name of the village's original founder and is answerable to its senior residents, including women who have produced children, and to the chief. A village prospers by producing children and sufficient food and by having access to water and other essential supplies for a healthy life.

Village life is never the same over time. Alliances shift under each headman, with marriages, with births and deaths, and as a result of conflicts and the daily comings, goings, and traumas that constitutes villagers' lives. The village environment also changes as economies fluctuate, ecologies alter, rainfall varies, and people interact, select, and modify certain components. Village needs for labor shift as its demographics change. Patrons are assumed "clever" (*cenjela*) as their active knowing enables them to keep pace with the contingencies of the surrounding ecology and politics. Generally, people gathered within a village are the descendants of those originally composing it, their children and spouses, as well as other individuals or minor lineages making alliances with its headman.

At the village level, a key to "staying well" occurs through keeping respectful relationships with others. A person demonstrates respect by accomplishing daily tasks assigned on the basis of gender and age, voicing regularized modes of greetings and inquiries about the health and the well-being of others, showing proper conduct to elders and peers, and mentoring dependents. "Staying with respect" means following diligently the activities and dictates of one's forbears and by focusing primarily on enhancing lineage interests. The prominence given to elders and ancestors ensured that some historical strategies are prominent in spirit and in practice even as their utilities erode within the crucibles of changing circumstances (figure 2.3).

Everyday and seemingly mundane activities are likewise a crucial part of remaining Bisa. Daily activities reflect many practices that both sustain and "gather people," for the critical cultural code in agency, power, and knowledge is "respect giving." The local idiom *goot milile* ("good life") describes people, who sustain "respectful" relationships, share news while "keeping secrets," which affirm personal and group boundaries.

FIGURE 2.3. "Showing respect." A village headman (center, white cap) turns to receive snuff "respectfully" delivered by his wife and daughter as a political party official (left with pen and pad) records his donation and comments. The other women and children keep their "respectful" distance by remaining in the background in the shade of another granary. (Photograph by author in 1978)

With "Respect" to Women's Roles

Women live respectfully by following the ways of their mothers, properly greeting others, showing appropriate conduct inside and outside the village, fulfilling their routine tasks of collecting and preparing food, gathering firewood, producing and taking care of children, brewing beer, washing clothes and utensils, and preparing, planting, weeding, harvesting, and storing crops, among a host of other mundane activities in the vicinity of their households. Young girls follow their mothers and learn their roles early by assisting with daily routines.

The labors of young girls and women are the foundation of a lineage's ability to "stay well"; at an early age, young girls are instructed, supervised, and relied upon for household routines. As they mature, they learn appropriate adult behavior and how to keep a husband through their puberty ceremonies (*cizungu*).[10] Few young girls are encouraged to attend and finish their primary education. Parents often offer their prepubescent daughters to older men to marry.

The recursive routines of two Valley Bisa women can be glimpsed through a selection of notes taken by a local observer during the agricultural season beginning in November 1988. Elder women, like Doris Kanyunga, then sixty years old, manage the activities of their nearby daughters and granddaughters, monitor happenings within village life, set standards, and mold opinions during the course of the day. The constant visiting, gossip, and sharing are essential features of village and household lives. Doris led a busy, grinding schedule as suggested by the snapshot of her in figure 2.4 and in the recorded vignettes of her activities and relationships during 1988–89.

Each of the following glimpses of village activities and interactions below are timed occasions that selected representive individuals were monitored and recorded by a research assistant during one day of each week during the 1988–89 agricultural season (November 1988–July 1989, see preface for methods). The individual's activities and interactions with other persons are briefly abstracted from these records to disclose the varied nature and content of villagers' daily lives during these months.

FIGURE 2.4. Doris Kanyunga grinding sorghum grains into flour between two stones prior to mixing the flour into boiling water to make porridge (*nsima*). Many such iterative tasks of food and beer preparation are the customary roles of women. (Photograph by author in March 1989)

29 November 1988: After hoeing [265 min] in her garden this morning, Doris takes a brief rest [at 1042 hrs] and talks about the striking appearance of hippos and makes fun of those who eat its flesh. She doesn't eat hippo, yet both her husband and granddaughter do. She instructs her daughter to bring her a *citenge* (cloth) from the house. She spends the rest of that day weeding her field [290 min], resting, listening, and entertaining guests [237 min], drawing water, washing plates, keeping household [120 min], preparing snuff [47 min], as well as finding and preparing a relish for food, eating [42 min] and child supervision [16 min].

6 December 1988: She greets [at 0659 hours] Ronald, a neighbor who has just returned from Mpika after several days of carrying another's luggage. As she continually supervises the activities of younger members around her, she inquires about how Ronald left his mother (in good health), and what he was paid [Zambian Kwacha ZK60] for this trip. Doris exclaims that inflationary prices in Zambia are "too high" and that "anyone who brings a dress or *citenge* from town really loves you." She asks about Ronald's father and learns that he has gone to Mpika to carry the hippo meat killed by a game guard last week. She states that the daughter of Mumbwe, who has come to grind sorghum grains, is a "very hard worker." Doris spends the rest of the day weeding her field [508 min], resting and child tending [102 min], borrowing implements from a neighbor [53 min], acquiring relish, firewood, eating [71 min], drawing water and doing domestic chores [36 min]. While back at her homestead again [at 1745 hours], she describes Mrs. Mupango as "stingy and selfish" for giving her so little maize when her husband [Doris's relative] had instructed his wife to provide more. Doris is concerned about her son, who is investigated by the game guards on a report that he had killed a buffalo. She alleges that her son only accompanied the hunter who killed a licensed buffalo for a teacher. She suspects a cousin of spreading the rumor and of having a "bad heart."

15 December 1988: Doris spends much of the morning in her field [361 min] and returns to her household after 1100 hrs. She checks the amount of grain in her bin and instructs her daughter to bring some sorghum. As Doris prepares a meal (154 min), she talks to a neighbor and says not to punish her [the neighbor's] daughter severely for losing a pot at the river, where she had gone to wash it. Doris also instructs her own son to invoke the ancestral spirits before going to hunt in the bush. She grinds snuff [41 min], visits and borrows something from her neighbor [60 min], eats with her daughter [13 min], bathes [15 min], and draws water and washes dishes [34 min]. As she rests and supervises her granddaughters [56 min], she talks about Mr. Peterson who has expressed a wish to marry her granddaughter. She refuses as Mr. Peterson is too old to marry her and the girl is a "classificatory daughter" [unmarriable relationship] to him. She comments, "Girls are rushing after men with money, seeking business and wealth, rather than marriage." She proceeds to attend a funeral nearby [42 min] and returns to additional household tasks.

20 February 1989: Doris greets her daughter-in-law [at 0613 hours] while loaning a pot to a neighbor. She speaks words of condolences about the death of Mrs. Mumbwe, expresses pity that her relatives and husband could not afford cloth to wrap the corpse for burial. Doris provides the cloth as she is concerned that people are talking nonsense about her contributions. Her sister-in-law comes by to collect her for weeding in a neighbor's field, to be reciprocated later in beer drinking. She spends over three hours at that task in the morning . . .

28 February 1989: Sick with malaria, Doris is suffering from body pains, vomiting and diarrhea. Throughout the day, a continuous stream of neighbors and relatives come to wish her well and in the afternoon, men gather to drink the beer she prepared prior to her sickness. While drinking these men discuss events and relationships. Among those wishing her well (at different times throughout the day) are a neighbor and her daughter-in-law, her sister-in-law, her brother (who brings local medicine for her sickness), her son-in-law (brings snuff), a game guard on his way to the chief's court (about his divorce), another sister-in-law, another daughter-in-law (who draws water and cooks *nsima* [porridge] with buffalo meat for her), her husband (who brings medicine from the clinic for her body pains), her older brother, the clinic officer, and the chief's wife.

A younger woman, twenty-three-year-old Rose Chiombo, has no formal education, and came from a small village some ten kilometers away to live in her husband's small settlement near the government school She married her husband (a cross-cousin) soon after his return from town. They spent almost a year in her village before her elders allowed her to join him in his settlement. Unlike Doris, with daughters and granddaughters, Rose has three small children and puts in long days as she builds relationships with other women in her new neighborhood. She is still learning the boundaries of their claims on nearby bush resources, such as firewood and wild plants for food. The glimpse of Rose, self-consciously cooking relish (figure 2.5), and day summaries of her other activities and thoughts evoke a consciousness of her world.

FIGURE 2.5. Rose Chiombo stirring a relish dish of bushmeat and vegetables, which she has prepared with porridge for her husband and small children. (Photograph by author December 1988)

5 December 1988: After chores around her household premises [39 min] and laboring in her garden [252 min], Rose returns home [at 1014 hours] to find some meat which her husband has brought in from the bush. She instructs her eldest daughter to gather live coals from a neighbor's fire as she takes the wrapped meat inside the house. She then expends the next 5 hours retrieving firewood from a hiding place [14 min], gathering relish [wild greens] and preparing, cooking, and eating a meal [317 min]. While pounding sorghum grains in the afternoon, she greets Mr. Chilunta, who is on his way to the chief's palace carrying a chicken. As he is from her home village, she suspects a funeral. A game guard's wife informs her that the blemish spots on her daughter's face come from eating bushbuck flesh. As she finishes grinding grains on her stone, she chats with three girls who have come to visit her children. She learns from her sister-in-law that a crocodile killed a young chap from her village, confirming her earlier suspicion that Chilunta's journey concerned a death. She responds to her husband's directions, proceeds to the border of her field, and returns carrying a large basket of meat. She serves the husband his meal of *nsima* and bushmeat while informing him about the funeral. After attending her husband, she proceeds to the river to bathe, to wash plates, to draw water [101 min] and to breast-feed her youngest child [51 min]. It is a long and full day.

22 December 1988: Rose spends most of the morning and early afternoon laboring in her garden [214 min], washing clothes and dishes at the river [179 min]. When she returns home, she sweeps the premises [28 min], prepares a meal and eats with her children [227 min] while taking care of her sick child [111 min]. Between 1407 and 1435 hours, as she prepares *mutelele* (a wild leafy vegetable), she says her husband hates it for he prefers *kandolo* (sweet potato—his code for bushmeat). As she breast-feeds her child, she talks about the brother of an elderly neighbor who has proposed to a young girl.

19 January 1989: Rose spends the morning sweeping her yard [26 min], working in her garden [202 min] and proceeds to the river to wash and draw water [32 min]. Returning home in early afternoon, she prepares food [403 min] and eats with her children [15 min]. While care-taking her children, she chats with a visiting neighbor, who is using her grinding stone [52 min]. With this person, she discusses her first grader's assignments, their husbands' refusal to help when wives are sick, their husbands' fondness for beating them, the lack of trust among neighbors, and recent divorces. She fetches firewood from some distance and complains that some neighbors are stealing her firewood forcing her to keep her wood in the house. "Men, who marry several wives," she says, "find it tough economically. Getting married to one of those puts a person in a tough situation."

24 February 1989: Rose spends most of this day at home preparing food [405 min], comparatively little time at the river and in her fields [129 min]. While accomplishing her chores, she muses about her husband's wish to marry a wife whose husband is dying. The wife's parents are against the idea, but Rose surmises that somebody must be encouraging her husband to marry this woman. When a neighbor drops by, they talk about husbands who deceive their wives by having affairs. Rose declares, "If I find one of these ladies with my husband, I will slash them."

3 March 1989: As she clears the weeds from her yard [34 min], a neighbor informs that the church bells last evening summoned the church choir for Mr. Frank's funeral. Rose expresses sympathy for the mourners, yet muses that "his death came from sleeping with careless women." As she arrives in her millet field [at 0645 hrs] she is still talking about the death. "Mr. Frank has been ill for so long for his sickness began when he divorced his second wife. She was responsible for transmitting the disease to him. If my husband gets married to this same woman, then he is putting himself into problems." She cautions her oldest daughter not to use the new path to the school as two buffalo have been spotted in a near-by bushy patch. After weeding in her field [220 min], Rose goes to the river [129 min], visits others [131 min], prepares food including collecting firewood [112 min] and family related activities [74 min].

These timed observations show some of the differences and similarities in the routine concerns of women. Women's status and circumstances may change abruptly depending on their relationships, their health, and the well-being of those around them, and whether there is a drought, flood, or extensive animal damage in their fields or to food bins. Women expect to bear children who will help them with daily tasks and assist as they age, lead good lives by valuing other people (particularly their husbands), and work hard to secure the welfare and prosperity of their husband's lineage. Although women may live in their husband's village, her children belong to her lineage, not that of her husband.

With "Respect" to Men's Roles

Young men living within a village follow deferentially the ways of their fathers and maternal uncles by expressing the proper greetings, showing respect to elders, performing tasks in the field and in the bush, and learning appropriate skills to enhance lineage prosperity. Boys attend the local school and expect to obtain employment to support their lineage's need for cash. "Staying with respect" means focusing on one's lineage, whose membership comes at birth from one's mother.

Joseph Farmwell, fifty-one years old in 1988, lives with his wife and daughters in a small settlement allied with another immigrant lineage with close ties to the chief. With the exception of a few years working in Ndola with his brother and occasional trips to Mpika, Joseph has remained at Nabwalya. Both he and his wife are subsistence agriculturalists with no formal education and few possessions. He raises chickens and pigeons to sell to purchase clothing and other household essentials. Three of his daughters are married, have children of their own, and live close by. Together these households cultivate adjacent plots as a corporate strategy against marauding animals. Joseph's activities are representative of subsistence cultivators whose diverse activities follow the timing and tempo of the rains.

> *18 November 1988:* For several weeks, Joseph, his wife, and daughters have prepared their fields. On this day, he clears bush for his field [400 min]. In addition, he collects firewood in the bush [63 min], goes fishing at the river [39 min], returns to his house in early afternoon to eat [40 min], to repair farm implements [144 min], to relax and converse with his friends [37 min]. While carving the handle for a hoe, Joseph mentions the importance of safeguarding the wild animals from outsiders and from local hunters. Eating roasted corn, he discusses with those recently returned from a diviner what they learned about those involved in a witchcraft case. His father comes to inform him that Joseph's brother is sick in Ndola. As he snuffs tobacco, the conversation turns to the good rains of last season and to the abundance of wild vegetables and rice then. When served his meal with a wild vegetable relish [at 1445 hours], he complains about the scarcity of meat. He instructs his son about caretaking bananas while removing a thorn from his son's foot, afterwards [at 1605 hours] he takes up his hoe and joins his wife planting seeds in their field.

24 November 1988: Joseph begins work early in his field [276 min]. While weeding, he greets his brother-in-law, [upon hearing gun shots] talks about a local merchant shooting a hippo and selling its meat on the plateau, complains about *lwiba* [a hardy "weed"] in his field, greets his grandfather who asks to take leaves from Joseph's sisal plant to make bird nooses, and exchanges news with an uncle about the location of the diviner. He takes his axe and spear [at 1015 hours] and proceeds into the bush to check his traps and snares [58 min] returning with some poles for repairing a shed. He puts his axe and spear away. In the settlement, he rests, roasts maize, and eats *nsima* with warthog meat [95 min]. He talks about shortages of animals and about the times he has escaped crocodile lunges at the river. In the afternoon, Joseph spins bird nooses from sisal fibers [to snare birds in the fields] and makes repairs to settlement structures [275 min], greets his brother-in-law and asks about their children, grinds snuff [34 min], talks about beer and drums as prestige items and the problems with his bicycle. Towards evening, he goes to the river, checks his fishing lines, bathes [37 min], returns to the village where he greets his brother-in-law, talks about wrist watches, and spins more bird nooses.

1 December 1988: Joseph begins work in his field [464 min] until rainfall sends him back to his homestead for shelter. He spends most of the day under the eaves of his house talking to friends and family [187 min], overseeing his younger children [14 min], taking a bath [29 min] and eating *nsima* with puku meat [38 min]. Conversations are about animals migrating to the higher, sandy ground since the rains began, about floods experienced before turning to the serious implications of stealing game and how some trappers seek to protect their interests. Joseph relates how a deceased man killed his own grandchildren because his son and a friend stole an impala caught in his snare. This man used magic (*bwanga*) to avenge the thieves and announced it publicly; yet his own son never confessed and fell victim along with his children. "Stealing animals from another's snares is dangerous business for the life lost may be one's own. Last week, I snared a puku," he continues, "and when I went into the bush to check my snares, I found my brother taking a dead puku away on his shoulders. I called him and later we shared the carcass." On another occasion, Joseph continues "I came across a snared zebra. Not knowing to whom it belonged, I put some thorn branches over the carcass to protect it from vultures. I returned to the village and announced what I had done. The owner responded, we went back together, butchered it, and I was given a share." Instructing his children, he continues, "It's good to ask rather than to steal something." The stories continue until nightfall about dangerous crocodiles and how Joseph got drunk on brewed spirits (*katata*) in Mpika, was lost, and had to be shown where he was staying.

13 February 1989: Joseph works (436 min) his field until noon rain interrupts and prompts him to seek shelter. He returns to weed in the later afternoon on urges from his wife. While in the fields, he hears about the mad woman who yesterday was caught by a crocodile at the river, how she held onto the roots of a tree, and was rescued by the men who speared the crocodile. He takes a bath in a near-by lagoon when he goes to check his snares (111 min) and later, takes his spear and accompanies his wife to the river while she washes and draws water (32 min). At home, he eats and talks with visitors (22 min) about finding honey in the woodlands and about a big crocodile that appears often at the bathing place. He sharpens his hoe and makes nooses for catching doves (84 min) before weeding his fields. There he greets his brother who has come from the bush with snares. They discuss where to deploy the snares in mopane woodland. Before leaving the field, he checks a large tree limb set to prevent hippo from visiting his field at night.

8 May 1989: Joseph releases pigeons from his dovecot, feeds them sorghum, and heads out to his field to scare birds [32 min]. He visits his sick uncle to discuss the beer, which his sister brewed in honor of their late mother. He reports that his mother appeared in a dream demanding beer from her granddaughter; the beer was brewed and everything done during his two days in Chalwe village [28 min]. He visits his "mother" (mother's younger sister) to talk about family affairs [39 min] and complains about dizziness from drinking so much beer earlier in the week. Once back home, he tells his wife that he is going for beer and spends the rest of the day around pots of beer in different homesteads [619 min] before returning home late at night. The conversations over beer are about the high costs of living and the low probability of securing employment in town, that hunters should organize, about yields of sorghum, about opposition to authority and how those meet with tragedies, and about shortages of meat and fish for relish. The drift of the conversation is that people earlier were never short of fish from the depressions along the river; consequently, only a few people used to hunt and wildlife flourished. Presently the river has stopped its flooding of the adjacent lagoons and forced many people to hunt thereby decreasing wildlife.

Mumbi Chalwe, twenty-six years old in 1988, is married to the daughter of a prominent local family. He completed grade seven at the local primary school, but his wife has no formal education. As his wife became progressively sick, they spent much of their time consulting with diviners and doctors about the causes of her sickness and about why they were incapable of having children. His wife has little energy to cultivate, and he has hung around looking for work, engaging as a carrier for loads to Mpika or picking up local jobs when possible. His situation is typical of many young men who, as "school leavers," wander around in limbo seeking employment or getting into trouble.

2 December 1988: Mumbi has just returned from a trip on foot to Mpika lasting 10 days. He carried buffalo meat for his cousin.[11] For this service, Mumbi's pay was ZK 110 [$13.75], which he used to purchase a pair of used trousers [$10] and a blouse for his wife. He spends most of this day resting and delivering [565 min] items (sugar, salt, soap) purchased for kin and friends. He accomplishes some chores around his house [12 min], visits his garden [16 min], bathes [65 min], goes to the clinic visiting a sick relative [54 min] and eats [50 min] *nsima* with warthog meat. During the day, his conversations are about the game guard's killing animals with government rifles for license holders, the lack of profit in selling small animals [as "all profit is consumed in carrier fees"], about magic so a witch cannot see one's house, a prescription for keeping pigeons, the drinking and smoking "daga" habits of his peers, about losing a hoe, and the shortage of meat relishes. He observes that many villages have split into smaller settlements because kin lack trust and do not cooperate. "People have always liked to eat meat," he says, "but now, instead of sharing it as in the past through common meals, most people consume it alone in their houses. Those not receiving meat may report the kill to the game guards so they can make arrests."

16 March 1989: Mumbi has just returned from another trip to Mpika and is considering changing his residence, as his wife's health has not improved. He rests for most of the day [301 min], visits his sick brother at the clinic [84 min], attends a funeral [252 min] as well as eats, bathes, and does other chores around his house [95 min]. At the funeral of an elderly man, who had fled from witch accusations and joined his son, Mumbi prepares the casket of sorghum stalks. He assists at the grave and listens to others talk about witchcraft. Their message is that one cannot escape from a determined witch just by shifting locations.

6 July 1989: Since his wife died at the hospital in Mpika, Mumbi waits to hear whether his in-laws will provide him another wife. He attends and assists a funeral [382 min], visits [315 min] and eats with his brother [49 min]. He muses about his chances of becoming a wildlife scout since English is an important prerequisite for employment. At the funeral, he listens as his elder brother tells stories about game guards listening for gunshots and then visiting villages at night to make arrests. His brother's fear comes through in his statement, "We shall all die because the guards are not following government rules." Later he asks his observer if he thinks the professor will write him an application for employment as a wildlife scout so he can go for training and secure a paying job.

The world for men appears more open, with more options, than that for women, whose role confines them at an early age to the protracted agricultural and household tasks. Formal education offers opportunities for some younger men, but it serves as a cultural chasm that divides them from their elders who have little or no education. Elders complain that younger men have become self-focused, materially acquisitive, disrespectful, and show little regard for the welfare of others. With no formal education, Joseph found it difficult to secure local employment and consequently continued to rely upon his labor—intensive agricultural practices with his wife and daughters, utilization of local networks for subsistence, and selling local products in Mpika for cash to meet his needs. On the other hand, Mumbi's local education met the employment criteria for a wildlife scout, and his marriage into an important local lineage helped his prospects. Of course, I wrote him the letter supporting his application as a wildlife scout and he obtained an appointment later that year. A man's local status depends upon his connections to women as well as to other men. In the next section, I examine a few basic components of Bisa cultural life, including some beneficial and contentious contexts.

"Wealth in People": Linking Men and Women within and beyond a Matrilineage

The ways in which wealth is valued, collected, displayed, and imagined are instruments of power and influence as well as products of cultural history. "Wealth in people" is a cultural strategy that well-positioned individuals employ to build "respect" and prosperity by accumulating the collective services of labor, knowledge, and skills of others through various alliances rather than investing exclusively in capital or land. Although this approach depends upon a variety of political, cultural, and environmental skills, it appears risk adverse under certain conditions as an appropriate adaptation for people living within chancy environmental and social surroundings (Guyer 1995; Guyer and Belinga 1995). Some scholars argue that "the good life" embodied in relationships with other people is at the center of much economic and social history within Africa (figure 2.6).

As a cultural system for expressing status and influence, "wealth in people" has a long history within Zambian societies (de Luna 2012). Its values continue today as men establish their own settlements by attracting dependents who reflect their patron's political and social pretenses (Richards 1939; Kapferer 1967; Moore and Vaughan 1994). Changes continually challenge such assemblages of people as they have struggled with the

FIGURE 2.6. Women taking a break to visit with one another as men drop by to socialize and sip the fermented sorghum beer the women prepared the previous week. Everyone uses the same straw to imbibe the beer. (Photograph by author, November 1988)

repetitive problems of controlling labor, maintaining social networks, and reproducing households, that have "appeared at different times in new forms, both conceptually and materially" yet their crafting of immediate responses have occurred within the context of previous arrangements. As Moore and Vaughan (1994: 165) write of the Bemba in Zambia's Northern Province, "the old and the new copied, replaced, and substituted for each other, but did so within a long historical process that was never complete or completed."

The "wealth in people" system depends on the rights of some people over the services and labor of others. For chiefs, such rights begin in the alliances struck between his or her ancestors and other clans and through continuing to build new coalitions. For commoners, the process begins in marriage and builds successively as individuals convert services, goods and cash into gifts, loans, support, and rapport to sustain and extend their alliances. The extent of an individual's grasp depends upon gender, experience, chance, and circumstance.

The Matrilineal Descent Group (Lineage)

Social relations and kinship are part of the daily substance of Bisa life and implicit in the details of how its members produce, distribute, and consume wealth. Degrees of kinship and marriage are the means through which men as political actors build their alliances and express, regulate, and interpret behavior. The basic bonds are with one's mother and her descent group (*mukowa wapacifulo*). Marriage is forbidden within this matrilineal descent group (*lupwa*). Broader ties recognize eight categories of kin: one's mother (*nyina*), her brothers (*mwizyo, bayama*) and sisters (*banyina*), her sisters' children (*bamunyina*) together with her children (*bepwa*). Also included are maternal grandparents (*bazikulu*) and all the children of one's sister's daughters (*bezikulu*).

All these primary kinship terms refer to specific groups of kin. For example, *nyina* (literally "mother") includes one's biological mother in addition to her female matrikin of comparable rank (mother's sisters or grandmother's daughter's daughter). Similarly, *bamwizyo* includes all *nyina*'s male siblings who are her child's main authority figures. Since the Bisa readily distinguish their natal mother from aunts, the point is that all class terms carry a cultural loading of moral obligations and meanings. Inherent in these associations is a binding reciprocity that locks kin together in "a dense and inescapable net of obligations." These relationships are not of personal choice and, depending where one is placed within the tapestry of kin, obligations are simultaneously "ill-defined and unbounded, and morally binding and indissoluble" (Crehan 1997: 102). In discussions, individuals are inseparable from their kinship location, which defines and often guages their activities. Conditioned by age and status, men generally have influence over women, as sisters and wives, as senior in lineage decisions and in public fora.

Each local lineage belongs to a larger umbrella group or clan (*mukowa*), the descendants of a presumed founding ancestress, whose members have become scattered. Each clan bears a name, usually of a natural object (such as *Muti*: tree; *Nzovu*: elephant; *Mbeba*: mouse; *Lungu*: seed; *Mvula*: rain; and *Mwansa* or *Nyendwa* for male or female genitalia, respectively). This widespread network is useful for travelers expecting hospitality from others encountered along their journeys.

For the Valley Bisa, a matrilineal descent group is the normative basis for determining an individual's social standing. New members are added by birth, and the dead are believed to retain an active interest in and influence over the group's fortunes. Thus a descent group exists in perpetuity with various secrets and traits, such as names, mannerisms, rituals, history, and sayings (membership memories and behaviors) distinguishing its members from its neighbors. Members share corporate rights over land and resources and are responsible for them to the chief. Elders venerate the memories of deceased members by soliciting their aid in decision-making, maintaining the group's integrity, and expanding its influence while these same processes sanction their own seniority.

Each descent group contends for women, children, labor, land, and all other necessities of "a good life." Competition and conflicts with some groups reinforce boundaries, while significant links with others enhance their reputations and may strengthen their advantages. Within the group, tensions and deep jealousies may develop over differences in wealth, power, and age. When these rifts surface, some members may be accused of witchcraft, forcing them to join other alliances or to form new villages or settlements. Relatives are generally among those usually suspected as the cause for the untimely deaths, accidents, and sicknesses of kin.

Marriages and Alliances

Initial matrilineage marriages frequently begin by betrothing a cross-cousin, the exchange of spouses between alternating generations, a time-honored strategy for intertwining lineages over time. Elders know the histories of those involved in these marital transactions and may use this knowledge in monitoring wealth and welfare as well as in resolving problems. Valley Bisa marriage customarily began for a young man with negotiations by his elders with his prospective wife's family ending with the groom's residence in wife's village, where he provides labor service during a specified time for his in-laws. Upon the successful fulfillment of his obligations.including an agreed upon payment, the husband may determine the next residence for this new household. This arrangement was the norm until the 1980s, when a local marriage more frequently involved a negotiated cash payment and no residential requirement. For a man, marriage is the crucial step for creating a following and expanding rights over others' services. Whereas some marriages may last a lifetime, they are fragile; serial divorces are common. In 1965–66, half of the civil cases heard in the local Nabwalya court were for divorce. In 1989–91, divorce cases were the overwhelming majority.

Men and women have different tasks and goals, and their interdependence and needs for each other's labor is not evenhanded. Men tend to emphasize certain objectives, such as clearing a field or building a house or structure, accomplishments that endure for a long period of time, a season or several years. An older man's supervision of lineage and its political alliances is continuous and demanding. Women's tasks generally are day-to-day and repetitive responsibilities—such as weeding and tending the crops, preparing food, drawing water, child welfare, and tending the settlement. Men need access to women's labor on a daily basis and require marriage and children to become responsible adults. Women can provide most of their own needs with only periodic access to men and their labor. Glimpses into the lives and gender roles of Nabwalya residents indicate the progressive nature of these changes (Marks 2014: 73–102).

A young man has connections to others through the lineages of his mother, father, and grandparents. These links persist as long as its members interact, serving as potential building blocks for young men who seek patrons or build their own following. Both men and women remain loyal to their descent groups, often living and working in the same or neighboring villages. Descent groups retain the land allocated to them by the chief as long as they cultivate it and care for its assets.

Inequalities of Gender and Age

Formal subordination of women to men appears as a local convention, yet there is increasing discord among younger educated women about this normative expectation. Subservience to one's male matrikin and to one's husband is different in kind and manners. In public life, a married woman supposedly follows her husband's directions and seeks to please him.[12] This relationship is one of economic interdependence, as men and women need each other's labor and proceeds. Through marriage, a woman provides the products of her toil and that of her dependent children to her husband and his lineage. Yet the wife's matrikin expect her to produce children, as her husband's sexual prowess determines if he supports his wife's goals. These contrasting mandates create a space in which some women may play the expectations of her matrikin against her husband's needs.[13] A woman's identity and security, together with those of her children, remain bound within the destinies of her matrikin. Likewise, a man's permanent affiliation is with his matrikin as they, rather than his children, expect to inherit his material belongings, such as a gun, sewing machine, bicycle, or axe.

In the recent past, there was little specialization as adult men and women were expected to acquire the knowledge and skills appropriate for their respective tasks relevant for their ages. The few exceptions were in hunting, blacksmithing, and skills picked up by men through their employment elsewhere, as well as divining and séance. Some women are also skilled in prophecy and spirituality. Other women are potters, know about local afflictions affecting women, and become known to men through their qualities of beer brews and cooking. More recently, in 2007, improved teaching within the local basic schools, together with the establishment of a secondary school and the founding of a Catholic parish and protestant churches, has expanded the outside affiliations and connections of many Valley Bisa, enhancing the life chances of some residents.

In addition to gender, one's age is an important variable on the local political and economic landscapes. Within a social context, all persons are sorted into definite elder/youth (or senior/junior) relationships according to age and gender as well as the historical reckoning between respective clans. Elders (*bakalamba*) were once the main decision-makers who embodied authority, wisdom, and experience. Yet no defined marker existed by which an individual achieved adult status or that of an elder; rather this standing was incremental with successive goals achieved as one moved through various life stages, beginning with marriage, establishing a household, having children and grandchildren, eventually assuming notoriety as a focal person or as a practitioner of a critical skill. Young people (baice) were earlier respectful of elders and generally followed their advice when consulted. This mentoring of the young by the old in villages was a casualty of the chaotic and traumatic 1980s and its replacement by required attendance at primary schools as well as the demographic increases of the previous decade.

The completion of primary education by many boys and comparatively fewer girls during the 1980s together with the local employment of the educated youth as wildlife scouts to enforce the game laws during the 1990s has worked to undermine the status of most elders. Local scouts, armed with modern weapons possess the authority to investigate and enforce the criminalization of customary entitlements in wildlife while their local employment provides means to marry and to possessions earlier without having to follow the older submissive rules.

Of Households, Villages and Settlements

Among the Valley Bisa, households are basic yet problematic units to study. In terms of study and economic organization, households appear organized more as units of consumption than of production. Spouses generally act autonomously, engaging in gender-specific activities to produce different goods and services elsewhere before eventually entering households. These differences are particularly noticeable in polygynous households where each wife manages her fields and kitchen and raises her own cash to purchase necessities to meet her needs and those of dependent children. Each spouse is responsible for their productions and income even as these assets are subject to a wide range of claims.

Although many households consist of a married couple and their dependent children, one cannot assume that "owners" (*mwine ng'anda*) are necessarily men. Generally, a household never has more than one adult male, yet female-headed households are still common. At Nabwalya in 1966, women led 33 percent of the households enumerated; among the same villages in 1988, adult women headed 23 percent of households. In 2001, female-headed households varied between 7 percent at Nabwalya to 14.6 percent in a village four kilometers away. The 1966 figure supports the inference that the married men in many households were elsewhere as migrant laborers. The later figures indicate the subsequent decline in migrant opportunities and the influx of men (and women) to the facilities around Nabwalya, and the stiff competition for temporary employment there. Polygynous households are widespread among middle-aged and some older men and remain aspirations for many younger.

The village (*muzi*) is the organization level above that of the household. As places for "staying well," villages wax and wane in size contingent upon the agency of its elders and the circumstances of its surrounding environmental assets. Few villages retain the names of their original founders generations ago or have remained generally in the same locations. Before the mid-1960s, many of the villages around the chief's palace of Nabwalya moved frequently as their soils and other assets diminished (Marks 1976: 21–23). As a hub of government services, famine relief, and employment, the permanency of the chief's palace currently attracts many Valley Bisa men and women from within the GMA and elsewhere. These immigrants make alliances with the chief or headmen, who in turn may suggest spaces to build and farm.

Witchcraft as the Dark Side of Kinship

The witch (*mulozi*) and nefarious witchcraft (*bulozi*) are omnipresent realities that cast threatening shadows over many personal interactions within small, residential groups. If kinship provides the tangible tangle of social knots embracing mutuality and positive welfare within these assemblages, the rhetoric of witchcraft exposes a darker, cruel side in which the supposedly inherent evil within individuals conspires maliciously to destroy the vital unity of the group. To explore witchcraft beliefs is to glimpse a bit of what life is like for those entrenched within the dense hegemony of kin, capricious circumstances, and meager resources. Witchcraft conjures powerful images of someone eviscerating the life substance from within communal lives, a vicious ideology for muddling through the failures and compromises assumed under dense kinship expectations.

A witch behaves inversely to normal expected human conduct. They are said to go about at night when everyone else is asleep, naked, to cause harm in the form of sickness, exploitation, or death, and to consume or exploit the dead while affecting the living or sleeping in ways not immediately obvious. Singularly or in groups, witches act for personal gains and may use intermediaries, such as hyenas, crocodiles, lions and even snakes to affect designated victims. Witches steal crops or wealth at night, even compelling their victims to transport these goods from their owner's bins, acquire the labor of kin at night to work their fields, even using the power of dead fetuses from "sisters" to capture game, or to kill detractors.

Most everyone uses prescriptions (*muti*) in a variety of contexts, from curative matters to adjuncts ensuring an objective. *Bwanga* is the term used for the narrower nefarious prescriptions to accomplish witches' missions. *Bwanga* is an inclusive term, assuming not only the substance but also the knowledge of and skill to deploy it destructively. People may employ *muti* and *bwanga* defensively as well as offensively. Only witches and *nganga,* the ritual specialist, know the indescribable aspects of *bwanga's* uses. A sick person or someone, who wants to know about an illness or death of a relative, consults with a *nganga* to divine answers to these afflictions from the dark side.

Most adult Valley Bisa men and women acquire the appropriate levels of skill and knowledge to carry out their basic tasks of living. They assume that regular skill becomes extraordinary and everyday fortune becomes enhanced through the possession and knowledge of the powers inherent in animal or plant parts. When these natural parts are combined in concoctions [prescriptions], the product contains magical properties inducing the desired goals. This form of magic, *muti,* is fairly common in use and is part of the repertoire of skilled practitioners, yet does not have the sinister connotations of *bwanga.* Local hunters typically employ a host of magical adjuncts (such that to find prey, to escape detection by wildlife scouts or dangerous animals), which are generally acquired to overcome a specific type of failure (not killing game after many forays, being wounded by a buffalo) for which the magic prescription is the antidote. Yet enviable achievements or a suspicious accumulation of undeserved wealth by someone

may cause jealous detractors to accuse him of using unconventional means such as *bwanga.* Hunters, who openly use magical adjuncts as prerogatives of their status and successes, are vulnerable targets for vilification as witches particularly by kin not favored in the distribution of meat, protection and in other behaviors in old age.

The charges of *bulozi* have changed shape and taken on different dynamics in the postcolonial world. An inversion of the loan term "poacher," pronounced locally as "boacher," was adopted by school youths discussing an elder's suspected activities in 1989. Their use of the term indicated a subtle, generational shift in discussions and accusations against some elder men's alleged exploitation and malfeasance. Every adult might whimsically accept the local designation as "poachers," yet these young people bet their reputations (and perhaps their well-being) upon their elders' inability to distinguish the sound heard of this term's the initial consonant. I asked the young man to describe his use of the term:

> "Boachers" are people who kill people. These people are recognized in many ways. Some have a lot of wealth through theft-using magic. They steal sorghum, maize and money from their relatives and friends. Some "boachers" are fat because they eat human flesh and also use their families at night to work their fields and to do pounding, grinding, and other hard work. We recognize some "boachers" by their abusive language during beer parties. When traveling, "boachers" use their kin to carry their luggage without their kin's awareness of what is happening. "Boachers" move with hyenas, jackals, antbears, and owls. That's why at night they do not lack transport. When young people are talking about the "boachers," they do not mention actual names—rather they use names of other things such as plants or trees or something resembling the subject of conversation. Therefore, older people cannot understand what we are saying.[14]

In 1989, elderly people complained that their world of prerogatives had turned upside down and that their grandchildren were rejecting their values and practices. With local employment favoring the young and educated, those so privileged respected their teachers as role models, who would manage their entry into a different world from that represented in the elders' material inadequacies. This visible "poverty" was that which primary school graduates were prompted in and expected to escape.

"One Finger Will Not Pinch a Louse"

The Valley Bisa instructively use proverbs (*mapinda*) as succinct reference guides to important issues. The proverb "One finger will not pinch a louse"[15] summarily emphasizes the importance of human integration as a group, rather than individually, to accomplish goals. The proverb can also be read variously as "A person never stands alone," "One person never knows everything," or "A single person alone is incapable of doing anything."

I use this local saying to emphasize two basic principles. The first is that the time-honored Valley Bisa identity is not that of an individual but that of a corporate entity. A person's behavior reflects his responsibilities within the broader contexts of a group. For the inhabitants of the Luangwa Valley, this assemblage has been the matrilineage. Woven by the threads of kinship and embedded in moral obligations, this social identity is very different from that of individualism within the modern world. Yet this corporate identity of the Valley Bisa appeared, in the early 2000s, to be unraveling as many joined larger non-kin-based alliances, particularly Pentecostal evangelical congregations and affiliated with other foreign groups. A second meaning of this proverb is that the materials from which I have woven the accountings in this book are not the products of a single author, as my name on its cover may suggest. This book is the product of many contributors and of different creative involvements. These others include inter alia the finances and affiliations with organizations that have made my roles possible, those providing the hospitality and means of getting me placed on this frontier, and a host of Valley Bisa residents who have scrutinized me observing them. Others contributed records, made observations or allowed them, provided generosity, participated in conversations or were helpful in other ways. All are contributors and subjects of this work and made it possible.

This chapter began with lineages and clans as the foundations for and political spheres by which chiefs and headmen continually gathered people around them. Loyalty to these alliances and "respect" for others were the keys to enduring relationships in which kinship was the moral fiber, safety net, and main recourse. The webs of kinship obligations were diffuse, open-ended, and

yet often so binding that individuals inevitably faced a number of competing claims to whatever material goods or fortune they managed to accumulate. Wide spaces existed between kinship expectations and individual decisions for kin resentment as well as for individual anxieties.

Unlike the dichotomized human and natural world of the modern Northern Worldview, the Valley Bisa appeared to assume that a seamless moral fabric bound them and the resources of their landscape together. Nothing exists that is indifferent to the morality of those within its human community, an assemblage of kin bound together by reciprocal webs of moral obligations encompassing the rest of a cosmic order. Events occurred because causality came with morality and intentionality rather than through chance. A group of kin may properly plant a field or an individual appropriately set a snare in the expectation that what they expect (ought) to happen and would occur. Yet wild animals or a drought may devastate the cultivated fields or a snared mammal may turn upon and severely wound its trapper. The space between what the "ought" of an expected outcome and what eventuates is where the suspected witch and his craft lurk as both causative and explanatory agent. Witches and their craft do not violate natural laws only moral prospects (Crehan 1997).

This proverb also highlights openness and uses of ethnographic materials, experiences, and explanations from different sources to explore different world frames. An open-ended and indeterminate nature has characterized my inquiries into subsistence and persistence, just as some Valley Bisa will admit to not knowing everything as they continue to live within their chosen spaces. In many ways, my understandings and conversations have been determined less by plan than through the contingencies and circumstances of "being there" with the Valley Bisa on and off over six decades. My "being there" also reflects the progress of my ciBisa language skills as well as an appreciation of the less subtle forms of language that friends used in communicating their experiences and ideas to me. My living and experiences with relevant individuals are mainly about how I came to know, understand, and now depict in writing some characteristics of this place and its particular peoples. The Bisa proverb also appraises the superficial "expertise" of those whose brief journeys only confirm their earlier held impressions or those, who appear on a scene with their models and plans to implement or see only those processes that conform to these visions.

The brief vignettes recorded at Nabwalya, during the agricultural season of 1988–89 and transcribed above, illustrate an openness and grounding in gathering information on the mundane activities and conversations of individuals by contextualizing them with personal concerns embedded within its local universe. Doris Kanyunga reprimands Mr. Peterson for wanting to marry her granddaughter, inquires about Ronald's mother's health, while she continues to demonstrate her role as a respectful wife to her daughters and grandchildren. In seeking to establish herself in her husband's village, Rose Chiombo works long days while rearing her dependent children and negotiating through the thicket of her husband's wishes for a second wife as she reads the cultural signs of what might be happening in her natal village. As he continues to sustain his family through sharing agricultural activities with his wife and their children, Joseph Farmwell also raises chickens to sell for cash, fishes for provisions as well as sets snares to provide animal protein for their meals. Though his stories, he instructs his children about the consequences of theft, shares with his kin, reports to his surviving "mother" and her kin about a recent journey, and respectfully participates with peers in conversations over beer. On the other hand, Mumbi Chalwe works his way through the youthful dilemmas of an ailing wife and finding employment by judiciously following the scripts of kinship as he listens and learns. Upon his wife's death, he hopes her lineage will provide a new wife and that he secures employment as a wildlife scout. For him, both hopes materialized.

Despite useful categories and handles, Valley Bisa life remains complex and many-dimensional as another interpretation of this section's proverb allows, "it's impossible to know [or record] everything." As the end of my initial studies was drawing to a close in 1967, I was left grappling with many obscure bits of accumulated information that appeared relevant. As a consequence, I spent my last few days at Nabwalya searching for ways or someone to help me integrate these fragments into my planned thesis. I realized that miles of land and ocean would separate me soon from my primary sources once I left to write my dissertation in a distant, cluttered, but barren space surrounded by notes and thoughts. So I compiled lists of confusing topics and puzzles hoping to find nuanced answers from impromptu conversations with Bisa elders. The response of one elderly man, perhaps irritated by my persistence to find him, my questions and anxious to get on with his own pressing tasks, suspended my queries soon after I began to ask them. He turned away abruptly to attend to his own tasks as he mumbled, "Machisa, you never will understand everything in a single day!" Recognizing eventually the wisdom in this voice has sustained my efforts over the decades to expand my network of alliances and frameworks and to listen to differences along with interpretations and their prospects. This meaning is also implicit in the song admonishing a new chief, the epigram opening this chapter that "a child is like an axe."

Notes

1. *Tangala elyo uli mwanabwalya; Wea wabubiza*
   Ciaalo cabene taci mekelwa wemwana wamubiza
   Yangu Somo ifyo aculile; yangu Somo ifyo ampengele
   Nabantu abengi balikakilwe bambi kufwilamo
   Nomba Somo, nimwe uleteka wilaba
   Necipyu, mwana Kazembe, kukumikomo watola wakobeka.
   Recorded by Florence Somo (1988).
2. Jaap van Velsen (1957: 83, 2) found that a lack of definition for "headmanship, land rights, the matrilineage, and, indeed with marriage itself," was a general feature of the Tonga institutions he studied. His fieldwork in Malawi turned up many differences between expected values and real practice. He also discovered that many concepts used by colonial administrators and anthropologists to understand Central African societies indicated a "definedness" he could not find in real life. Martin Chanock (1998: 25–47) discusses similar issues with African law and anthropologists. Helen Tilley (2011) describes the role of anthropologists and others in the development of the British Empire in Africa.
3. For example Kate Crehan (1997: 53–75) shows how the Kaonde in Western Zambia became a "tribe" and how this label was used in the complex exchanges between parties of unequal power, such as by colonial anthropologists at the Rhodes-Livingstone Institute, the colonial officials stationed within North-Western Province, and Afri-

cans, whom the "Kaonde tribe" supposedly located and described. Crehan notes that "one of the great silences" under British hegemony of indirect rule was the denial of the colonial state's ultimate authority and its devastating impacts on the autonomy of the societies that it dominated. Within this silence, she locates her analysis of the ways the colonial state supposedly controlled, the ways anthropologists sustained their profession, and the different ways Africans themselves appropriated colonial concepts. See also Martin Chanock (1998) and Henrietta Moore and Megan Vaughan (1994).

4. In her discussion of how vocabulary restricts analysis and understanding, Crehan (1997: 75) employs the analogy of comparing maps as representations of physical realities to the categories employed by cultural observers to depict social realities. The descriptions and analysis in this book to chart the recent history of Valley Bisa life and relations to their surroundings make no pretense of avoiding a similar predicament. Such an admission is instructive as well as humbling, particularly if one seeks to contextualize the ideas and objectives of others, including those of the colonials and wildlife conservationists, whose employment and deployment did not encompass the experiences and friendships I have gained among all parties through my field studies.
5. Adding to this confusion is that many Africans use the English word "tribe" to describe their own affiliation and identities. Yet their use of the term does not imply the same fixed and invariable sense associated with its uses by foreigners.
6. Personal transcript of forum chaired by Senior Chief Kopa at Nabwalya, in my field notes for 23–24 June 1967 and in the unpublished manuscript (Marks and Marks, nd). A synopsis of the the fora before and the day of the Senior Chief's arrival appear in chapter 10 of this book.
7. At independence in 1964, the new Zambian government sought to weaken chiefly authority in order to prevent the development of regional tribalism amid hopes of creating a strong centralized and independent state. Rural chiefs had been an important component of colonial policies as a means to "divide and rule." In 1982, the government reversed its policy and promoted traditional ceremonies as a way to reassert government and chiefly authority along with "traditional values." Because of the long interregnum and disputes over the Nabwalya succession, the Valley Bisa ceremony *Malaila* was not performed until the coronation of the new chief on 15 September 1990. This ceremony is not held every year. In wildlife-rich GMAs, this occasion allows a reigning chief to "squeeze" government

bureaucrats and safari operators for supplies of meat, grains, and other necessities as well as to host a large communal celebration from which many residents and outsiders benefit.

8. The traditional ceremony held each year at his palace on the plateau by His Royal Highness, the Honourable Senior Bisa Chief Kopa on the plateau west of Mpika.
9. Quotes from the prepared speeches delivered at the ceremony.
10. Although prominent in most villages in the 1960s and 1970s, these public ceremonies have since become infrequent.
11. In 1988–89, Nabwalya residents, while transporting bushmeat for sale on the plateau who could show when challenged by wildlife scouts that they had previously purchased licenses to hunt for the specific species in their possession, together with a permit for hunting in the Munyamadzi GMA, were generally not considered as criminals or as having committed a prosecutable offense.
12. In some local songs, a code that symbolized the sexual behavior for wives was domesticated chickens, while that for husbands was wild guineafowl or bushmeat. Examples include: *Nankoko wacizumpa niwe walanga bana ati balya tebantu bobe* [literally translated as "The mother hen has foolishly taught her chicks that those people are not relatives"] and *Balume batana nekanga, naine ndebatana mesekese ndecoca chinyo chandi* [literally "As my husband has refused to give me his guineafowl, I will refuse to give him what most men need or expect"].
13. Close monitoring of a husband's sexual powers follows his marriage. He is expected to abstain from sex during the last months of his wife's pregnancy. Difficulties during a wife's delivery maybe ascribed to the husband's infidelity, but all parties must confess to any improprieties and be purified with special medicines to restore normative relations. A husband's responsibilities include keeping his wife properly clothed and fed.
14. My field notes during March 1989.
15. *Munwe umo tausala nda* (Bisa proverb).

Chapter 3

Never an Isolated Place Suspended in A-Historic Space

The surveyors have finished the animals in our land
For this reason, they must return to their homes and eat vegetables
Animals are like flowers in our place
That's why they must return to their home and eat vegetables.[1]
Valley Bisa verse (recorded 1989)

For Europeans, the central Luangwa Valley was a difficult place to reach throughout most of the nineteenth and twentieth centuries, and as such, it held for them a certain mystique. The valley's remoteness and isolation conjured images of "wilderness," "adventure," and "challenge"—values that readily translate into saleable commodities for the leisure traffic of northern worlds. Some of these aspects brought me here as a young graduate student in the 1960s and still remain as promoted pictures in the markets of the international tourist and safari industries. Yet this chapter reveals that those inhabiting the Luangwa landscape were never so isolated; rather, they were always connected to other people and regional and global events, and as such have responded in turn. Neither national legal boundaries, imposed on the ground, nor the institutional mindsets prescribed for its residents reflect the ambiguous, contested, and open realities of human life within this valley.

Conflicts between outsiders and residents over the valley's resources, particularly its wildlife wealth, continue today, yet with more substantial and subtle twists. Recent clashes over entitlements and aesthetics, between nonconsumptive and other users, and who those users should even be have become increasingly blatant and vehement. Yet the struggles over the tangible aspects of these resources are often individual skirmishes compared to the wider, deeper chasms of distrust and despair over land claims. The irony is that as the physical distance between valley inhabitants and outsiders has decreased, the figurative gaps in metaphors and minds have widened.

Although mining and gem explorations were widespread during colonial times, the epigraph at the beginning of the chapter refers to government-sponsored foreign explorations for oil and gas throughout the Luangwa Valley during the energy crisis of the 1970s. Under federal auspices, a contractor bulldozed parallel tracks across state and communal lands, enabling men with mobile units and equipment to test for subterranean resources by detonating explosives. While the results of these surveys remain classified, the explorations left long-term scars on the landscape and bequeathed atypical exposures within local memories. These rootless operators offered short-term opportunities and interactions with outsiders as well as markets for local products and services. The verse above is the reflection of a young woman balancing the tantalizing wealth of these strangers with the services allegedly sought through her.

Differences between local and international purveyors of the valley's resources surfaced in the political economy of earlier centuries. With the cessation of the slave and ivory trades during colonial times, wildlife became the focus of contention. I summarize the wildlife cycles of the twentieth century and examine the ways the colonial and Zambian states have sought to curtail local uses of wildlife while seeking to enhance their own. As agriculturalists and tenants on this landscape, the Valley Bisa annually pay a high tariff in life, limb, and property through their competitions with wild animals. Beneath this physical price is the more difficult-to-discern mental cost of being marginalized as dependents upon discriminant and inefficient government services. State boundaries imposed on the ground to separate nonconsumptive from resource users, along with the disparate cultural webs in the minds of foreigners and residents, remain contentious, porous, and continually challenged and confronting in obvious and covert ways.

Images of Isolation and Remoteness

Recent Images of Inaccessibility

Prior to 1960, the only way to get to Nabwalya village was on foot, a journey that under good conditions took a minimum of five days from Mpika. From Mwanya, a chief's village in Zambia's Eastern Province and west of the Luangwa River, an expedition on foot required a minimum of three days to reach Nabwalya. For decades from an administrative view, the remoteness represented in these time requirements for its officials to reach Chief Nabwalya's village had become a persistent problem and needed a resolution. Finding a suitable vehicular passage from Mpika through the Muchinga Escarpment, surveying and building a seasonal road took several years during the late 1950s. During the 1960 dry season, the first government vehicle ground its way down a precarious roadbed of switchback stony curves and rickety pole laced bridges over deep canyons, triple clutched the steep inclines of the escarpment, slid down its rocky foothills and ridges, forded the clear Mutinondo River though an opening in dense reeds, crawled along the edges of the uneven, solid surfaces of black stiff clays, sand, and pebbles along the northern edge of Chifungwe Plain, entered and exited the steep gulches of numerous then dry stream beds, each sculptured anew after each downpour in the wet season, passed swiftly through the flat, sandy stretches of mopane woodland, descended the brittle rocks, clays, and sands of Ngala hill onto a level patch of riverine savanna, made a sharp left turn to arrive at Chief Nabwalya's palace nearby the Munyamadzi River. The initial trip was not continuous as it took nearly a week as officials incurred many diversions, repairs, and unexpected delays as they explored the best transit across the valley floor. Using this established bush track, a four-wheeled drive vehicle could accomplish the outward journey to Nabwalya within a single day, but for subsequent decades the route remained passable only during the dry seasons.

The Lumpa uprising in Northern and Eastern Provinces was an African indigenous religious and political threat during the transition from colonial rule to that of Zambia. Prime Minister Kenneth Kaunda declared a state of emergency, deployed state troops to suppress the riots, killing hundreds and arresting its spiritual leader, Lenshina. During the years prior to independence and the Lumpa uprising in 1964, district officials felt vulnerable when making the long mandated foot journeys into the valley and chafed at having to endure the arduous and lengthy foot journeys to this marginal section within Mpika District. From a post-colonial perspective, the road connecting Nabwalya in the valley with district (Mpika) and Northern Provincial (Kasama) centers on the plateau placed communication and oversight significantly more within the administration's orbit. Yet the arrival of vehicular transport was more controversial locally. Some residents welcomed the opportunities that came with it, for it promised access for development and opportunities to create local wealth. Others preferred to keep their distance from external surveillance and welcomed the obscurity of their own socioeconomic relationships.

A heavily loaded Land Rover could traverse the 140 kilometers from Mpika to Nabwalya in ten hours during the dry season (June–October) of 1966. With the return of the rains in November through May, the only feasible way to travel there was by foot. During our stay in 1966–67, my wife and I used both means of travel as we learned the landscapes and tuned in to the rhythms of residents. Many times we felt removed and isolated—feelings reinforced each evening as we huddled with our local neighbors around a small radio to listen and discuss broadcasts about Rhodesia's Unilateral Declaration of Independence and other world events.

Since then, the options for getting to Nabwalya have increased for wayfarers, contingent upon their connections. In 2006, travel by foot or bicycle remained the mode for most residents during all seasons and likely for everyone during the rains. A few salesmen on bicycles "hawk" light supplies and used clothes as well as convey dried fish, chickens, and local products to outside markets. Government officials, missionaries, safari operators and wealthy outsiders travel by motor vehicle and plane. Since the 1980s, the government has occasionally used helicopters for wildlife surveillance, to deliver famine relief, and to collect ballots during elections. Encouraged by officials as a "self-help" scheme, the community completed a dry-season landing strip for fixed-wing aircraft near the chief's palace in 1991. Light planes used this strip for anti-poaching surveys and, with improvements the next year, larger aircraft landed and delivered famine relief. During the dry season of 2006, two airstrips within the GMA buzzed daily with take-offs and landings as safari operators brought in clients and fresh supplies to their camps. These airstrips caused the flight time between Nabwalya and Lusaka, Zambia's capital, to shrink to within two hours, but only for those with wealth, seeking adventure, or affiliated with powerful organizations used them.

Residents have petitioned repeatedly for an all-weather road to provide an adequate link to the plateau and to Mpika.[2] Lack of this amenity remains a major constraint and greatly increases costs for all constructions and developments. Until 2014, well-connected outsiders with hunting interests have outmaneuvered residents in this request; as powerful strangers, they are convinced that a good road would quickly deplete the wildlife and their easy access to it within the central valley.[3]

Despite outsiders' impressions and the area's geographical constraints, residents within the central Luangwa Valley never belonged to isolated communities; on the contrary, they have always been connected to powerful personalities and groups elsewhere. Even today, the flows of people, goods, and information interlock this valley and region with the larger world. Wildife in this valley has been a major asset attracting outsiders' interests for enhancing their own wealth through commerce for centuries. Recently its foreign values, promoted in various forms of recreational tourism, attract a different class of foreigners who seek to preserve its supposedly pristine qualities as "remote wilderness" as these features appear to have dissipated elsewhere throughout Africa. We next examine the relative richness of this magnet for strangers within the past century as well as its benefits and dangers for its residents more recently.

Abundant Wildlife and Living with "Problem Animals"

During the late nineteenth century, trading or raiding parties as well as European caravans subsisted on scattered wildlife through or across the Luangwa Valley. Although these travellers expected to trade some of their goods for agricultural products, they found that the small bands of suspicious residents, hiding in the thickets, or those living within stockade villages, possessed insufficient crops to share or trade. Unsuspected skirmishes caused suspicion with strangers, and outbreaks of the rinderpest epizootic during the 1890s reduced some wildlife populations.

As the wildlife within the central Luangwa Valley were the lure attracting the attention of outsiders and residents alike, the abundance trends for these wild animal populations is comparatively well known for the twentieth century. Archival accounts of administrators, records of travelers and hunters, counts by wildlife experts and local hunters, together with the recollection of local residents, form the basis for profiling these wildlife trends (Marks 1976; Freehling and Marks 1998).

Wildlife Trends under Government Management Policies

From the accounts of early administrators stationed at Mpika, most game species could be found scattered throughout the district. Already known for its wildlife, the central valley attracted European sportsmen and adventurers until the outbreak of sleeping sickness and the First World War. These European hunters came seeking escapades on a colonial frontier and natural history targets from within its diverse environments.[4] Among others, C. H. Stigand (1913), Dennis Lyell (1910, 1924), and Owen Letcher (1911, 1913) sought trophies for their collections and for profit in a mythic place they seemingly knew from their readings of the earlier exploits of Cornwallis Harris (1852), Gordon Cumming (1857), Frederick Selous (1881) and Rider Haggard (1885) on the southern African frontier. Letcher (1911: 255) described the Luangwa Valley in the style of Rider Haggard as a place where "nature unfettered and unrestrained is there on all sides" and as the "land of wild beasts and uncivilized men, where the vegetation has run riot, and the great mammals which the Creator put on this earth roam free and wild." The fauna together with a longing for the free and unrestrained life attracted these sportsmen. These valley habitats were suited for wild beasts and "savage" people as these collectors stalked the antithesis of their restrained society and confirmed its myths. Some more wealthy adventurers were the bane of resident BSAC officers, who feared retribution if officials did not graciously host them on the hunting fields or fined them for game law violations.[5] A few hunters lost their lives during their exploits and are interred in the European graveyard at Mpika, including a New Zealand hunter (Twigg); a soldier, hunter, and prospector (Mills); and an elephant control officer, allegedly upon the killing of his 350th elephant (Ross).

For many years during the 1910s and 1920s, sleeping sickness and tropical diseases closed large portions of this valley to Europeans except prospectors and those on administrative duties. These restrictions left the taking of wildlife largely to valley residents. There were few firearms as most men used bows, poisoned arrows, spears, game pits, and snares (see chapters 5 and 8). As the human population remained scattered and small, wildlife increased, dispersed, and recovered from rinderpest.

When the Colonial Office assumed control of Northern Rhodesia in 1924 from BSAC, it allowed Africans to purchase muzzle-loading guns. The Colonial Office commissioned a territory-wide fauna survey in 1930–31 led by Colonel C. R. S. Pitman, a retired military officer and game warden in Uganda. Pitman indicated that Mpika contained "more game in greater variety" than any other district save possibly Namwala. Mirroring what he had heard and others had presumably witnessed, Pitman reported that during the previous twenty-five years, most game within the Protectorate had decreased by as much as 75 percent. Reflecting his sentiments and those of his contacts during the consultancy, he blamed Africans' hunting with muzzle-loaders for this drastic wildlife drop. Elephant and buffalo showed significant increases, however, a result he attributed to these species being beyond the capability of African hunters using muzzle-loaders (Pitman 1934; Marks 2002).

The general consensus is that wildlife populations in the central valley peaked in the late 1960s, about the time of my first year of residency at Nabwalya in 1966–67. My records indicate that wildlife was indeed abundant, diverse, and visible. In the early mornings and late evenings, I saw and encountered many species close to the villages. Wildlife counts varied seasonally from about one animal encountered per minute (April 1967) to one every five minutes (July 1967) while walking out and returning to the villages during forays. The Valley Bisa population at the time was small at Nabwalya, with people living in well-defined villages centered around their granaries in the dry season and in scattered households surrounding lineage fields during the wet, growing season (Marks 1973a, b). These formats provided some protection for their crops while local hunters and state-employed elephant control guards reinforced these perimeters against persistent wildlife damages.

Wildlife around Nabwalya has decreased significantly from the benchmark numbers that local hunters and I observed in the 1960s. Reasons for these declines include the downward turn in the national economy beginning in the mid-1970s, which created a scarcity of jobs in town for migrants from the valley and reduced the means by which lineage men could generate wealth

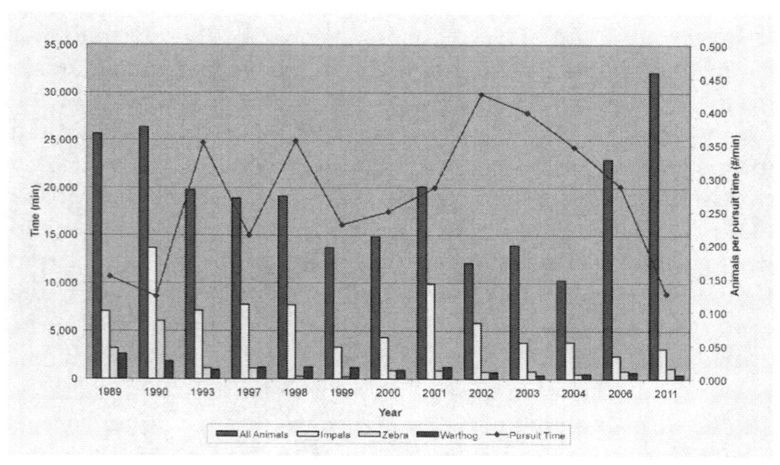

FIGURE 3.1. Pursuit time by a local hunter and number of prey observed in minutes. Numbers of smaller species (impala, zebra, warthogs) observed, various years 1989–2011. Numbers of animals observed per unit of pursuit time in 1989 and 1990 were for all months; other years' observations were made during six months of the dry seasons. See Marks (1994a) for methodology and relevant details; modified after Marks (2014: 106).

beyond exploiting rural natural resources. Other factors include frequent droughts and floods, competition between people and wildlife for space and fertile habitats, and unlicensed off-takes of wildlife by all parties, including officials. Since the beginning of the "community-based" wildlife program in 1988, immigrants settling around the chief's palace seeking jobs, collecting famine relief, and searching for protection from wildlife depredations elsewhere have placed greater demands on the resources of the adjacent habitats, and in doing so increasingly disturb wildlife (Marks 1994a, 1996).

In comparison with many other African landscapes, Nabwalya remains comparatively plentiful in wildlife. In the 2006 dry season, wildlife counts around Nabwalya showed recent increases in buffalo and elephant numbers and encounters, slight increases in zebra and wildebeest, and fewer encounters with impala and warthog. Every morning, signs of buffalo and elephants appeared within the villages and adjacent fields. These increases for the larger mammals and decreases in the smaller species suggest that current enforcement practices may be having an effect, as local hunters have shifted their means from guns to snares and targets from buffalo to smaller species. Although an occasional buffalo may still be snared, it is more difficult to hide and dispose of its meat without attracting the wildlife scouts' attention (Gibson and Marks 1995; Marks 2014). Currently, local hunters legally kill a small number of buffalo according to the few GMA-licensed residential quotas given each year, while more nonresident Zambians take many more buffalo on national licenses, GMA permits, and special licenses issued by the minister of tourism. Foreigners on seven- to twenty-one-day safaris are given a buffalo quota twice that for residents, as well as sizeable quotas of other mammal species.[6]

Living with "Problem Animals"

Living with wildlife, particularly large mammals and predators, has high costs for residents. Human conflicts with wild animals can end in tragic loss, such as the killing or maiming of relatives as well as the destruction of their properties (crops, destruction of granaries or houses). All wildlife incursions on human lives and welfare are highly charged representations of state repression, particularly since state agents have disarmed most local residents and made them dependent for protection upon an inattentive and often recalcitrant cadre of wildlife police officers. A sixty-two-year-old man in 2006, responding to a general question about his experiences under the wildlife program, stated,

> In February, elephants finished my maize field in a single night, I reported it to ZAWA [Zambia Wildlife Authority] officers and they never followed up. In May, buffalo grazed my sorghum field. I reported it to ZAWA officers who never set their feet in my field to check on the damage. In the event where they do kill an animal in one's field, the scouts are the ones taking the lion's share [of the meat] to the detriment of the farmer. Even when there is an elephant "crush" [person killed by an elephant], ZAWA scouts cannot respond immediately to the tragedy. They first have to consult the regional warden, who then he must consult with headquarters. So in light of this doing, does this mean that elephants are more important than human beings?[7]

Even before 2006, a common complaint was that ZAWA and community scouts had been negligent for years in responding to individual and community requests for protection against crop-raiding mammals. Occasionally firing into the air over raiding animals in fields at night has only a temporary effect, if any, and scouts' repeated failures to respond to destructive animals appear reprehensible given the disarming of residents, who earlier had recognized rights to protect their properties. Scouts' normative responses to local appeals for help are that a game animal killed in the fields has no value other than as temporary meat and as an "individual benefit," whereas the same animal killed by a safari client is worth a lot of money and is therefore more of a "community asset." This scout response is a reversal of lineage and community sanctions during earlier decades when lineage hunters killed animals for "community" welfare and for protection, but were scorned if they killed game to sell or "wasted powder" to scare intruding wildlife away. If persistent marauding episodes or local tragedies with wildlife are known, reported, and scouts are pushed to respond, these episodes are simply summarized conveniently in wildlife reports as scout responses and observations on residents' insolence.

The tragic death of elderly Doris Kanyunga, killed by a rogue elephant on the night of 25 April 1998, illustrates the dreadful dimensions of these engagements. Kanyunga was respected, as mentioned in the previous chapter. The lifetime resident was married to a worker within the chief's compound, and she was selected to participate in the activity survey during 1988–89.

> The elephant was grazing (at 0300 hours) in a sorghum field along the Muchinka-Nabwalya road near Ngala and Kapili Ndozi. When the owner of that field heard and scared it off, the elephant ran into an adjacent field where Roger and his young wife [Doris's daughter] were sitting by the fire. When they heard and saw the beast, the couple fled towards Doris' house on the opposite edge of the same field. In their rush, a "brother" in a nearby hut joined them. The elephant stopped briefly at Roger's field house, tore off the roof, and scattered the hot ashes before continuing its rage. Doris was just emerging from her house when she saw the elephant and started running for a safer place. Unfortunately, the elephant could detect her movements despite there being no moonlight. It ran faster, caught her with its ivory and flung her against a tree trunk. Neighbors heard her mutter *"nafwa"* ("I am dying") just before they heard a thud as her body hit a tree trunk. The elephant screamed sharply and went off into the bush. Early next morning (at 0500 hours), neighbors went to check and found her scattered remains. They picked up the pieces and placed them together for burial.
>
> The day was sad because people had not begun yet to harvest their sorghum fields and they realized the elephant was still about. Looking at her remains, they found it hard to realize that one of the lively and important people in their community was now dead. Everyone reached the funeral home with a chilled body because of her untimely, dangerous death.
>
> The chief was very annoyed with the wildlife scouts because they had failed to respond to the numerous reports of crop damage and threats to human life. The [Wildlife] Unit Leader had told his scouts that whoever would kill an elephant, no matter what damage to crops or to persons, would serve a prison sentence alone and that he would not defend them. The chief chased the wildlife scouts away from the funeral house and told them to hunt that elephant. They left at once with their guns. Scouts were not happy because the Unit Leader [an outsider] had caused all this by always threatening them each time they saw a dangerous animal.
>
> Around 1500 hours, the scouts returned and announced that they had killed the elephant from the herd. It had some blood on its ivories. People rushed there to where the carcass lay, but some refused to get any meat because they believed that an animal

that has killed a person is the same as a murderer and eating its flesh the same as consuming a dead person. I also couldn't get a share with reason that I respected human life more than that of an animal.

On 26 April, Doris Kanyunga was buried, her intestines carried to the grave in a pot. All the people who had fields around the Ngala area were in fear [of additional devastations by wild animals] and harvested their sorghum prematurely. That's why most of them have a poor harvest this year.[8]

Through sharing space and even brief moments with other people, one begins to grasp the flowering of life under difficult circumstances as well as to sympathize with the difficulties of others living under imposed regulatory institutions and shackled by inadequate leadership and training. One might also learn about daughters encouraged, due to thwarted parental ambition, to further their education instead of marrying elderly men, and of young sons exploited by untrained teachers, their minds depressed through rote memorization rather than liberated and enabled to cope in a very changing world elsewhere.

Outsiders conventionally perceive of human-animal conflicts as local predicaments arising from inappropriate behavior by either the humans or the animals. These events become stripped of all cultural context under the official category of "Problem Animal Control (PAC)." Yet victims of problem animals with whom one has shared space, time, and conversation are much more than mere statistics. Months after I learned of Doris's tragic demise, I also mourned her death, for I knew that her husband, her family, and all her neighbors missed her routines, opinions, and energy. Unfortunately, deaths from wild animals within this GMA are rather common occurrences.

Each year, some Valley Bisa are maimed or killed by wild animals. During the twenty-three-year period between 1990 and 2013, I recorded 133 fatal encounters with wild animals, some of which were not officially recorded (table 3.1).[9] Of that number, crocodiles killed the most (seventy-three) people, while elephants killed twenty-five, buffalo killed nineteen, lions eleven, and hyenas two. Poisonous snakes killed three individuals. In addition, wild animals maimed 162 people. Crocodiles wounded ninety-six people, mostly women and children while they collected water or washed at the rivers. If surprised or cornered along the rivers, hippos can also become deadly. The comparative low number of residents wounded by this species (nine) may appear to contradict its conventional reputation as one of Africa's worst human killers. Although the Valley Bisa have a reputation for not consuming hippo flesh, many hippos were killed during the famine years of the 1970s and later in the 1990s, under a government quota to exchange on the plateau for grains (see chapter 7). Consequently, within the GMA's perennial rivers, hippos are not as abundant as they were in the 1960s. Buffalo maimed thirty-five people, mainly men and boys hunting or collecting in the bush, while lions and hyenas, prowling around villages at night, wounded twenty people. People assaulted by elephants usually do not survive beyond the charge, but two individuals did escape with minor injuries in 1912–13.

The state, with its interests in stocks of large wild mammals, has progressively restricted the entitlements of GMA residents as its agents promote their relevance as protectors and conservers of wildlife. Among their supposed activities are annual assessments of each wild population of game species in order to calculate the "surplus numbers of animals" beyond those needed to reproduce adequate numbers to sustain subsequent generations. This theoretical yearly assessment is the basis for establishing the annual quota of animals to match the game licenses issued for sales to hunters, thereby granting these buyers the legal right to pursue a specific species for recreation, for meat or for other purposes. These license revenues together with other hunting concession fees become the means to support government and community efforts to sustain wildlife, suitable habitats, and to enforce management rules. Yet under real world circumstances, these theoretical calculations are difficult as wild animals are difficult to count, and the reproductive success of any wild animal is contingent upon many human and environmental factors (*inter alia* disease, climate, other predators and "poachers"). As of 2015, Zambian wildlife officials, even with considerable donor and private supports have yet to achieve their iterative goals of "sustainable wildlife conservation" or had produced appropriate evidence for "improved Valley Bisa general welfare," at least within the Munyamadzi GMA (Marks 2014). Because such public claims have a long textual history, various state documents and statements are prime

TABLE 3.1. Numbers of persons killed or maimed by wild animals, Munyamadzi Game Management Area, Central Luangwa Valley, Zambia, 1990–2013. (Modified from Marks 2014: 65)

Killed/ Wounded	By Animal	1990–1	1992–3	1994–5	1996–7	1998–9	2000–1	2002–3	2004–5	2006–7	2008–9	2010–1	2012–3	Totals
Killed	Elephant		1	1	4	1		3	3	4		2	6	25
Wounded	Elephant												2	2
Wounded	Hippo						2	1	1		2	2	1	9
Killed	Buffalo		1	1				2	1	13			1	19
Wounded	Buffalo	3	3	2	2			3	4	12	4	2		35
Killed	Lion			1	2			2	3*	1		1	1	11
Wounded	Lion		1		2		1		7	1			2	14
Killed	Hyena	2												2
Wounded	Hyena	2	3		1									6
Killed	Crocodile	5	10	6	7	8	3	5	9	13	2	2	3	73
Wounded	Crocodile	22	10	5	9	12	1	10	8	10	5	2	2	96
Killed	Snake										2	1		3

Sources: Letters and personal notes, clinic and Wildlife Records, and local diaries (modified after Marks 2014: 65).
* girl killed by leopard

sources to match and evaluate these wildlife policies with assessments of their outcomes, which might lead to their subliminal and "intended" purpose.

The Regional Political Economy and the Closing Nooses of Colonial, National, and International Agendas

Earlier Political Economy and Perceptions of Valley Resources and Users

As Brian Morris (2000) suggests for Malawi, the political and economic history of the central Luangwa Valley makes little sense without reference to the precolonial exploitation of human and natural resources. The landscape throughout Southeastern Africa continually shifted during that time as warlords, traders, and residents depended upon slavery and trades in animal and other products. Land was plentiful but labor remained in short supply. Well-connected people used others to pay their credits and injuries as well as to obtain valuable properties such as muzzle-loading guns, textiles, beads, and copper. They exploited, captured or purchased slaves for labor, services, and transporting trade goods to and from the coast. Residents assimilated new strategies and technologies from invasive or transient outsiders for warfare and hunting larger mammals. Strategically placed residents applied new techniques to obtain ivory, horn, skins, meat, slaves, and important regional trade goods. Each exploitive activity had its players, clients, and survivors; detractors either became victims or went into hiding. According to some historians, this landscape was a "rising tide of violence" brought about by traders and warlords competing to supply escalating international demands (Birmingham 1976: 300; Roberts 1973; Alpers 1975; A. Isaacman and Rosenthal 1984).

During the final decades of the nineteenth century, the control of persons increased the number of conquered, captive and enslaved; the retention of those conquered and captive was always at risk between belligerent parties. British Central Africa (BSA) commissioner Harry Johnston described in 1894 the "constant hunting of man by man [that] keeps the whole country in a state of unrest" to dramatize his struggle to garner resources and men to administer the newly acquired realm. Yet, the nature of this place, as "one of the finest hunting grounds in the world," attracted Johnston's interest. He made "discoveries" of several large mammals (bushpig and okapi), that still bear his name in their Latinized scientific identities, and noteworthy contributions to regional natural history (Johnston 1897).[10]

Beyond gradually stopping the slave trade, the BSAC exported in 1893 about 1912 tons of ivory, then the struggling territory's most valuable export. Johnston expected this production would increase as his company acquired the stockpiles that Arab traders managed to slip past his agents. Furthermore, he pondered the product's future by offering a regulatory solution to conserve elephants as a source of revenues.

> I have come to the conclusion that provided the Brussels Act is enforced and guns and gunpowder kept from the natives, especially the Arabs, and Europeans only are allowed to shoot elephants by taking out a license, the elephant is likely to exist with us for all time, and yet supply a sufficiency of ivory for trade. The fact is we should leave the bulk of elephant killing to those natives who kill elephant by trap and spear. They do not perpetuate anything like the same destruction as the natives armed with guns who indiscriminately shoot every elephant they come across.[11]

Pitched for "peace, order, and good government," his edict required the registration of all guns in African ("native") hands and the purchase of an annual license. Any African not complying was subject to a fine or imprisonment plus the forfeiture of his weapon. Premised upon racial and cultural discriminations, this mantra to further the empire and save "the game" gathered momentum within developing colonial structures.[12]

Starting with the BSAC in the 1890s, every successive government has expressed alarm over the numbers of firearms in African hands and sought to control these weapons. Perhaps feeling more secure after the conclusion of World War I, the protectorate administration permitted Africans to invest their migrant wages in muzzle-loading guns, which by then were considered obsolete for prolonged warfare. Subsequently, administrators and others couched their disquiet about firearms in terms of Africans lacking suitable hunting constraints and its overall devastating effects on wildlife. The same worries about rural residents' impact on wildlife persist today in Zambia and elsewhere within conservation and tourist circles. I describe the technical reversion to snares as a local response to the state's wildlife regulations that have physically and symbolically become, to employ a trapper's metaphor, "closing nooses" upon some rural village identities and impinge even more on customary livelihoods. Whereas the snaring of wildlife and other ways (game pits, downfalls, poisoned arrows) were indigenous means for taking wildlife, they were all declared illegal, "cruel," and criminal offences by the colonial and the subsequent state. This illegal designation had little effect on eliminating these practices as I witnessed their clandestine uses during most of my stays in the Luangwa Valley.

Although local snaring of mammals decreased as guns became the dominant form for taking wildlife in the 1930s, its reappearance as a major means corresponded with enhanced local enforcement against gun hunting and the collection of "illegal" guns during the 1990s. The revival of snaring, and to a lesser extent other customary means, at this juncture is significant as it is a reassertion of local entrepreneurship that challenges the state's distant authority and management of a perceived local endowment. As was gun hunting, snaring proactively provides animal protein, adds some protection against depredation and property loss as it reaffirms a stealthy space to cover the autonomy of local actors. As these practices rapidly diminish wildlife quantities, which the state seeks to increase and its foreign allies value, snaring may also become "a cultural protest," an indicator of a centralized resource management system gone awry, failed promises, and continuing threats to local lives.

Firearms under the British South Africa Company

Well-placed Valley Bisa chiefs and others knew about muzzle-loading guns from their use in the slave trade and elephant hunting during the nineteenth century, and a few had acquired them. As European countries rearmed with more rapid-fire and precision guns, large quantities of muzzle-loaders became available as trade items and were given as favors along the trade routes throughout South-Central Africa. Prior to British domination in this region, muzzle-loading guns proved decisive elements in skirmishes and became privileged and prestigious possessions.

Always anxious over the possibilities of a rebellion and aware of their own vulnerabilities, company administrators disarmed Africans and dealt severely with those hiding their weapons. The territory itself required a minimum of foreign administrators, who depended upon local African agents to transmit and enforce colonial demands. The colonials living under this fragile structure remained acutely sensitive to any indication of African unrest and apprehensive of whatever form it might take (Yorke 1990).

Administrative control of firearms remained a vexing issue and was the topic of many deputations with chiefs. Beginning around 1902, Mr. Kennelly, the BSAC officer stationed at Nabwalya, sent policemen to collect all guns with instructions to arrest anyone they saw wielding one. The BSAC administrator in 1908 responding to demands for firearms from Bisa chiefs (including Nabwalya), granted each a Snider rifle and twenty rounds of ammunition per year for self-defense and for crop protection. He also "allowed all natives so long as they behaved" permission "to kill any game in their gardens by any means of traps, snares, pits [without pointed and poisoned stakes], bows, arrows, spears etc." He threatened to revoke any of these permissions should any African use forbidden means against elephants. Embedded in these official "concessions" was the restless ghost of his administrative predecessor Commissioner Harry Johnston, who worried about the ruthlessness of Africans with firearms. In 1913, the company state charged Chief Saili, then a recognized Valley Ng'ona chief, with failure to report a gun in the possession of his counselor. The court found the chief guilty and punished him by withholding his monthly subsidy for two years.[13]

At the beginning of the World War I, administrators withdrew all guns loaned to chiefs and headmen, promising to return them upon the war's conclusion should the administrator consider their requests.[14] In June 1923, Africans "were given permission to carry muzzle-loading guns" upon purchasing an annual license and registering their guns with a magistrate.[15] In 1924, the British government formally took over the protectorate's administration and inherited an increasing African demand for firearms.

Firearms under a Protectorate of the British Government

At an Eastern Provincial governor's meeting in 1929, Valley Bisa chief Kambwili, who lived directly across the Luangwa River from Nabwalya, emphasized that "although natives were encouraged to buy muzzle-loading guns, these were more or less useless against elephant," which continued to destroy their gardens. He requested new rifles, as "the Martini Henry rifles, at present lent them by the *boma,* were old and unserviceable." Another chief agreed that these loaned rifles "repeatedly misfired . . . and asked for more reliable ones." The governor responded that he intended "to make it as easy as possible for any native to own a muzzle-loading gun and powder" and that he would try and replace the worn rifles "provided they were procurable from government stocks."[16]

When Captain Pitman traveled throughout Northern Rhodesia in 1931 to make his faunal survey, he noted an increase in gun ownership by Africans in Serenje District from 14 in 1918 to 1,130 by the end of 1928, and up to 1,400 by April 1931. "With few exceptions," he wrote that officials and settlers, including missionaries, expressed their displeasure to him over how the Africans were allowed, "practically without restrictions," to procure these guns. Pitman quotes the district commissioner (DC) at Mpika that, under prevailing conditions with large numbers of muzzle-loading guns in "native hands," any attempt at successful game preservation was impossible. In Mpika District, the DC reported possession of these guns "monthly . . . increasing by leaps and bounds."[17]

Pitman's (1934) recommendations for curbing the "utterly improvident and irresponsible" Africans' hunting reflected earlier administrative pejoratives and resonated within the prevailing norms of indirect rule. His proposal for the creation of a game department included the development of native authorities (chiefs) as colonial confederates providing its incentives would turn Africans into submissive subjects. Registering firearms and requiring the purchase of annual game licenses were essential ingredients for discouraging African "licentiousness" in hunting, and he insisted that chiefs be given authority as well as accommodate chief's "traditional" prerogatives. He emphasized that a prudent administration should sustain its subject's meat supply, particularly in tsetse fly areas lacking domestic stock. Pitman's other endorsements included the creation of two large game sanctuaries in the Luangwa Valley and the retention of the Valley Bisa homeland separating them.

Once it sanctioned gun ownership for Africans, the colonial administration was reluctant to reverse its decision despite complaints by officials and European residents. Despite attempts to control gun ownership through taxation and registration, muzzle-loaders increased as local blacksmiths mastered the technology and produced unregistered numbers of these firearms. Residents in the Munyamadzi Corridor of Mpika District officially possessed 176 muzzle-loaders in 1953. That number increased by 1962 to 347 muzzle-loaders, 7 shotguns, and 2 rifles (Marks 1976: 73).

Mpika game ranger Eustice Poles employed his perverse humor to explain the "amazing increase" he observed in hyena numbers during his third Luangwa Valley tour (dated 28 July 1947). He notes that, whereas all other species were decreasing, only hyenas had benefited from the proliferation of the "destructive muzzle-loaders." His conclusion was deduced from "the large number of young, inexperienced native hunters one meets these days, when every sophisticated native seems to trot with his father's [*sic*] gun." Since much of the game shot was lost and eventually died, he knew that these carcasses, especially the bone and gristle, kept hyenas fit long after the flesh was removed.

Building and broadening support as part of pre-independence maneuvers, Zambian nationalists encouraged their rural constituents to disregard and disobey any colonial regulations on firearms and hunting.[18] Consequently, political disturbances together with falling copper prices in the late 1950s forced drastic cuts in the European game staff (by some 40 percent) and the placement of the Department of Game and Fisheries under the oversight of the Ministry of Native Affairs. Dr. F. F. Darling, a British consultant, expressed the views of many Europeans by saying that "the general feeling of the natives is that game is wide open, and that there is nobody to stop them doing what they like with it" (Northern Rhodesia Government 1960: 30).

Firearms and Wildlife Enforcement under Zambian Agency

The newly independent government in 1964 reorganized the Department of Game and Fisheries under the Ministry of Lands and Natural Resources and greatly expanded its plans and staff, including expatriates. The Luangwa Valley became the focus of a five-year United Nations Development Project on wildlife conservation that ended in 1973 and launched decades of international funding and consultancies by foreign experts. In 1972–73, Zambia's numerous game reserves became national parks with permanent guard posts on their peripheries and inside the adjacent GMAs. Hunting within GMAs came increasingly under the surveillance and regulation of the new Department of National Parks and Wildlife Service (NPWS).

Beginning in the mid-1970s, NPWS was one of the last national agencies to become staffed by Zambians in its senior ranks. When copper prices fell again in world markets later that decade, this department possessed little equipment and its budget was greatly diminished. During this period, the department depended upon coordinated "village sweeps" in conjunction with the military and police to uncover illegal activities and unregistered firearms. Villagers greatly detested these unannounced raids because the authorities interrupted normal routines, confiscated weapons at will, harassed and imprisoned men on the slightest pretense, pressured women for local information, and searched houses at will. Outside funding, expertise, and equipment enabled many of these enforcement efforts, including a local initiative in South Luangwa, Save the Rhino Trust (Bowden 1982; Antrobus 1983; Leader-Williams and Albon 1988; Leader-Williams et al. 1990).

In the late 1980s, northern donors sought to combine development assistance with conservation under the mantra of a "community-based" Natural Resource Management program (CBNRM) throughout southern and eastern Africa. Zambia's program, Administrative Management Design for Game Management Areas (ADMADE), emphasized wildlife, placed much of the onus for enforcement on local village scouts, supervised through national and district leaders, and provided a return of some safari hunting revenues to a local committee for approved uses to build schools, wells, and clinics and to pay for local labor. The state created a new institution, the Wildlife Management Authority, at the district level, chaired by the district governor and staffed by the wildlife warden and other district officers. Its mandate was to supervise developments in and conservation management of the GMAs within its territory. Each GMA had its own Wildlife Management Sub-Authority, chaired by a chief and staffed by the wildlife unit leader, resident civil servants, and a few appointed headmen. They were responsible for collecting revenues from ADMADE and accounting for the money's dispersal among predestinated community development projects, as well as for wildlife protection. Besides recruiting and training local villagers as wildlife scouts, chiefs were encouraged to enforce wildlife laws and reduce the number of illegal muzzle-loaders and snares. The initiating activities of ADMADE program were standardized workshop formats, yet were promoted in the media by national and international staffs as local dynamic leadership responding before receptive rural audiences. Earlier media reports reflected mainly positive results, but they were often based on flimsy or shallow data for marketing global appeals for funds and support. In 1989 and 1990, a newly appointed Chief Nabwalya became embroiled in a confrontation that reveals some of these tangles and the national scales of its connections. The chief describes the wrangle in a letter to the NPWS director:

> During the last year, plans were made as to how the local people who own the private, unlicensed, self-made muzzle-loading guns would give in the guns to minimize poaching in the Munyamadzi area. The first plan was to list down the names of the people who possessed the guns and go to them to ask them to surrender the guns, but I saw this was not a good method. It was inefficient, few people or guns would be surrendered and not in a peaceful way. I then devised a method of persuading by holding several meetings in both [political] wards to explain about the importance of conservation of animals and the benefits that would come to the area in terms of development of schools, clinics, roads, wells, and because more tourists could be attracted to come to see the animals and hence bring in more money.
>
> I started meetings in Chifungwe Ward and explained that conservation would bring long-term benefits to us. I persuaded them to surrender the guns and the people were assured that their names would not be listed and they would not be arrested. And of course, the people listened and they began to surrender their guns

> in both wards. Now about 16 guns were surrendered in Muchinga Ward.
>
> These guns were waiting to be taken to the Governor [in Mpika] and to the Warden, Bangweulu Command, but surprisingly enough Game Guards from Mfuwe [in Eastern Province] came to Muchinga Ward between 20–22 February 1990. They approached the ward chairman of Muchinga Ward and said that His Excellency the President had sent them to collect the guns and now we wished to collect the guns.
>
> I want to find out if the guards were sent by the President [Dr. Kaunda]. We want the 16 guns back to Mpika Boma, so that the Governor can see them, see our work. Game Guards from M'fuwe come to trouble people here. They came and got a licensed muzzle-loading gun of a certain headman last year and the gun is still with them. We want it back. Why are these Game Guards from M'fuwe (sic) working like this? Muchinga ward Chairman told them to see me, or the Unit Leader, before collecting the guns, but they wouldn't heed. They came to collect the guns of which they didn't know the arrangement or organization about and even otherwise frustrating the initiated plan, destroying our initiative because we want so many people as possible to surrender the guns peacefully, in the persuasive manner.[19]

This local initiative fell victim to two internationally sponsored conservation projects competing for control of and benefits from the same turf. The Luangwa Integrated Resource and Development Project (LIRDP), headquartered in Chipata (Eastern Province), chaired by the president of Zambia and funded by Norwegian Aid (NORAD), employed these game guards from Mfuwe. Their mission centered on the South Luangwa National Park and Eastern Province, yet its directors construed its mandate more broadly to allow patrols into the Northern Province, including the Munyamadzi GMA and on the adjacent plateau. None of these latter sites derived any financial benefits from the LIRDP project. The Munyamadzi GMA was technically the providence of NPWS and ADMADE, headquartered near Lusaka, and initially funded by USAID through the World Wildlife Fund, with additional operating revenues from safari hunting (Gibson 1999; see also chapter 10).

Like many areas in the Luangwa Valley, the Munyamadzi GMA became contested terrain for competing national and international interests, whose authority depended on financial and political clout within the national arena. I do not know if the chief's letter on pages 142–43 received an answer. Yet in 1995, the chief received a "meritorious award" from the Honorable Minister of Tourism, a prize processed through the Southern African Wildlife Trust of Washington, DC, using funds administered under the African Elephant Conservation Act through the U.S. Fish and Wildlife Service. The award commemorated the chief's accomplishment in getting his people to surrender their illegal firearms. The Mfuwe scouts received publicity and other remunerations in their own province for their successful raid and capture of this cache of unregistered weapons.

In addition to amnesties, government and safari operatives have sponsored other initiatives to reduce the numbers of poachers and weapons within the central Luangwa Valley, including cash prizes to scout units capturing weapons, for convictions of illegal hunters, as incentives to undertake a specific number of bush patrols per month, for educating and reforming "renowned poachers," and for exchanging bags of grain during famines for guns and bundles of snares. The physical, emotional, and financial cost of these activities upon local residents are significant, but not well known. When I administered a questionnaire throughout the Munyamadzi GMA in 2006, 38 percent of 460 village respondents told us that either they or a close relative had been arrested recently for a wildlife violation. Arrests rates varied by location from 51 percent to 19 percent, with the highest percentage being in the villages around the chief's palace, the site of most scout activities. The most frequent offences were for the possession of meat from buffalo (34 percent) and impala (28 percent) (Marks 2014). These percentages are high for any resident population and had not decreased since the beginning of the program. We might expect a decrease in arrests if residents were benefiting from the new incentives under ADMADE or if arrests were a sufficient deterrent to residents. Enforcement has been successful, if measured by the infrequency of gunshots heard since 1990, but residents have reverted to earlier methods of securing the local "game."

The State's Attempts to Curb Snaring

If at one time the colonial state felt some responsibility to protect its rural residents and their property from marauding beasts, neither NPWS nor ZAWA appear quick to draw its bead on truculent animals.[20] In 2006, when I asked a wildlife scout if he had any complaints about his job, he replied, "As scouts we don't complain because people come to us when they experience crop damage and tell us, "Your animals have eaten from my garden—so we are now like the 'owners of animals.'"[21] Residents despise scouts particularly when they kill wildlife illegally for their own needs, blatantly refuse to share this bounty, or upon killing a crop raider, require residents to take the meat to their camp. Common local knowledge is that some scouts snare on the side, as do some residents.

Snaring is a common way to put meat in a relish dish, and the proper placement of snares, along game trails leading into fields, protects crops. Snares were once constructed of sturdy vines, but they can now be built from readily available steel wire or cable. Both snaring and problem animals were on the agenda of the ADMADE Leadership Conference for Northern Province held at Mpika in April 1994. This convocation brought together chiefs, government and political officials, and technical advisors to develop action plans and resolutions for the province's GMAs. One of the tasks assigned to the committees was "the elimination of the practice of snaring." As chairmen within their respective GMAs, chiefs agreed to accept a penalty should any professional hunter find snared wildlife species anywhere within their territories. This penalty, equal to a thirty percent reduction of the thirty-five percent (community) ADMADE share, derived from the license fees for that particular species during a safari hunting year and would be transferred into the GMA's account for wildlife protection. The chief warden developed a procedure wherein animals "that pose a direct threat to the lives and property within a GMA [i.e. 'Problem Animals']" would be dealt with speedily. He reported that wardens were required to ask safari hunters to pay additional fees to kill these menacing animals for the "communities."[22]

This resolution on snaring, which the chiefs signed, was primarily a promotional tactic to attract external funds as it lacked traction on the ground. I found only one reference to this provincial resolution within local documents. In the May 1995 minutes, Chief Nabwalya, reporting to a local assembly about a recent meeting in Mpika, ended by cautioning that he would consider "door-to-door searches" if residents continued to snare wildlife, because the "community" would lose a lot of money if any safari company found a snared animal.[23] This rebuke did not dampen the practice of snaring, although it may have dimmed awareness of its prevalence.

In September 2000, a professional hunter brought a complaint through the community liaison officer (CLO) at Nabwalya that snaring had reached "an alarming state." This hunter had killed a snared lioness near Kanele Wildlife Camp and reported that he had seen many animals elsewhere entangled in wires. Given the origin of this complaint, the newly appointed Community Resource Board (CRB), the replacement under ZAWA for the former Wildlife Sub-Authority) decided they must address this issue. The CLO visited every village cluster to warn that if snaring did not stop, the chief would permit the dreaded "village sweeps" (household searches) and that the CRB would not employ anyone from areas continuing to snare. Villagers responded that snaring continued because there was "no[t] any sources of protein as people can't go on same diet [potato leaves and okra] every day". Other reasons for this "epidemic of snaring" were the discriminatory practices in the allocation of wildlife licenses, the distribution of employment, and the [mis]allocation of direct benefits from community funds. Some of the Village Area Groups (VAGs) complained they had received no benefits compared with those around the chief's palace.

The CLO recommended that the CRB make available dried sardines and beans at reasonable prices, that meat from safari hunting be given to residents, that soccer balls given to schools would keep young men busy competing within the villages, that more frequent meetings would help to keep villagers "on the right track," and that headmen should always be on the lookout for people setting wires. Supplementary solutions included the alleviation of inequalities and the recruitment of additional volunteer scouts from different areas. Local complaints underscored the needs for communication and for the redistribution of protein, commodities, and wealth.[24]

While present within the GMAs during the dry season, white professional hunters and their clients (for an additional price) took over much of the operations against marauding animals. When safari operations closed for the season, residents complained that the scouts lacked resolve to protect lives, property, and crops during the rains. They requested compensation for their losses, yet preferred personal permission to kill the rogue animals raiding their fields and the rights to dispose of its meat.

Boundary Problems on the Ground and in the Mind

Once the colonial government officially accepted many of Pitman's recommendations, it appointed T. G. L. Vaughan-Jones as an acting game warden. In 1938, Vaughan-Jones synthesized a memorandum for his game department based upon Pitman's report and the 1933 International Convention for the Protection of the Fauna and Flora of Africa (Vaughan-Jones 1938). Both report and protocol became the foundation for a combined Department of Game and Tsetse Control in 1942. Within the game estate, Vaughan-Jones proposed the creation of both reserves, where wildlife (fauna) had aesthetic, recreational, and scientific values, and controlled areas (CA), where wildlife had economic and exploitative worth. The latter were primarily on tribal trust lands where residents could hunt for food under their respective native authorities and served as "buffer zones" around reserves. Nonresidents could hunt in CAs under permit and with payment to authorities. Later these areas became first- or second-class CAs; the former contained more game and, if strategically located, was subject more to department regulations. Under the 1941 Game Ordinance, the Munyamadzi Corridor between the two game reserves became one of the first CAs and later a significant first-class CA (Astle 1999).[25]

Following the Pitman report in the 1930s, authorities proposed extensive game reserves for the Luangwa Valley, including portions of four districts in three provinces: Eastern Province

(Petauke District), Northern Province (Mpika and Chinsali Districts), and Central Province (Serenje District). Establishing these extensive boundaries on paper was easy compared to the ongoing skirmishes on the ground to define and defend them. Some boundary markers shifted in sections where their presence was contentious because of ecological and political changes. T. S. L. Fox-Pitt, Mpika's district commissioner, negotiated his district's boundaries and advised his provincial commissioner that any elephant sanctuary must be far away from villages, adding that "no scheme is going to be of much value if it is made by feelings unsympathetic to the natives."[26] Some designated sections of the two Luangwa Valley game reserves contained small settlements whose residents were encouraged to relocate. Some of them resettled along the Mupamdazi River under the Nabwalya Native Authority. These new boundaries became contentious later as good land became scarce with increasing numbers of residents grappling with progressively restrictive wildlife regulations, yet with few realistic local options.

The Beginning of Some Differences and Contentious Issues

Pitman recommended that while most of the "unoccupied" land west of the Luangwa River to the escarpment should become game sanctuary, a corridor of land including the Munyamadzi and Mupamadzi Rivers, where most Valley Bisa resided, should remain as native trust land. This counsel has had far-reaching consequences for residents, who have seen their opportunities shrink in relation to those in many other rural areas, but also for government, which has yet to devolve the institutions or determine the means for resolving this human dilemma created by an earlier European consultant.

Under indirect rule and left largely to their own devices, the Valley Bisa developed their own strategies to cope with increasing numbers of wild animals, localized famines, and outside pressures. These procedures included purchasing muzzle-loading guns to protect against marauding mammals and to secure animal protein, as well as soliciting government elephant control agents to protect their crops. To survive in times of famine when their cultivated sorghums failed, the Valley Bisa built upon indigenous knowledge of place for tapping into the varieties of edible indigenous plants and the gathering of two varieties of wild rice, exchanging labor for food, encouraging and utilizing wide networks for support when needed, and exchanging wild meats for grains. Valley residents also produced mats, baskets, tobacco, fish, chickens, and bushmeat (primarily hippo) for trade and sale elsewhere. Local lineages directed most of their young men to become laborers throughout Southern Africa, to supply their relatives in the valley with cash and manufactured goods. With reference to official oversight, this valley economy differed noticeably from the rest of Mpika District, particularly because getting there required foot caravans that took weeks and wildlife numbers were increasing rapidly.

Successive district commissioners commented on this distinctness for some looked forward to their official visits in the valley as a romantic interlude in the heart of "merrie ole Africa." As D. C. Treadwell wrote about his official *ulendo* [foot safaris with carriers] in 1947: "This area, far removed from civilization, and devoid of educational facilities, is yet peculiarly happy, and is one of the more interesting of the Native Authorities areas of Mpika District."[27] Yet such patronizing sentiments never entered the minds and field notes of most agents of the Department of Game and Tsetse Control (DGTC), who considered the hapless residents within the Munyamadzi Corridor as persistent culprits "poaching" on their department's turf and to determined to do what they could to undo "Pitman's legacy."

An early advocate of this position was Maj. W. Eustice Poles, the European game ranger stationed at Mpika in 1946 until the mid-1950s. He seemed to have arrived on post with this objective in mind and wrote loathingly about valley residents, particularly its Chief Nabwalya, whom he distrusted as a nemesis. In 1947, Major Poles followed Colonel Pitman's 1932 journey into the valley during the same season to make comparisons. He saw less game and determined that conditions had worsened since Pitman's visit. Poles wrote in 1947 that

> One of Col. Pitman's strong contentions is that a meat-loving African agricultural community and game cannot exist together. I believe that this is very true and therefore find it difficult to reconcile these views with his recommendation . . . that the Munyamadzi

> area should be allowed to remain under native occupation. . . . The meat-loving African is ever insistent for game destruction on his behalf and is never satisfied. I predict that the Munyamadzi Corridor, which even now is responsible for considerable wastage of game from the Reserves, will prove a very real danger to the future welfare of game and the government's policy of game conservation in the Luangwa Valley. In the long run, the interests of human beings must take precedence over those of game . . . the African's voice is becoming ever more insistent.[28]

Successive generations of European as well as Zambian wildlife officials have found themselves dealing with this perennial "Bisa problem" and using their pejorative vocabularies and other means to get their points across. For some, the adversary was "unrestrained" hunting ("poaching") that kept wildlife from reaching its potential numbers ("carrying capacity"); for others, it was a truculent chief failing to follow instructions and discipline game violators; still others employed a vocabulary focused on the "primitiveness," "backwardness," "ignorance," and "irrationality" of its residents. Some wildlife officials still use similar cultural slurs altered to fit current discourse, to express their management dilemmas with the corridor's inhabitants.

As Astle (1999) claims, the colonial administration may have realigned the boundaries of the South Luangwa Game Reserve taking in villagers' comments and concerns; at other times, it expanded *de facto* its own claims without any consultations. As the official lines demarcating this park's northern boundary were defined on legal paper as the course of the Mupamadzi River to its junction with Luangwa River at one point in time, the former river's subsequent perennial pulses of water through its channel banks of alluvial sand have become erratic as well as persistent problems in human relations there. The human dimensions of these shifting sands near the confluence of these two rivers illustrate how physical factors may trump shifty bureaucratic protocols.

The original definition of this northern boundary in 1938 was the course of the Mupamadzi River to its confluence with the Luangwa River. Subsequently, under heavy rainfall, the river shifted its course some distance further south to the Chibembe stream before entering the Luangwa River. This alteration isolated several Valley Bisa villagers from water and their fishery. As the

Game Department remained adamant about retaining the original boundary, this persistence spawned contentious correspondence between various government offices.[29] The department always suspected that these villagers were consistently trespassing into the reserve, assumed "notorious poachers" among its headmen, and fought against constructing wells to compensate for their lack of water. Its officers hoped to reach an eventual consensus with other government agencies to resettle these residents and eventually join the two reserves. The Mpika game ranger opposed the DC's plans for building wells and wrote his provincial biologist that "the more Gov't. (sic) funds sunk in the area, particularly in the form of permanent amenities will later add to the difficulty should it be desired to clear this area of its population and embrace it in the Luangwa Valley Game Reserve."[30]

In championing local residents' rights against Game Department ambitions, the DC at Mpika granted permission for residents to enter the reserve for water, supported the digging of wells, and remained against the removal of the affected villages, as it possessed no evidence for the alleged "poaching." He wrote his superior provincial commissioner in May 1952, attaching an extract from the Game Department, which "shows an unequivocal desire to obtain the removal of the people altogether from the Mupamadzi River." He noted that "if the proposed wells are a success, the Bisa Tribal Council may consider the proposal to re-define the boundary more favorably" and that if the DGTC assured residents that they had no desire to remove them, this would help alleviate "the natural anxiety of the Bisa."[31]

Department plans to resettle Nabwalya residents and to join the two Luangwa Valley game reserves in a united national park were discussed in correspondence prior and more openly at the Southern Area Development Team when it met in April 1954 at Mpika. The department hoped for a compromise by creating a Nabwalya game reserve around Chifungwe Plain with Mupamadzi residents resettled along the upper Munyamadzi River near to the escarpment. This plan would connect the reserves along the Luangwa River. Other colonial agents remained divided on this plan. Both Chief Nabwalya and the Bisa tribal council adamantly refused this proposal as they remembered their loss of tribal lands earlier in wars with the Bemba and of the extensive tracts alienated by the government for the game reserves.[32] This setback did not deter the Game Department from resurfacing their objectives for the Munyamadzi GMA after Zambian independence and new allocations to increase their staff and expertise.

In 1966, the newly independent government brought in two wildlife biologists under the UN's Expanded Program of Technical Assistance (EPTA) and placed them under the newly energized Department of Game and Fisheries. Their mission was to survey the Luangwa Valley and draft a proposal for international assistance. The consultants did not analyze other options for economic development; rather, they assumed that wildlife conservation was a necessity everywhere in the valley as their objective. The consultants took up the mantra of their departmental hosts by crafting in their report an imminent "crisis narrative" for the central Luangwa that included the removal of the Valley Bisa. After reviewing materials at their disposal, these experts wrote that the valley suffered from long-term "improper land-use practices," which would increase in proportion to its human population and bring only protracted conflict and mounting costs. For the benefit of "all Zambians," they reported that the valley should be designated as game estate rather than developed for agriculture. The Munyamadzi Corridor became integral to their proposal and provided additional biological legitimacy to the department's claims to demonstrate the value of wildlife's capacities to deliver state revenues (Dodds and Patton 1968; Astle 1999: 81–82).

As for the Valley Bisa, the EPTA team described the Valley People as "a people with simple aspirations and simple needs who respond only to those aspects of their environment upon which life depends." "Corridor people," they scripted in paternalistic and period professional prose, "are not now capable of judging what may be best in the future for their progeny" and prescribed for them a greater future "outside their present environment" (Dodds and Patton 1968: 125). As elsewhere, the Bisa's remaining isolation would lead to "the more progressive young people" leaving and "the people in the Corridor would remain, as they are now, the less articulate, less ambitious and less able, but in growing number" (p. 124). The EPTA team was given to understand by their host ministry that government "was committed in principle to the resettlement of the entire Munyamadzi

Corridor population" (p. 124). This possibility for displacement persisted for years, but never happened. Only in the late 1970s did the government reconsider and provide funding for further human developments in the Munyamadzi GMA.

The Luangwa Valley Conservation and Development Project under the United Nations Development Program lasted until 1973. When the government declared the South Luangwa National Park by statutory instrument in 1972 under the National Parks and Wildlife Act of 1968, its boundaries included the Chifungwe Plain, a large chunk of the Munyamadzi Corridor. Dodds and Patton (1968: 46) described Chifungwe Plain as "one of the most productive elephant habitats in the valley" needing immediate protection and conservation. Astle (1999: 40) noted that the inclusion of Chifungwe Plain in the national park was accomplished "without the detailed discussions with local villagers, which had characterized the determination of the boundaries of the original Reserve in Colonial times."

The alienation of this additional Valley Bisa land, including earlier much of that within the Luangwa Valley game reserves, occurred under the long tenure of Chief Nabwalya, who died in 1984. The legislated park boundaries between the South Luangwa National Park and the GMA along the lower Mupamadzi River and near the chief's palace [that included the excised Chifungwe Plain] remained ill-defined on the ground. Although some residents continued to trespass into the park and confronted occasional wildlife patrols, the loudest protests for these ill-defined boundaries came from safari operators accused by NPWS personnel of incursions into the park or of "unethical" luring of animals (lions) out of it.[33] This appropriation of additional land by the state continued as a thorn in local memory but seemed to erupt suddenly in an incident under the new chief, appointed in 1990.

In 2002, a diligent wildlife officer from Mfuwe unilaterally undertook to permanently mark the eastern boundary of Chifungwe Plain and the Munyamadzi GMA. This officer did so (apparently inaccurately) and without consulting the chief (Blackson Somo), local residents, or even the game warden at Mpika. This intrusion, without showing "proper respect," deeply grieved and angered the chief and added to other burning issues within the inexperienced Zambia Wildlife Authority (the successor to the

NPWS) staff. The chief held meetings and wrote letters of complaint to the Zambian president, as well as to ZAWA, the minister of tourism, the Constitution Review Commission, and several international organizations. After attending a meeting with the aggrieved chief, the sympathetic new area warden, who had earlier served as a unit leader under ADMADE within the GMA, described the situation to his superiors in ZAWA.

> You may wish to know that the situation in Nabwalya is very bad. Chief Nabwalya and his subjects are very bitter about this unfortunate development. The Honourable Chief [*sic*] has even gone to the extent of saying ZAWA should remove all the wildlife species from his area to avoid similar problems in the future. This I repeat is a very unfortunate development although it might sound an exaggeration to you. The deputy minister also expressed similar concerns about the same issue and he is very upset about the whole issue . . . I also still stand that it was not done in good faith, even if there could have been some substance of truth in what they did. It is totally unfair. My fears are that such actions are going to render our working relations with the local communities difficult if we do not involve them in such sensitive issues.[34]

The issue remained unresolved and intensified subsequently. When I returned to Nabwalya in 2006, the headline in the *Post* read, "Chief Nabwalya Tells ZAWA to Leave His Area" (the Zambian national *Post*, June 16, 2006; Marks 2014: 30). The chief alleged "that ZAWA had brought more problems by turning part of his land in[to] . . . South Luangwa National Park. . . . They should remove the beacons, if they refuse to leave that would be the beginning of war. We will look after our own animals." The chief claimed that the state had brought about poaching and had been unable to control the situation, and the authority should be given back to chiefs.

In the questionnaire used to assess the Zambian "community-based" wildlife program during 2006, we asked residents for comments on their reputation as the "place of wild animals" and about a plausible decision to resettle a village in order to create space for wildlife within the GMA. Most respondents (82–94 percent, depending on village location) agreed with the statement, "Nabwalya is known by most outside people as a place of wild animals," yet an adamant group strongly voiced the opinion that residents made Nabwalya important, that the animals had followed the people into the valley because the animals depended on people and still lived around them. People were the initial colonizers and owners. Regarding the second question, the majority responded that the chief, as "owner of the land," might possess the authority to move a village, but only if a better site was agreed upon by those affected. A few were against resettling people for any reason. In concluding his interview with me, an elderly Nabwalya resident summarized his opinion by integrating identity, place, land, and culture in the following telling way:

> Animals are now much freer than we are. They have more protection and rights. We are restricted and mere objects in our own land. Many government leaders, who passionately enjoy the revenues from wild animals, wish us evacuated from the valley to allow Honourable Animals to walk, reproduce, and graze freely. We will die for and in our land. I trust you understand that this lack of access to game meat is at the root of most politics here and the source of our scarcity of relish today." (Marks 2014: 48)

Contextualizing these Materials

All organizations, lineages, churches, governmental institutions and private agencies are products of "human culture"—an abstract synopsis of a group's learned behaviors and the presumptive ideas that seem to inform its members' observed or written performances. Outside of their own groups or structures, individuals may find another group's ideas and behaviors difficult and even challenging to comprehend. In the face of new exposures, spectators may reinforce their own cultural boundaries or remain porous selectively to new developments that appear advantageous under circumstances. If rural residents have held tenaciously onto their culture, identities and relationships while responding nonchalantly to those seeking to transform them, these communities do change, but not necessarily in the manners envisioned by the wannabe change agents. In this book's introduction, the external threats and reprimands selectively focused on community residents during the "poached" elephant episodes seemed never to end until temporarily suspended when a suitable local "scapegoat" was identified to protect those in power. Those in authority, both within the valley community and among the inscrutable elsewhere, needed security should government agents press charges or need explanations. In frontiers or in rural areas, the playing fields of power are never level or equal. Encounters are never simple relationships, for what is at stake are the intertwined threads of history, power, identity, livelihood, and sometimes destinies of involved parties. An individual's brief observations and instantaneous written interpretations, whether insider or outsider, build upon their group's consensual cultural foundations based on the person's place within their group's ranked structure. As one group discusses development or practices their way, their audience may build subtle resistance and buffer their silences and reactions off-stage. I have raised this feature in arguing that the resumption of snaring and other serendipitous strategies among residents appears of this nature. Residents may have acquired the language of development while erstwhile scrambling to maintain their existence and identities in their changing homeland. No group is ever homogeneous or consistent, as both residents and wildlife officers have changed subsequent to the time frames presented in this study; the tragedy is that both groups appear moving in different directions with reference to the GMA's asset of biodiversity.

The department's dilemma appears structurally constrained by its military deportment, its bureaucratic nature, and that the staff seem to lack basic social skills in dealing with other people. The agency's national legislative mandate insists that its staff focus on, and is accountable for, protecting wild animals, enforcing the game laws, and revenue generation. While assuming "the walk" of their colonial predecessors, many Zambian wildlife managers have absorbed also its cultural "talk" as their activities are embedded in their employment derived, directly and indirectly, from foreign ("northern") values, supports, technologies, and funds. Successive generations of wildlife officers have found themselves struggling unilaterally and unsuccessfully with their self-defined and truncated goal of subduing the ghost of their supposed "Valley Bisa problem" rather than negotiating and accepting its imaginative and mutual challenges as creative opportunities for building a future, including their own. Despite the department's articulation of progressive, scientific, and democratic ideals to promote human developments with the ideology of wildlife "conservation," its actions afield remain grounded by its legislative mandate, organizational structures, staff training, and historic legacy.

Whereas rural individuals might flaunt the agency's strict oversight through snaring and the "illegal taking" of an occasional mammal, the office of the chief, and its structural ideological position as "owner of the land," appears to remain the main collective, effective cultural and political counter to outside attempts to take or (re)make their landscape. In the following chapters, I assess this local legacy and its cultural framework for regulating social relations and perhaps offering new ideas useful for crafting meaningful relations with wildlife within this challenging environment.

Notes

1. *Basalufeya bapwenama muconde bweleni; cibanga nkonde naimena*
   Nama maluba yaciaalo muconde bweleni; cibanga nkonde naimena.
2. During the 1960s, some Valley Bisa men working on the Copperbelt formed a political group to lobby against government resettlement attempts and established a fund intended for constructing a road. These funds were offered to the public works department but were allegedly refused because road building, planning and funding fell under the department's purview. Petitions for an all-weather road and for other developments began in the 1970s, particularly after the government decided against plans to resettle the Valley Bisa. Surveyed by me in 1978, most Nabwalya households named road construction as their main developmental need.
3. Wildlife abundance is often a casualty when road construction occurs into new areas. Given weak government support and lack of commitment for conservation, wildlife decreases in the central Luangwa is plausible, especially since gangs of commercial hunters from the plateau continue to make illegal kills of large mammals there despite governmental patrols.
4. These notables and visitors are listed in Mpika District Notebook (vol. 2) in the Zambian National Archives, Lusaka.
5. For example, Sir Harry Johnston's letter to Sir Percy Anderson in 1893 reproduced in Hanna (1956: 266–67). Sir Percy Anderson was a prominent actor in the British Foreign Office (Louis 1966).

6. The specified 2004 wildlife quotas for residents was 35 animals (including 9 buffalo, 15 impala and 6 warthog) and for safaris 135 animals (including 18 buffalo, 23 impala, and 12 warthog). These totals do not include the licensed quota for traditional ceremonies (6 animals, including 2 buffalo), as well as unknown national and special licenses. ZAWA drastically diminished local quotas in subsequent years.
7. A resident in Paison Village, interview with author, 22 June 2006.
8. Based on an account submitted by Elvis Kampamba, a long-time resident, certified teacher, and eventually head teacher of Nabwalya Primary and Secondary Schools. A few details came from additional sources.
9. This table is an accumulation of information from several sources, including the Nabwalya Rural Health Clinic, Kanele Wildlife Camp, local diaries, records, and personal notes. Since most records are listed by name, village, sex, and dates as well as describe the outcome of attacks, my list minimizes double listings. These figures are minimal as circumstances often dictate what statistics are recorded and many casualties are known only informally.
10. Quotes from Commissioner Johnston's *Report on the Eastern Portion of British Central Africa,* 31 March 1894, UK Public Record Office FO2/66.
11. Johnston's Report listed in endnote 10. Johnston's contention was that Africans with spears "naturally selected bull elephants with good ivory" as they realized that cows with calves were dangerous.
12. These draft regulations (dated 24 January 1989) were cited as "The Gun Licences Regulations, 1894." They were specific as to race (Europeans and Indians were excluded) and applied to "every description of fire-arm." UK Public Record Office: PRO FO 2/65.
13. Zambian National Archives (ZNA) Mpika District Notebook, 4/1, vol. 2, p. 385, under Inspection Notes; KSD 4/2Records of Magistrate's Court, Rex vs. Saidi, Case 17/1912, May 7/1913.
14. ZNA Mpika District Notebook under Ndabas: ZNA Lundazi District Notebook KT 3/1, p. 230.
15. ZNA Lundazi District Notebook.KST 3/1 Lundazi District lists twenty muzzle-loading guns registered in 1923, seven in 1924, seven in 1925, four in 1926, eight in 1927, and three in 1928. The records show a few shotguns listed with the muzzle-loaders. This district is directly east, across the Luangwa River, from Nabwalya.
16. ZNA Lundazi District Notebook, 17 October 1929 Ndaba with the provincial governor, provincial and district commissioners and between 4–500 natives.

17. Serenje Distict is immediately south of Mpika District and contains both plateau and valley terrain. Comments attributed to the district commissioner at Mpika are found on page 58 of Pitman's 1931 game notes; figures for the numbers of muzzle-loaders were copied by Pitman from the Serenje District Notebook and are found on page 66; and the quote about those guns in African hands is on page 92. Colonel Pitman's game notes are stored in the library and archives of the Natural History Museum, London.
18. Reference for this assertion is a post independence conference I attended given by Dr. Kaunda, Zambia's first president, on 6 October 1966 for UNZA students (see Marks 1984: 105).
19. This letter addressed to the director, National Parks and Wildlife Service, Chilanga, and to the warden, Bangweulu Command, Mpika, by the Honourable Chief Nabwalya, was dated 3 March 1990. The chief's main audience was neither the NPWS nor the warden, but the governor, who was chair of the Munyamadzi Wildlife Authority. Neither the chief nor his local subjects respected the existing warden, whose underground poaching in support of his personal enterprises was commonly known and discussed. There were advantages for the chief's association with this governor, who belonged to United National Independence Party (UNIP), the political party that had gained Zambian independence and that had appointed and confirmed him as a chief.
20. Protection of life and property, especially on native trust lands, was an important objective of the elephant control work that began in the 1930s under colonial rule and continued for a number of years after Zambia obtained independence in 1964. There were provisions under the wildlife law for individuals to kill animals in self-defense and in defense of their properties, provided that officials were notified of these activities within a certain time frame. This right, or privilege, was not publically continued under the ADMADE program.
21. Interview with author, Kanele Wildlife Camp, 15 June 2006.
22. Resolutions from Northern Province ADMADE Leadership Conference, held at Mpika's International Guest House, 23–29 April 1994.
23. Minutes of the meetings held by "The Honourable (sic) Chief Nabwalya with his subjects at their respective villages, 8–13 May 1995," page 3.
24. Quarterly and Monthly Reports for Community Liaison Officer (Blackwell Banda) to his coordinator during 2000–2002; copies in author's collection.

25. William Astle was employed by the Game Department from 1965 to 1973 and returned to the Luangwa Valley later to work on his book. His history of wildlife conservation within the Luangwa Valley is based upon archival materials, his employment as a staff officer, and later and by informal contacts with the NPWS staff.
26. Zambian National Archives (ZNA): (49) SEC6/226, folio 202/3/B/37. Astle (1999) reviews what is available of these early discussions about boundaries, including some areas where there was little, if any, consultation with residents. He could find no records of negotiations with Chief Nabwalya. My recollections are that the Game Department always harbored the hope of finding some way of joining the northern and southern sections of the Luangwa Valley Game Reserves and were reluctant to engage in other possibilities.
27. Zambian National Archives (ZNA): SEC 2/838 Mpika Tour Report, August–September 1947,
28. Eustice E. Poles, typed Report 1/1947: Report of a Tour of the Munyamadzi Controlled Area, 236 miles on foot in 16 days, 3–21 February 1947 (excerpts from report in author's possession).
29. ZNA 44, 46.48 SEC 6 contains relevant correspondence. These disputes are mentioned in Astle (op cit. pp. 39–40). I also cite additional correspondence, copies of which are in my possession.
30. Letter from Game Ranger Mpika to Biologist Kasama` dated 22 February 1952, marked confidential, Munyamadzi Controlled Area. Northern Rhodesia, Department of Game and Tsetse Control, Mpika.
31. Letter from the office of the district commissioner, Mpika, to the provincial commissioner, Kasama (REF 184A/52[715]), 27 May 1952. Residents did not follow the DC's rules to fetch only water inside the reserve. Under pressure from the game ranger, he subsequently withdrew his permission.
32. District commissioner, Mpika, to chairman, Southern Area Development Team, Kasama, referenced Munyamadzi Valley, 27 February 1954; provincial game officer, Kasama, to game ranger, Mpika, 25 January 1954; game ranger's office, Mpika, to provincial game officer, Kasama, 15 March 1954, referenced Munyamadzi Corridor—Future Policy, marked confidential; Mpika District Notebook ZNA 4/1, vol. 2, under Records of Discussions, 8 December 1954.
33. The professional hunters involved informed me of these accusations in 1993, 1997, and 2006.
34. I visited Nabwalya shortly after this event and was shown copies of these letters. I was also taken to the Mupamadzi area and shown where land allegedly was removed from the GMA.

Section II

On the Quest for Local Sustainability

Chapter 4

A Cultural Grid

Making Sense of the Natural World

When you go into the bush, you go for everything[1]
Bisa Proverb (1966)

What westerners call "nature" is hardly natural at all, for, as William Cronon (1996: 25) notes, nature is a "profoundly human construction" and so enmeshed in our assumptions and values that in discussing "nature" we expose a lot about our personal dispositions. In this chapter, I will examine ways that some Valley Bisa used their cultural imaginations to describe life and the events around them. Like us, these men and women often use familiar terms and many of the same categories to describe themselves as well as their surroundings. The perception of animals and plants recorded here for the Bisa is more akin to that of practicing residents on the English or American countrysides during the sixteenth and seventeenth centuries than to current urbanites, adventure-seekers, or academics (Thomas 1983; Ritvo 1997; Anderson 2004).

The Valley Bisa did not separate themselves customarily from other forms of life, nor did they typically objectify "things." They did not have an indigenous word to match the northern abstractions for "nature," "environment," or even the more recent "biodiversity."[2] The Bisa noun *nchende* sometimes translates into English as "place" or "environment," yet its vernacular meaning includes both people and the assets that sustain them in a particular place. The basic presumption was that the human and the "natural" worlds were parts of a seamless whole.[3] Both were integral parts of a landscape: the productive bush responded to and was responsible for sustaining village inhabitants in the same manner that it remained the abode of the Bisa's ancestral spirits and therefore extensions of themselves. No domain of "nature" existed outside the morality of the human community, for reciprocal obligations extended outward and embraced other forms of life and spirit. Yet causality embedded in moral principles and human intentionality still remain basic factors as to why "good" as well as "bad" things still happen to and among the living. Bad things happen because someone, somewhere, has violated ethical expectations, not natural, mechanical principles. As Elizabeth Colson (2006) notes for the Zambian Tonga, people need an explanation for disorder rather than order.

I begin this chapter with a series of events I witnessed during a two-night bush encampment in 1973. This experience provides the context for grounding some local categories and meaning for an event at one point in time. Themes, such as the categorical spaces of bush and village, as well as the efficacy of colors, witchcraft, and spiritual agency, serve as explanatory threads connecting and interlacing these cultural domains. Residents may acknowledge the more persuasive cultural threads, yet their plausible connections and interpretations within specialized domains, such as hunting, are restricted to the few practitioners who act creatively with them. Among hunters at Nabwalya, this understanding reflected the traditions of their lineage, their ages and training, their experiences. Even their personalities, and their wisdom changed as they became progressively familiar with wild animals and exposed to other ideas. Based on metaphors and figurative reasoning, these ideas enable individuals to scan a scene or event as they simultaneously employ their culture's classifying and personal or experienced meaning threads to interpret it. Such cultural procedures require knowledge of folk categories deeply embedded in language, identities, and images (Johnson 1987; Morris 1998: chapter 3; Levi-Strauss 1966).[4] Although this chapter's epigraph suggests the bush rewards those venturing within it, common sense based in experience demands that one should proceed there prepared for the unexpected.

While "mammals" may have an autonomous reality of their own, people recognize and relate to them through cultural images. Valley Bisa concepts that I translate loosely, for example, as "wild animal," "mammal," or even "wildlife" often embody subjective and cultural elements not conforming to northern notions considered as specific and "objective" categories. Local hunters' categories reveal what their prey is like, as well as the behavior that hunters expect during encounters with them. Their domains express local interests, needs, and uses, which for individual hunters may further encapsulate their personal fears, histories, and experiences with prey. Traditionally, hunters addressed resource management issues tangentially through beliefs in capricious spirits, observable and culturally mediated rituals and prescriptions, and by following normative processes in the distribution of its proceeds. In their interpretations of unobserved agents and observable agency, lineage elders monitored the compliance of subordinates within this "spiritual" system as they imaginatively structured the cultural means to restore legitimacy and order in times of crisis, uncertainty, or in the aftermath of suspected slights. This homocentric system presumed that if hunters remained true to their protocols, the "spirits" overseeing the prosperity of prey and community welfare would ensure the future of both.

Making some Connections: An Encampment in the Bush

In August 1973, I accompanied three local hunters and an understudy for three days on an encampment in the bush (*malala*). Each hunter belonged to a separate matrilineage, yet all were related to each other in the intergenerational and classificatory way that reflects generations of intermarriages and alliances among neighbors. Chibale, the elder hunter, was seventy years old, a noted village headman, and long-term counselor to the reigning chief. Chibale's grandson, the teenaged Lankson, accompanied him as a personal assistant and understudy. Chibale was not just an elder; he was also the classificatory grandfather of the two other younger men. Both Lowlenti and Hapi had spent some time under his hegemony as their headman, and they often borrowed his muzzle-loading guns. Lowlenti had even selected a cross-cousin in Chibale's village as his first wife, and he had made his first primary residence there. Hapi's mother's sister (whom Hapi addressed as "mother") resided in Chibale's village, and Hapi had also married and divorced a woman living there. For these and other reasons, Chibale was the person to whom we all deferred to interpret events as our journey proceeded. He assumed ritual responsibility for the group and acted as our guardian as we moved further from village life.[5]

On the first night, we encamped in a thicket[6] close to a water hole some fifteen kilometers north of the Munyamadzi River. We found few animals here, so early next morning we backtracked along the edge of the eroded foothills and positioned ourselves for the second evening on a ridge with an overview of the surrounding terrain, including a deep water hole beneath a spreading ebony tree.[7] My companions pointed out that in earlier times, spear hunters sat in this large tree awaiting the appearance of elephants and other game. Spearmen would wait, perched on a limb, until an animal moved directly below before launching a poisoned spear between the front shoulder blades. Later that afternoon, three elephants approached this water hole; they sensed our presence, trumpeted, and retreated.

Soon after our arrival on this promontory, Hapi demonstrated his mastery over warthog curiosity and behavior. Despite the incessant chatter of a nearby honey guide[8] enthusiastically announcing our presence, Hapi used the sunlight reflecting on his gun barrel to entice an inquisitive warthog increasingly closer. Once it was within range, he teased it to shift from a frontal to a broadside stance, thereby providing a larger target. We spent the heat of that afternoon butchering, curing, and drying warthog meat.

Toward evening, we withdrew to a commanding site on the ridge. The younger men cut mopane branches and placed these and other dried vegetation along the boundary of our designated sleeping space. As it was a full moon, these branches would provide warning should a hyena or other specific dangerous animal (*cizwango*) seek us harm. We deliberately created a "human space" within the surrounding alien bush while Chibale, as our guardian, went alone to collect "medicines" (*muti*[9]) to add yet another layer of protection. At sunset, Chibale proceeded east and found a mopane tree with a slight hollow. In this hole, he placed a mixture of dry grass and sticks for "protection" and to hold a smoldering fire. He then went west, south, and north, finding trees in each direction in which to burn his "medicines." Returning to us, Chibale placed two branches from a shrubby thorn tree on the fire. When I asked what he was doing, he replied only that this prescription would "prevent fierce animals from coming to harm us while we slept."[10]

Within this protected space, Chibale and Lankson slept on their backs with their heads pointing west, their faces in the direction of sunrise. In contrast, the rest of us slept stretched in parallel north-south lines, with Hapi and Lowlenti on either side of me. In the early morning hours, Hapi dreamed of a porcupine that, undetected, invaded our encampment and stomped on his back. This vivid dream so startled him that he awoke and sought an explanation. He first consulted his peer, Lowlenti, before elaborating on its details for Chibale. We spent the rest of that night vigilantly apprehensive, for Chibale interpreted the porcupine dream as an ominous sign (*cibanda*), suggesting that something bad, which we had yet to experience or witness, would occur. When nothing happened that night, he decided that we would return to the village, expecting that we would learn of its meaning there.

In the morning as we followed the line of eroded foothills back to the village, we became aware of intermittent growls and aggressive noises some distance to our right. Unable to determine the nature of the sounds, the younger men suggested we investigate. As we ventured slowly forward, my companions appeared more curious than apprehensive. The lead, Lowlenti, had an axe in his hands, while Hapi came last with a muzzle-loading gun unprepared for firing.[11] As we rounded a slight rise, we suddenly came upon three lionesses and six cubs in the process of disemboweling a freshly killed zebra. The hunters shouted strong imperatives in the Afrikaans [South African Dutch] and, surprisingly, perhaps only to me, the lions quickly abandoned the carcass and

FIGURE 4.1. An unanticipated gift and omen to interpret an evening's dream. Returning from an overnight encampment, we found a zebra freshly killed by a pride of lionesses and their cubs. Headman Chibale, with his walking stick and spear, stands to the left and directs his grandson, Lankson, where and how to retrieve a small portion from the carcass with an axe. Lowlenti is standing on the right. (Photograph by the author, August 1973)

retreated further up the slope out of sight. Immediately, Chibale instructed Hapi that this event was the one forecast in his dream. Hapi's guardian spirit had foretold a positive event (by metonym) instead of a figurative and dangerous omen as originally interpreted by Chibale. The key to determining this resolution was that the zebra and the porcupine belonged to the same "striped animal" (*vizemba*) category. The hunters took a token piece of meat from the zebra's flank, leaving the rest for the accommodating lions, and we proceeded on our way back to the village. Along the way, Hapi allowed that Chibale's explanation had satisfied him, for if we had not come upon the dead zebra, his dream would have continued to haunt him.[12]

In the following days, these events provided grist for numerous conversations about how residents characterized mammals in both practical and symbolic ways and about local ways of "knowing." The following threads and thoughts recognize that the formality of such an exercise discounts the fluidity of expressions as well as subtle difficulties in translating their meanings.

A Cultural Grid for Sorting Life's Forms: Cornucopias of "Goods" and "Bads"

People do not simply adjust to their environments; they create and recreate themselves through their activities and through their visions of what they see and use. Many rural African societies locate themselves in space via the contrasting, oppositional domains of "village" and "bush" (Croll and Parkin 1992; Morris 1998; Colson 2006). People's thoughts about their environs and assets are not separate catalogues, despite the foreign methods I engaged in to clarify their coherence and meaning, but reveal complex abstract and differently positioned philosophical worldviews. When people classify, they proceed to mark distinctions whose meanings explicate human experiences and lives. Such distinctions are provisional and transient interpretations of underlying values, as conditional perspectives change with time and through circumstance. Changes materialize as individuals give voice to their practices of coping, experimenting, and living while hesitatingly expressing these outcomes through seemingly timeless cultural vocabularies.

The framework that Chibale articulated while preparing and protecting our encampment in the bush made sense then for his audience. Later that evening as listeners, we instantly deferred to his immediate ominous interpretation of Hapi's dream. Likewise, early the next morning, we unilaterally accepted Chibale's more positive reassessment and analysis upon the unexpected provision of a zebra carcass. We had experienced something unexpected and mysterious in Chibale's ability to make meaningful connections between two states (a dream and an encounter), emphasize their similarities (porcupine and zebra) even as it was presented to us through an unexpected agency (lions). It was as if some unforeseen force had revealed its presence and thereby effectively changed our moods instantly from despair to positivity. Yet these interpretative frames were not immutable except perhaps for this elder's generation, who had invested in these

guideposts from experience over much of their lives. By 2006, I encountered few men with Chibale's experience, or others who would characterize their world with his same authority and verve, or find a group of such attentive, vulnerable clients.

The Village and Bush as Metonymic Spaces

For the Valley Bisa, a homestead or village (*muzi*) is a space constructed, ordered, and managed for most human activities—intimacy, sleeping, socializing, storing a harvest, instructing youth, preparing and consuming food, drinking beer, and celebrating. It is associated with what is human, known, cultivated, and cultured. During the daytime, the village descriptively becomes "hot" (*-kaba'*), packed and bristling with activity and life. The grounds are cleared of vegetation and swept clean of debris each morning by women or their children. "Village" is a metaphor for ordered human life, the way it should be lived according to the many expectations of those who constitute a group. In their routines and imagery, residents go about their daily grinds as cultivators, hunters, traders, diviners, food preparers, migrant laborers, fishers, as well as perform their lineage roles as grandmother, mother, uncle, daughter, father, nephew, or headman.

At the edge of every village homestead is a "rubbish heap" (*cizhala*), a symbolic valve of depressed space, through which refuse materials are discarded to begin their slow eventual incorporation into the neighboring bush. Yet a village's garbage contains esoteric and restorative properties useful for concocting potions or in rituals to rebuild former capacities or to dispose of misfortunes. Hunters use these places to wash and restore their guns, symbolically disposing of the disruptive forces influencing their weapons.

The contrasting sphere of the "village" is that of the "bush" (*conde, mpanga*). This space is devoid of human residence, yet out of it comes many items and foods that sustain village life and nourish its residents. The bush is only partly known, especially the nearby shrubs where most people spend some time each day and from where they clear and cultivate fields and gather wild plants and other resources. Knowledge of the bush depends upon one's activities and goals, as it is simultaneously the shadowy world of dangerous powers as well as contains vital and regenerative energies for both men and women. The far bush is a metaphor for disorder, dangerous and mysterious powers, the site of confusing transformations. It is the abode and domain of wild animals, spirits, and witches. This landscape is known by few, mostly men, who go there to hunt, fish, gather other products, or just pass through. These individuals, like Chibale, are aware of the dangers present within this unfamiliar land as well as the available powers to sustain and protect human life. Chibale was mindful of our visible "hotness" and remained apprehensive of our vulnerabilities within this alien and descriptively "cold" (*-talala*) environment. As elders had before him, he knew how to protect human bodies, spaces, activities, and accomplices while remaining constantly vigilant to the whims of the "spirits," which might appear in unexpected ways.

For nearby Malawi as in rural Zambia, Brian Morris (1998: 121) shows many of the complementary dualisms implicit in these contrasting spaces:

Village	**Bush**
agriculture	hunting
wet season	dry season
matrilineal kin group	affinal males
(blood)	(semen)
living humans	spirits of the dead
domestic animals	wild animals

Morris suggests that "dialectical oppositions" in the social practices and ritual lives of many rural Malawians reflect an ambivalence of communal fears and respect toward the uncertain bush and its products. In terms of the bush's relationship to their primary dependence on agriculture, Malawians are aware that its denizens are hostile and destructive of human welfare and enterprise. On the other hand, the "bush" is the source of life-sustaining powers and materials, the ultimate guarantor of the human order. Wild mammals typify these dangerous forces, such as during instances when one suddenly encounters a dangerous wild beast or when a herd of elephant devastates a field.

Yet mammals are a treasured source of meat, as well as a source of activating particles for important charms (*cizimba*) associated with vitality and the spirits. As an essential source of fertility for a matrilineage, in-marrying men are associated with the uncertain domain of the "bush," or elsewhere, yet contribute significantly to the reproduction of kin groups and to the continuing (re)production of the social order itself.

Whereas the "bush" produces many "goods," it also harbors unexpected dangers or "bads." In the bush, villagers bury their dead, and unsanctioned sexual activity occurs there out of bounds. The bush is a source of powerful prescriptions to recapture reproductive powers or to destroy an "other." In a comparable way, "village" can become the operational space for the minds of witches, whose immoral activities at night with their substances and animal familiars destroy human relations and feast on its wreckage. Symbolically, the well-groomed grounds of villages as well as the tangles and thickets of the nearby bush are where "good" and "bad" things, respectively, happen to people.

Such contrasting domains may help to structure the oppositions explicit in some rituals and beliefs, but as Morris cautions, a simple grid of oppositions cannot adequately encompass the cultural or creative realities of human lives. The deeply rooted and interconnected "forest of symbols" (Turner 1967) that continues to energize rural lives builds upon the conceptual legacies of specialists' imaginations as they explore ways to capture human circumstances and experiences. Realistic interpretations and translations are not captured in a day, a year, maybe even a lifetime, for their relevance may shift unexpectedly with time and in different situations. In 1973, what "worked" for Chibale's expositions seemed to have little resonance for subsequent generations, who had experienced different slices of life and times. In coping, many of them had extended their ideological boundaries increasingly outward, incorporating new threads of meaning and more distant contacts. In 2006, a Pentecostal convert would base his conversation literally on a biblical text, largely memorized and contrasted to the errant ways of thinking and acting of immediate predecessors. The converted's perceived remove from earlier ways would serve as a marker for the progressive life they were trying to live (Marks 2014).

A Utilitarian and Familial Taxonomy

Visible, wildlife is an exceptionally situated subject for thinking about human society. Like people elsewhere, the Valley Bisa organize and sort life into various groupings based on assumed and assigned similarities. Analogous to humans, mammals have agency, awareness, and intentionality; they possess knowledge with some assumed sensitivity to local cultural sentiments, such as sanctions, shame, and pity, in humans. All mammals have intrinsic value and possess "inherent powers, properties and potentialities that are independent of humans" Morris 2000: 36). Mammals may exist on their own, yet their recognition by humans (*bantu*), as "elephants" (*nzovu),* "lions" (*nkalamo*), or "buffalo" (*mboo*), are products of language and social representations within a particular place and time. These cultural constructions shape values toward these entities as well as build expectations about them.

The Valley Bisa refer to any edible wild mammal as *nama* (or *nyama*), a cognate term widespread in Africa that typically may exclude predators and many smaller mammals. Schoffeleers (1968: 406) suggests that besides "edible quadruped," *nyama* also refers to a transcendental quality in "the spirit or power released by the blood of a slain person or animal" (cited in Morris 1998: 146). This mystical force, associated with the vengeful nature and potency of unspecified "others," is prominent in beliefs about killing larger mammals, particularly the lion, elephant, and eland (the largest antelope). As a general term for suitable prey, *nama* is the opposite of *cizwango,* the term for any wild animal (including reptiles as well as mammals) considered potentially hostile or predisposed as harmful to humans. These categories refer respectively to the beneficial (as meat) as well as the negative (harmful, hostile) relationships of animals with humans. The Valley Bisa use these terms flexibly and circumstantially. Buffalo (*mboo*) is prototypically *nama* as its meat is eaten by nearly everybody. Yet a specific buffalo demonstrating dangerous, aggressive behavior becomes *cizwango,* and people take care to avoid it, if possible.[13]

Hunters distinguish larger males from females and from younger animals. The categories describing classes of elephants are more detailed than those for other species. *Nzovu* is the term for elephant, *nkungulu* a large male, *nyinanzovu* a cow, *cipembe* an immature with no tusks. Adult elephants with aberrant tusks are also categorized: *tondo* (*nyungwa*) is an adult with no tusks, *cibuluma* an elephant with one tusk, and *sante* an elephant with four tusks. A person feeling vulnerable to elephants uses none of these terms, but rather its nickname (code), *munyepe.* Supposedly, such codes remain secrets from the animals; otherwise, they might appear suddenly, as if summoned. Other dangerous mammals also have code names such as *pundangala* or *mundu* for lion and *cigwele* for hippo.

Broader practical categories lump species based on size, possession of claws or hooves, and presence of skin or fur. *Nama nkulu* (large animals) include elephant, rhino, buffalo, and hippo; *nama mpele* (medium-sized) include most antelopes and zebra; *tunama tunono* (smaller species) include the puku, impala, bushbuck, warthog, grysbok, and other small animals. On the basis of their possession of claws (*nama zyamabutu*), lions, leopards, and hyenas are grouped with wild dog, civet cat, mongoose, and mopane squirrel; although some of the latter may be further differentiated by having the diminutive prefix attached (*kanama kamabutu*). Hoofed mammals (*nama ya mabondo*) include antelopes and buffalo, while hippo, elephant, and rhino are *nama zyambombo* (with toes). Furry animals (*nama yamasako*) include most of the antelopes, predators, and smaller species, such as the bush baby and mongoose. Animals with edible skins (*nama ziankanda* or *mpapa*) include bushpig, warthog, porcupine, elephant, and rhino. These last two groups indicate the usefulness of skins in addition to that of meat (Marks 1976 [2005]: 92–99).

Some hunters also sort species into groups based upon a triad of basic colors—red, black, or white. In this scheme, most large mammals fall into either the black (rhino, elephant, buffalo, warthog) or red (lion, hippo, and most khaki antelopes such as the puku, impala, and roan). White mammals, including albinos, are rare and when present serve as "protectors" for other species with which they associate. A few, like the porcupine in Hapi's dream, display two prominent colors and constitute another significant category of marked animals, *vizemba.* Given that the context of a dream is often inconclusive or ambivalent, it may cause angst until appropriately connected to an event. This was the source of

Hapi's apprehension upon Chibale's initial labeling of his dream as inauspicious (*cibanda*) rather than as auspicious (*mupashi*).

The *vizemba* group (mammals with stripes) includes the kudu, eland, zebra, and bushbuck, as well as the porcupine. Some people prohibit the consumption of flesh from this group for people prone to suffer certain types of sickness (skin rashes, diarrhea) or convulsions. A neighbor attributed a rash that appeared on Rose Chiombo's daughter's face to bushbuck flesh that she had consumed, illustrating this connection (see chapter 2, entry for 5 December 1988; pp. 107–108). If consulted, diviners may also prescribe such abstentions. The reigning chief (in the 1970s) did not eat zebra, yet others of his lineage did so. The largest member of the *vizemba* group, the eland, has more menacing associations, including the plausible death of its killer, onlookers, or kin.[14]

The general term *mutundu* (singular, *mitundu* (plural) meaning "kind" or "variety") refers to members of a specific species as well as to affiliations among humans. The Valley Bisa assume an apical affiliation among animal groups as among themselves. Elephants and pythons are primordial or apical types for mammals and snakes, respectively. Their size and strength give them dominance, infer descent from, and imply "respect" from other species. Other features associate them with other species.

> The elephant is the "mother of all animals" and we call it *zimwe-zimwe* (giant or monster) because of its size. An elephant has no fear of people or for any other animal on earth. It does a lot of damage to people's crops and activities; people fear it because it is difficult to kill. It uses its feet, tusks, trunk, and weight to kill people. When it moves, there is no sound because of its padded feet. It belongs in the same family group as the warthog, bushpig, hippo, and the rhino as they have similar body features—similar skin, little hair, teeth that hang outside, a big head and the same tails that stand straight up as they run from enemies. Each screams loudly when angry or threatened. Their skins are gray (blackish) from bathing in muddy water. The tusks of the warthog, elephant, bushpig and the hippo and the horn of the rhino feature in trade and in certain traditional medicines. Their tails are in high demand as potions. All these animals have the same meat texture while, as the [metaphorical] "mother" of all animals, elephant flesh also includes that of the fishes. Because of this incorporation, some people do not enjoy consuming elephant meat because it may contain a flesh prohibited to them.

The python (*luzato*) is the "mother of all snakes" because it is the biggest snake in Zambia. When the python reproduces, it bears many species (sic) that scatter. If bitten by a python, a person becomes immune from other snakebites for this "mother snake" leaves a smell that other snakes fear. If bitten by mistake, the python's saliva neutralizes all other poisons as it controls other snakes. Pythons possess the colors and venom of all snakes just as elephants contain the flesh of all other mammals. People use python fat for treating some diseases such as earache, severe burns, and reproductive problems. It is poisonous and one should not swallow it.[15]

Mammals belong in groups (lineages, *ulupwa*), with the largest considered the ancestor for the group. Antelopes and buffalo comprise such a subsidiary grouping with the latter its apical ancestor "as it is bigger than the rest and can be near them without any problems." Predators belong in another lineage. The lion is dominant in this group, kills its prey by itself, and roars to frighten all other animals in the bush.[16] As do their human counterparts, predators use "medicines" (charms) to enhance their abilities.

The lion has hunting medicine that it hides under its armpit. When it kills an animal, the lion dips this medicine into the blood of its victim to increase its power for catching additional animals. This magic is round in shape and is composed of the fur of all animals caught before. When the lion feeds, it hides its medicine some distance from the kill site. When people disturb it, the lion runs away and collects its medicine without people knowing where it is. If the lion finds people where it has stashed its medicine or people try to follow it, the lion threatens them "heavily" so that they run away and the lion can collect its charms. As a lion is shot and dying, it throws its medicine away where people cannot see and use it. If they find this medicine, hunters use it to enable them to kill easily all the animals that the lion has.[17]

Like the lion, other predators have distinguishing characteristics and exhibit behaviors with magical properties sought by certain people to achieve their objectives. Their acknowledged purposes are wealth or welfare yet they may have sinister designs as well.

Although powerful, the hyena makes very few catches because it fears other animals. It moves with lions to obtain leftovers. The lion is aware of the hyena and leaves something for it. The hyena sleeps during the day and dreams where the lion has made a catch, so that as soon as it wakes up, it can proceed straight to where the lion is hunting or catching prey. The hyena's dreaming is exact and that is why many people seek the brain of a dead hyena so that they can dream what is going to happen, or what is happening where they are not. Those who use snares also seek hyena brains. These brains enable a trapper to know, through their dreams, if his snare has caught an animal and if wildlife scouts might be waiting there to arrest him. This potion enables the trapper to avoid capture.

The cheetah makes its own catch and is very fast. No animal can challenge it except the elephant, which is more powerful. The cheetah is rarely seen by people and because of this some people use its parts in medicine so that they are not easily seen and do not tire readily while working at tasks.

We observe the leopard rarely during the daytime. When it catches an animal, the leopard takes meat into a tree because leopards do not share. Traditionally, people respect leopards because leopards usually avoid people. When such attacks occur, people say there is something wrong with the person attacked such as misbehaving by making love with a woman in the bush or grabbing the leopard's food or cubs. When a person kills a leopard, he cannot enter the village without using cleansing medicine.[18]

People recognize elephants, lions, and buffalo through cultural representations. The descriptions in the previous examples imply power, food, competition, material substance, danger, and relatedness, all attributes based on long-term association and a history of sharing the same landscape. Living with the larger mammals is hazardous and some of their activities clash with human purposes. Under these conditions, protecting human lives, properties, and sustenance demands continuous vigilance.

Bisa classification of mammals has a strong pragmatic and utilitarian bent, emphasizing the use-values of meat together with other products and the management of adversity rather than "dominion" over "nature." Elvis Kampamba notes that women classify animals differently from most men, for they rarely deal with wild mammals except to preserve or to cook their flesh.

When women encounter wild animals, these animals know that the women are not harmful as sometimes women themselves run away from these animals. Women cannot easily kill animals that they might need for meat. If people see a woman with some meat, they readily know that some man was involved in killing and shared it with her. The only meat that women have access to

are chickens and ducks, which are domesticated. Women mainly classify animals according to accessibility, edibility, fear, and medicine.[19]

Folktales recognize a fundamental similarity between people and wild animals. Stories distort the differences between people and animals whereby some animals become more human or people become more like animals to emphasize important social values. Additionally wild animals and their ascribed cultural attributes feature prominently in numerous proverbs and riddles to express metaphorically both positive and negative attitudes and value. Brian Morris (1998) provides examples of such stories and proverbs.

Animals also play important roles in rituals and religious life. In the divining process, a ritual hunt (*lutembo*) is a useful experience in determining whether the human perpetuator is a member of the lineage or an outsider. In this process, a hunter is deployed to make a decision for the group, which is determined by the sex of the mammal he kills in the bush. If he returns with a female, then the source of the affliction or the witch is declared within the matrilineage; if he kills a male, then someone outside these bounds (an in-marrying man or woman perhaps) is culpable. Some wild mammals are powerful by virtue of the plants they consume or by the nature of their symbolic uses. Warthog tails, which stand erect when the animal runs, supposedly promote potency in men; the nerve tusks of elephants allegedly reduce the same esteemed masculine virtues. Other mammals are given more nuanced consideration depending on specific interpretations and local circumstances. These readings incur human agency linked into that of occult worlds. This domain is full of power and mystery in relation to the spirits and to materials known to only few people.

Shape-Shifters, Spirits, and Animal Familiars

Some animals may not be what they appear: some people and spirits supposedly have the capacity to transform (*kuzanduka*) themselves into animals, achieved through the use of special potions. Animals alone do not have this capacity; only spirits, chiefs, some hunters, witches, and diviners have this power. A dead chief's spirit embodying a lion is generally harmless and protective—an omen, when seen or heard, to remind people to respect the "owner of the land." Witches are believed to control the activities of some lions through magic. These animals are dangerous and detrimental to human lives and properties. They are sent specifically by witches to attack certain individuals. A third type is said to exhibit the confidence of place and a secure identity. It is this gift of self-confidence and leadership that most people admire in "bests" or frontrunners, that are sought by politicians. Each of these types of lion forms has a specific name, exhibits particular behaviors, occurs in particular places, and requires appropriate ways to cope (Marks 1984: 1–2). Besides lions, witches may also control, among other dangerous animals, elephants, buffalo, and crocodiles.

The divinity (Lesa, Mulenga) may appear in various animal forms, such as snakes (python, puff adder), to manifest its presence or concern. Mostly, the divinity is distant, creative, and sometimes destructive, wielding powers beyond the influence of humans. Mulenga occasionally is interpreted as showing concern for residents when it allegedly kills groups of buffalo as they come to drink at the rivers. The drowning of many buffalo is looked upon as an unexpected "gift" of a deity in response to their hunger for buffalo meat.

Historically, the relationships between people and their ancestral spirits are reciprocal, as the spirits are construed as having close and intimate connections with their descendants. People show fear and respect to their ancestral spirits as a complementary part of their moral community by showing deference through requests for ancestral guidance, by observing customary rules, by displaying appropriate behaviors, by performing rituals, and by remembering names. As guardians of the local community and custodians of normative kin relations, spirits reciprocate by ensuring food (such as game animals and good crop yields); by providing special knowledge about the future, cures, or charms through dreams; and by their presumed responsibility for human fecundity and reproduction.

A witch's practices are always assumed to be destructive. As Chibale did in his initial interpretation of *mupazhi* (good spirit) or *cibanda* (bad spirit), classifications were assigned through the

intuitions of the moment. These designations are spontaneous interpretations based in whether the expectation is present or uncertain. The morning sighting of the tangible lion-killed zebra enabled Chibale immediately to change his initial interpretation of his grandson's dream, of uncertain (suspicious?) connections of a porcupine the night before, from a fearful to a beneficial finding. By designating certain events as witchcraft, people are able to explain contingent events in everyday happenings by linking human agency with supernatural powers. Envy and greed are key human motivators connecting events to the actions of supernatural forces, and Chibale (as a headman) decided to return to the village earlier than expected anticipating that the referent in the dream might appear there.

Witches are associated with animals in many ways. People suspect a witch when they observe the untimely appearance or behavior of dangerous animals or when traumatic events happen to otherwise decent people. Long-term grudges, circumstances, and relationships intertwine in the accounts of people accused of witchcraft and assume credibility and focus on marginal individuals in times of social strife.

> Before George was born, an in-law slapped George's only maternal uncle, who died shortly thereafter. Relatives linked the uncle's death to this show of disrespect and informed young George that this incident was the reason why he had no guardian or sponsor. Later in life, George's bitterness and hostility led others to suspect him of using magic (bundles of small sticks like matches) to concoct deadly animals for use against his enemies. His in-laws alleged that George "concocted" a leopard to attack people living in their village, causing several deaths eventually forcing them to huddle in separate gendered groups in separate houses at night. One of George's in-laws managed eventually to kill this leopard with a spear, the singular weapon with power to kill it.
>
> Later George reputedly made a lion using the same magical processes and deployed it toward more members in the same lineage. The spearman killing this lion confronted and accused George of nefarious motives. In the ensuing accusations and postures, George threatened to kill this "in-law," while the latter allegedly reciprocated by summoning a lightning bolt against George. These counter exchanges closed the conversations at the time, yet as its memory lingers, people will resurrect these conversations to contextualize future disputes between the two lineages.[20]

Whereas I recorded these events initially in the 1970s, George and his created lions remained vivid memories decades later. Their occurrence was inscribed publically in the landscape as stories of George's animal familiars were recalled as one passed nearby to where the events allegedly took place. It should be known that these types of lions were dangerous to their owners as well. Those who summoned the animals had to follow prescribed protocols when their "commissioned animal" returned from its mission, otherwise the owner could become its target. Reputedly, George ended his own life by becoming the victim of his own creation.

Prey Attributes and Goals of Local Hunters

All humans intend to kill the wild animals they hunt; yet the methods, motivations, and social tendencies of specific groups are culturally different and variable. Here I am specifically concerned with the local hunters around Nabwalya, initially witnessed in 1966 and the next generation of practitioners during the subsequent two decades, and their cultural incentives and motivations. During these decades, hunting was a pursuit actively practiced by comparatively few resident men and stage-managed by fewer patrons. Embedded in its cultural and spiritual contexts, Valley Bisa hunting was more than just predation or mere food gathering (Richards, 1939; Turner 1967; Scudder 1962; Stephanisyn 1964; Marks 1976; Morris 1998).

First, this activity occurred largely, yet never exclusively, for subsistence purposes.[21] Its fundamental premise was that the bushmeat produced was for sharing among lineage members within allocations sanctioned by elders within an ideology of ancestral monitoring and assistance. While a hunter's skill was important, this attribute alone was insufficient, as the hunter was vulnerable to many unseen dangers lurking in the bush. Other risks lurked in the village even when the hunter was successful and made a kill. Certain kin anticipated certain portions (chest, ribs, and fat) as a gift of "respect," acknowledging both the enabling spirits and as a statement of their relationships. In the recent past, chiefs also expected tribute from the larger mammals.

Second, certain mammals possessed ritual potency. A hunter approached these species deferentially using powerful charms that allegedly shielded him and his kin from spiritual reprisals from the "ghosts" of the prey. Prior to a hunt, protection began with prayers, sometimes with offerings, to the spirits (usually the individual's specific guardian spirit) asking for fortune and favor. Beyond the recitation of supplications, a hunter may carry "enhancing hunt prescriptions," incised on their bodies or carried with them, to provoke their success. Success in the bush was attributed partially to a hunter's possession of "prescriptions," partially to protective surveillance of a guardian spirit, and partially to the protective monitoring of lineage spirits, expected to respond favorably if descendants demonstrate appropriate behaviors during a foray. An elder hunter who had killed the most powerful prey (lion, eland, and elephant) told me that slaying powerful creatures was similar to homicide, in that the killer must properly cleanse his body and, through the enactment of publically sanctioned rituals, "lay the deceased's spirit" in order to protect both himself and his kin from harmful retribution. Hunters were mindful of the tastes and dietary prohibitions of residents within their villages; consequently, they directed their forays toward buffalo, whose flesh had few personal prohibitions and whose pursuit, if confined to those in herds, was not considered generally precarious or vindictive.

Third, hunting always included the use of prescriptions, or charms, whether acknowledged or not. These charms were to ensure success, protect the hunter against witchcraft and misfortunes, and neutralize any harmful effects brought about by a prey's spiritual essence. Other indigenous technologies, such as snares and pits, shared similar prescriptions and processes. Many charms and curatives contained portions of certain animals believed to possess proven influence. Hunters rubbed these medicines into incisions on their bodies, soaked them in solutions for washing, and wore or wrapped them around their necks or limbs. A hunter's persistent success might come with a social cost if his detractors persistently accused him of abusing his powers or of practicing sorcery. Some of these accusations emerge in chapter 5 as three generations of Nabwalya hunters share their life experiences.

Finally, a host of rules separated the ritual and procreative domains of human sexuality and that of hunting. A pregnant or menstruating woman was alleged to mystically affect the efficacy of a hunter's weapon, and her condition might even render him vulnerable while in the bush. Therefore, hunters kept their weapons out of contact with such women and abstained from intercourse the evening prior to hunting. This ban also applied to any accompanying man, and its effect was to keep the hunters "cool" and focused on their objective of pursuing and returning with prey. Contact with pregnant or menstruating women or with any sexual behavior made individuals conceptually "hot," a condition enabling maligned spirits to distinguish their presence in a metaphorically "cool" bush. An aborted fetus from one's sister was a particularly powerful and sinister potion allegedly used in certain "secret" or suspect charms for killing game, and was also an acknowledged ingredient in forms of witchcraft that could devastate a lineage.

Hunters could not please their relatives' and clients' expectations at all times. As with any small group of people existing face-to-face in a small space, envy, strife, rumor, and hindrance are the custodians constantly testing the social demeanor and reputation of individuals. With their apparent repute for killing, hunters are suitable subjects for village commentators. When a hunter made a kill, he faced a dilemma whether its carcass was sufficiently large to accommodate everyone whom might learn of it. With the domains of the bush and village "infested" (one hunter's favorite expression) with malevolent spirits and personal whims, only few hunters managed to outmaneuver their detractors perpetually.

Local hunting was a social identity predominantly managed by lineage elders, who claimed to oversee the welfare of their group's members. Elders controlled guns and the knowledge of the ingredients for prescriptions and the rituals sanctioning the successes of their younger charges. Moreover, elders selectively chose young men to become hunters by interpreting their dreams as an ancestral summons to protect and enhance local lineage welfare. Some of those destined to become hunters might spend a few years as wage laborers, but their short ventures in town were insufficient to acquire the capital for purchasing a muzzle-loader. If owners of guns returned to town as migrant workers, they placed their weapons in the custody of lineage elders as placeholders symbolizing the absentee's loyalty and care for other kin during their absences. These gun owners expected to use their guns along with other accumulated wealth as patrons if and when they eventually retired to the village. Lineage elders managed the "placeholder gun" by loaning them to younger clients and vetting their youthful ambitions as local hunters rather than as migrants. Most boys and young men were encouraged to become part of the migratory workforce in the developing towns and return iteratively with cash and other wealth for the benefit of the lineage. Villager welfare depended upon cash and social networks elsewhere more so than just protection and meat at home.

Patrons as managers of community resources focused on human relations rather than on materials or products per se. The bush was "common property", not in the tragic sense of "open access" but one that the elders managed through their interpretations of human relations, beliefs in ancestral spirits, and allocation of their group's human power (Marks 2005: 244–45). As long as normative coherence existed in village life, wildlife and other bush benefits were expected in "place" (*nchende*), and there was plenty of wildlife around until the national economy deteriorated in the mid-1970s, prompting the commercial wildlife slaughters. These widespread slaughters of wild animals were largely the work of large commercial groups from the adjacent plateau and directed at the larger mammals such as elephant, hippo, and buffalo. The few hunters at Nabwalya were peripheral to these larger commercial gangs, participated opportunistically to make ends meet and secure necessities during the rapid inflationary years.[22]

Ambivalent Attitudes toward Wildlife

The perspectives and attitudes that members of any given society hold toward animals are never monolithic or homogeneous despite the broad brushes of conformity conveyed in generalizations and descriptions. Thoughts and their expressions are always in flux, subject to new experiences and exposures to other people and their notions. In his classic ethnography on *The Power of Animals*, Brian Morris (1998: 2) notes that the neighboring Malawians, like "all human societies have diverse, multifaceted, often contradictory attitudes towards the natural world, especially towards mammals." His comments also apply to their neighboring Zambians, including the Valley Bisa, whose attitudes are equally complex and inconsistent. My generalizations reflect mainly those experiences with lineages among whom I have spent the majority of time while at Nabwalya. These Valley Bisa have a long history of conflicts and sharing their environs, competing for scarce resources, and protecting their lives and properties in the shadow of large and dangerous mammals. More recently, they have found themselves dependent increasingly upon the bureaucratic patronage of a weak state, which enforces distant legislative acts that handicap their dealings and relations with troublesome wildlife.

Wildlife remains significant in local social life as a source of food and "prescriptions," as a subject for proverbs, and as a spiritual medium. In the distant past, wildlife was also an important resource for clothing, fat, and trade goods. As primary characters in oral traditions, animals feature as sources of inspiration and instruction, as well as in divination and in religious life. Yet some wild animals are fundamentally aggressive, hostile to and competitive with humans. People monitor continually for these animals as their unexpected presence is a daily and nightly risk. When there were fewer people, relatives managed these contingencies under the hegemony of a headman, who provided subsistence and protection against wildlife marauders.

Under most circumstances, Valley Bisa attitudes remain ambivalent as their cultural orientation has remained utilitarian, practical, and homocentric. When the interests of the state were more distant, such vagueness allowed individuals and kin groups flexibility in handling their contingencies regarding wildlife. Yet the drift in local attitudes has become more compelling and critical as people's perceptions of their vulnerabilities have shifted and their numbers have increased. After some seventeen years under "community-based" wildlife management, residents in 2006 possessed few legal firearms, wildlife products were "criminalized" as unlicensed state property in most local hands, and legal licenses were reserved for high-paying foreign and national licensees. The wildlife scouts have become the *de facto* "owners of wildlife" as well as

unresponsive to local complaints about crop-destructive animals. Many of those interviewed in 2006 had experienced beatings and arrests or had one or more of their close relatives apprehended by the wildlife scouts. These intensive interviews then throughout the Munyamadzi GMA showed that its residents knew that they had paid a hefty personal price under these new wildlife regulations, that most were antagonistic toward the wildlife agency and its inflexible policies, and saw wildlife as the main reason for their own lack of development. Some preferred to see all wildlife vanquished if that would improve their standard of living to that of other Zambians. Then they anticipated that they might be accepted as citizens, rather than anachronisms (Marks and Mipashi Associates 2008).

Belief and Behavior: The Impacts of Time, Space, and Circumstance

Good health, bountiful harvests, an ordered world, and amicable relationships are taken for granted. What require explanations are the upheavals brought about by droughts, by capricious catastrophes, by human suffering and pain, by social distrust, and by outsiders. Earlier valley generations, with their more parochial visions and truncated boundaries of experience, grounded their explanations largely in human error or in human intentions. Human error or forgetfulness required the identification and reestablishment of harmony with an offended relative or spirit. The causes behind these intended afflictions were immoral and often embedded in accusations of witchcraft toward a certain individual (Colson 2006: 204).

People do not exist long, if ever, within an anthropologist's proverbial present tense, since the moment of fieldwork usually represents only a single, short frame within the continuous runs of unfolding events. Just as the experiences and exposures of past generations literally and figuratively impact their landscapes of mind and behavioral space, topographies of thoughts and places of recent generations also take on new dimensions and meanings reflecting their expanded horizons. A recent generation of adults has found a much more calamitous, uncertain, and challenging world than their parents did. The frequencies of droughts, floods, and wildlife beset their subsistence activities, while some customary resource entitlements and practices under earlier policies have become "out of bounds" by the state. Poorly staffed primary schools and slack teachers make most graduates' employment opportunities and the fulfillment of their parents' dreams of eventual support, implausible in outside labor markets. Residents currently compete vigorously for the few seasonal jobs at home or exploit wild resources surreptitiously as auspicious surveillance and circumstances allow. With their general lack of education and early socialization into their mother's routines, young women increasingly become the pawns of poorer families and older men. Younger men and women suspect their elders of conspiring against them, stealing their energies for nefarious purposes.

While some in the elder generation hold tenaciously to earlier grounds of meaning, many of the young within the "fractured" communities of the 2000s have shifted their grounds to a new "spirituality." Most residents, as of 2008, face larger economic and political issues, including climate and environmental changes, over which they have little control. This shift in perception of vulnerabilities establishes realignments in the boundaries and values between self and community, between Bisa-ness and other identities, between human and nonhuman life, as well as between good and evil. Local worldviews have become, in Elizabeth Colson's words (2006), a "medley of ideas and practices"—more heterogeneous, more speculative, and, for many, even more conspiratorial and confrontational. Their different experiences set new boundaries for the metaphors of known (village) and unknown (bush) forces, which extend outward into the regional towns, cities, and even international arenas of donors and organizations. Ambiguity and complexity reign within all these scales.

The image of the witch has morphed and extended to include distant strangers—the *banyama* ("zombies") and the "body snatchers." Groups of strangers, unrelated and illicit persons, find vulnerable individuals, often at night, and allegedly kidnap them for tasks or dismember them for blood, heart, and body parts for unspecified international corporations. Known individuals vanish, are never seen again, allegedly drained of their life forces to serve unknown others. These foreign evildoers are not discov-

erable through the neighborhood witch finder or diviner, who remains the medium for discovering the causes of more localized failures (White 2000; Colson 2006). People continually experiment with the flexibilities of their boundaries and metaphors; some of these explanations that earlier had moorings in provincial meanings now appear irrelevant and impotent. At issue in this questioning are ideas about the locus of power and control that some things or events exercise over people, that some people have over others, and that some other people may have over certain things (Croll and Parker 1992: 11–36).

Since the late 1970s and the decline in the national economy, the fear of witchcraft has escalated in Zambia as it has in many parts of rural and urban Southern Africa (Comaroff and Comaroff 1992; Crehan 1997). The inflating economy largely closed the outside market to unskilled labor and forced many young men to remain in the villages, increasing the differences between haves and have-nots as well as increasing dependencies based on age and gender. Along with the curtailment of many former government services in rural areas, particularly those of health, rates of mortality and sicknesses increased concomitant with the rapid spread of HIV-AIDS. In the central Luangwa, rural people have become increasingly dependent upon international charities such as those provided by established and fundamentalist churches as well as nongovernmental organizations (NGOs), including safari operations.

Earlier generations were mainly concerned with what their lineage and ancestors believed and how they behaved rather than about their distant neighbors' convictions. Their world of provincial rituals and practices made proselytizing unnecessary. Today many residents voice the formalized and standardized creeds of western churches, particularly those of the more fundamentalist sects, just as others repeat the formalized tracts promulgated through conservation workshops. These outside links emphasize individualism and an ideology that crafts individuals, who believe that difficulties arise as products of weak belief and trust in God (Pfeiffer 2005). Whereas Christian beliefs of the established churches (Catholic, United Church of Zambia) may have challenged the role of the ancestors and the older rituals, those of the newer fundamentalist churches (Pentecostal, Apostolic, Faith

Mission) demonize and stigmatize these former beliefs and practices as indicators of Satanic activities.

For the younger generation, the shift from local to more regional and global affiliations began during the difficult times of the 1990s with fracturing lineages and increasing individualization, along with the conversion of many residents to Pentecostal faiths. A young man from some distance away, who had recently settled among his in-laws near Nabwalya, spoke nonchalantly about this change during an interview with me in June 2006. He claimed no formal affiliation with any church.

> Some time ago, people believed in spirits for their deliverance, but today this is not the case. God was in the spirits but today's spirits have been lost and people pray to God. People don't trust spirits now and spirits don't trust in people. Earlier, people would come with luck from home when relatives complained they needed "relish." Sometimes hunters did not even use guns as animals died on their own and the hunter's role was just to coordinate the spirits and the relatives. The hunter praised the ancestors on behalf of their relatives; that is why they respected and maintained their reputations. A hunter would not make love when he goes out hunting or else he would become vulnerable and caught by an animal, especially leopards, buffalo, elephants or even bites from dangerous snakes (sic). This time hunting is done like a game and people don't kill and they don't prepare for spirits. Spirits don't give animals anymore because people have in the first place ignored the spirits.

I find it difficult to assess the durability of these recent conversions as they underscore individual associations with employment or with a more recent and wider network of connections within a distant world (Ellis and Haar 2004; Colson 2006).

Land alienation through the state's unilateral extension of national park boundaries, entitlements, and endowments in wildlife, including how local assets are used and by whom, remain important issues of identity and politics for the Valley Bisa. In responding to specific survey questions in 2006, residents recognized the strong national image of abundant wildlife associated with Nabwalya and wished for a return to thinking about the more practical, mundane benefits of consumption rather than the abstract, exclusive economic commodity that wildlife had become under the state's policies (Marks 2014). In expressing a negative response to whether they received benefits from just tolerating wildlife, responders referenced a more beneficial, remembered past during which they openly consumed bushmeat, and compared it with the state's suppression of such intakes today.

> *[Man, thirty-seven years old:]* My people no longer kill animals; only scouts kill them now. That's why we are now eating bush vegetables. In the past, bush vegetables were eaten rarely, only to change the diet; but now we can't eat meat. We are eating animal food, "bush vegetables."

> *[Woman, fifty-six years old:]* The Community Resources Board has stopped us from having our game meat by employing many scouts and putting up so many camps around us. We are now "coughing wild greens" (*mutelele*). We never "coughed" meat, it only made us healthy and fat.[23]

The appropriate constituent of the "relish dish" (bushmeat), typically eaten with consistent lumps of daily gruel (*nsima*), has earlier precedence. In 1966–67, observations and a survey of local food prohibitions at Nabwalya showed a high level of local compliance with a local norm against consuming the flesh of hippo and baboon; both species were fairly common then, and everyone knew where to find them. A likely cultural explanation for this abstention then was that these mammals had taken on metonymical cover terms for their more powerful neighbors (especially the Bemba and Ngoni), who in the recent past were belligerent aggressive enemies and now, as resident civil servants (schoolteachers, game guards, court clerks, and tax assessors) at Nabwalya, ate the flesh of both species. These officials acted superior to local residents, discussed local "primitiveness" in harsh terms as they sat in judgment, and controlled access to state and outside services, especially for local children. At that time, when asked for reasons for this prohibition, they invariably replied about a fear of "leprosy" or gave another innocuous answer (Marks 1976: 78–79; 99–102).

The interpretation of bushmeat as being "god-sent" shows a different political sensitivity and moral acknowledgment not encountered previously. In recent years, villagers have found dead immature and adult buffalo left behind and drowned by the large herds drinking at the river. These finds are welcome as most villagers no longer have ready access to bushmeat, and these buffalo carcasses can provide a feast for many people similar to when they could be killed earlier. Wildlife scouts acknowledged the social significance of these occurrences and make distinctions, between scavenging a predator kill (to save lions) and scavenging drowned buffalo (as crocodile fodder?), on condition that scouts are called to "certify" the drownings prior to any butchery. A young woman interpreted the importance of such a grouping of buffalo deaths in 2006:

> *Mulenga wampanga* is a spirit that kills animals when the people complain that they need relish. This spirit only kills buffalo because they are plentiful and occur in groups large enough to feed many people. I appreciate that these wild animals die because they become relish and the only way that we have meat is because game guards don't arrest and do not kill meat for people even after the animals attack our field.[24]

Public opposition to the state wildlife management restrictions surface in the continuing arrests of local "poachers," in the increasing incidents of snaring and the impounding of snares, in the verbal and occasional physical battles between scouts and other residents, in the distrust of government promises and projects, in reluctant attendance at government-called meetings, and in the proliferation of standardized answers to outsider inquiries. As in other marginal wildlife areas globally, such opposition against centralized state management may have a longer history (Vitebsky 1992: 223–246; 2006).

Cultural Boundaries: On Searching for Their Limits

The presumed agency of witches and spirits still invigorates local conversations about unsubstantiated events and unknown motives of kin and strangers. These events include hunter kills of animals as well as when a wild animal kills or mains a relative or neighbor. Besides the visible boundaries of gender and age, lives are not lived in reference to static compendia or according to the limits suggested in lists of inclusive domains. Most ideas about agency and motivations shift in significance as well as content depending on the experience, reputations, exposures, and claims of individuals and groups, who confront and may change their dimensions and presentations through time. These transformations occur as individuals and as a group, with reference to their networks and even detractors, weave the more enduring threads of conventional wisdom into their explanatory cloths and perpetuate the local stories that capture their essence and standing within the context of current events.

To show how various categorical threads come together to explain a sequence of events that unfolded over four months in 2006, I depict some personal observations with information volunteered then or later in response to my questions. The story ties together strands about the prevalence of witchcraft, beliefs about crocodile and elephant motivations, elderly mothers, errant sons, finding a place to settle, limited choices, the inheritance of guns and the management of their uses, unresponsive civil servants, and extralegal forays for wildlife and their consequences. I begin by mentioning the circumstances of the three occasions that I personally observed the events and either sought context then or later. My placement was significant for what was learned by first listening and later by asking appropriate questions of knowledgeable persons. Second, I provide some background to the events. This context suggests some of the interactions and conditions faced by ordinary villagers and the conventional wisdom about causes and of personal options. Finally, I asked for additional events during my absence and before my departure from the GMA in mid-August 2006.

Personal Observations

My awareness of these events began the day after my arrival at Nabwalya on June 6, 2006, to conduct a survey throughout the GMA. The next day the chief and I visited the villages in the vicinity of his palace to make courtesy calls and to familiarize me with the landscape and its changes since my last trip. We visited a headman, who had been particularly helpful in my earlier studies, yet was suffering from the late stages of what he described as "cancer" on his pelvis. I was distressed to see him suffering as he asked for medicines, which I did not possess. A local friend was to write me a few months later about my friend's painful death. I listened as he and others discussed a particular widowed woman from a village four kilometers up the Munyamadzi River who was intent on leaving that village and had sought permission from the chief for a space near the palace. The chief had granted her request, but this headman had refused her request for space in his village and was relating his reasons to the chief.

A week later, I was on my way from Kanele Wildlife Camp, walking along a path that went through several villages. In one of these villages, I passed an elderly woman bitterly exchanging words with another headman about a space that she allegedly said the chief had allocated to her. He was adamantly refusing her request and claims. This was my first encounter with this elderly woman, whom I did not meet, but whose circumstances I would learn about later.

A month later, before leaving with my local team to spend weeks interviewing in other villages, I was in the chief's reception hut awaiting my summons for lunch. The same woman, accompanied by a local "sister," who was refused land by the headmen came seeking an audience with the chief. She presented him an oral petition to which he and a counselor responded. I listened and, after her departure, asked the chief about what I had observed and overheard. He kindly elaborated upon this story, its actors, their motivations, and the circumstances.

After finishing my study and before departing in mid-August, I inquired from other residents information on what had happened in the interim and what they knew about these events.

Cultural Context

Beginning around 2000, residents in Chalwe Village, situated some four kilometers west of Ngala Hill along the Munyamadzi River, experienced years of devastating calamities to their persons and welfare. Over the course of a decade, a crocodile (or crocodiles) caught and drowned three adult women and children as they washed and collected water from the river. As some of these victims were related, residents in the village determined that the attacks might have been initiated by a wizard and reached a tentative consensus about who might be the cause for these deaths. Concurrently, a few individuals staying in marginal households and guarding agricultural fields prior to harvest experienced repeated devastating elephant attacks to their sorghums. These fields belonged to a small lineage, an elderly woman and another with grandchildren, who possessed no guns or other ways to protect their property and who had made overtures about joining the larger village of Chalwe nearby. These elderly cultivators began thinking of the elephant traumas in terms of sorcery and witchcraft since their fields were the only ones in the vicinity repeatedly devastated. Both villagers and those of the smaller lineage knew *what* damage crocodiles and elephants were capable of, and yet their concerns were centered on the cultural *why* (are these tragedies happening to us now?) and the *how* of its means. Both groups petitioned wildlife scouts on numerous occasions for retribution. Every visit to Kanele Wildlife Camp ended with the scouts acknowledging having no ammunition and that they would appear soon to follow up the case. Yet these scouts always had greater priorities and reputedly never showed.

Having faced several rejections for help for assistance with these animals from the scouts, a delegation of Chalwe residents visited a well-known witch finder living in Chief Chitungulu's area (east across the Luangwa River) to establish the reasons for these animal "afflictions" and their options. After being briefed about their circumstances and relationships, the witch finder reputedly implicated several individuals possessing crocodile magic and other potions capable of jeopardizing other's lives. Ngulube's[25] mother, an elderly widow in the small lineage, was among those indicated through his cultural séances.

When the delegation returned to Chalwe, they told Ngulube's mother about what they had learned and accused her of being the witch, who had caused their misfortunes with crocodiles as well as her marginal group's elephant problems. Subsequently, she made plans to resettle in a village around the chief's palace, a space she considered safe under the circumstances. The chief granted permission and indicated possible sites for her contingent upon the headman's agreement. Yet both headmen in the villages where she wished to settle used "strong, bad words" when she visited them and refused to yield her a place. Each headman told her separately that, as a "suspected witch," she possessed occult powers to disturb the environment as well as village relations and would therefore destroy the integrity of their villages. They warned that several residents were ready to leave their village upon learning of her past reputation. Upon hearing these words, Ngulube's mother retreated and changed her mind, deciding instead to relocate elsewhere, the village of her son's in-laws.

Ngulube had gone to live with his wife's parents (part of the dowry arrangements) in another chieftaincy, nearer to the Muchinga Escarpment and close to the Luangwa North National Park. When Ngulube discussed his mother's decision with his in-laws, they accepted her resettlement proposal and indicated for him the place where he could build her new house. Before Ngulube started work clearing a new homestead adjacent to his, he arranged with some commercial meat hunters from the plateau to show them where they could hunt and find game in the valley. His intent was that he would use the cash sales from this bushmeat expedition to pay his in-laws and neighbors for building his mother's house. Composed of two hunters with rifles and twenty-five carriers, the hunting party, accompanied by Ngulube, descended the escarpment to hunt in North Luangwa National Park. They killed a buffalo, impala, and warthog, but scouts at a nearby wildlife camp heard the shots and found the group drying meat. The scouts surprised these "poachers" and captured seven men, including Ngulube. The scouts were surprised when Ngulube claimed Nabwalya as his natal village, as he was captured about 100 kilometers from that place. The arrested men were beaten and taken to the magistrate's court in Mpika. The hunters were sentenced to three years in jail (or a fine of ZK 300,000, then about $100 US) and the carriers to one and a half years imprisonment at hard labor (or a fine of ZK 150,000 each). Ngulube was one of the captured carriers and was sentenced and sent to prison.

When Ngulube's mother heard about her son's imprisonment, her duty became raising the money to release her son from prison. Her late brother had inherited a muzzle-loading gun belonging to the lineage (*mfuti wapacifulo,* nominally not one for sale). When the brother died, the gun went to the chief's palace for safe keeping instead of to Ngulube, his nephew. Local opinion was that this uncle knew that Ngulube was "possessed" on occasion by "mad moments" and that had he inherited the gun that he would use the weapon for "poaching," likely losing the gun and jeopardizing the lineage's security.

When I saw Ngulube's mother during the second occasion, the mother had come to petition the chief to release her late brother's gun so she could sell it. Listening to the conversation, her intent was to sell the gun on the black market so she could pay for Ngulube's release from prison. The chief refused her request and upon my question to him, said, "there is a mistake in this kind of thinking." I took this comment to mean that the disposition of this gun was not something this woman could or should be considering. Three months later, Ngulube remained in prison. I was told that his mother had raised some money, but remained indecisive with few allies and no one to take the money and her petition to the magistrate.

In the next five chapters, I begin by sharing the life histories for three consecutive generations of local hunters, who lived largely in the shadows of the chief's palace at Nabwalya. We learn of their rough-and tumble worlds and how they accommodated the rapidly changing political and economic worlds around them. In the following chapter, we accompany an articulate young hunter on his forays after game during two different seasons as the terrain evokes memories of ancestors and recent events and as we learn about how he interprets the bush signs, sights, and sounds in designing tactics for finding prey. From these and other forays, we learn about the behavior and cultural attributes of large and small mammals and their roles in provisioning descent groups.

Notes

1. *Mwenda mpanga endela vyonse* (recorded valley version);*Conde afuta abamwendamo,* (White Fathers' Bemba-English Dictionary 1954: 140 (ciBisa version); *Ukayenda m'chire wayendera zonse* (Skjonsberg 1989: 207).
2. *Fingwa wa Leza,* "made by God," is the term used in reference to discussions of natural assets (mainly wildlife) under the community programs (2006).
3. I write here in the past tense as a sea change occurred in the late 1990s and 2000s. Many of these ideas recorded in conversations

during earlier decades no longer appeared tenable by the children and younger literate adults during my conversations in 2006.

4. My ideas follow those of Mark Johnson's "semantics of understanding." According to his theory, people derive meanings from embodied experience and from the imagined and learned structures for understanding their experiences. His analysis suggests a distinction between the classification that orders a specific domain (such as that for mammals) and the ritual classification (metaphors, metonymy, symbols) that cross-cut and link several domains.
5. More on the hunting careers of Chibale and Hapi, including the differences in their generational experiences, are elaborated in chapter 5.
6. *Combretum obovatum* is a common shrub that forms thickets. This straggling bush has long spiny stems and dense branches that arch to the ground around its periphery; on the ground near its branching trunks is a space wide enough to accommodate several sleeping men. This thicket provides a barrier against wild animals during the night while Chibale's medicines offered an additional layer of spiritual protection.
7. *Diospyros mespiliformis* (African *ebony*) is a tree whose fruits and leaves are eaten by many species of mammals, birds, and reptiles.
8. Honey guide (*Indicator indicator)*: a robin-sized bird chatters incessantly to attract attention. If actively searching for game, local hunters generally consider them a nuisance, as the birds alert other animals of an impending danger. On other occasions, the birds live up to their name and lead people to honey.
9. *Muti* (*miti* plural) is the generic and vernacular term for tree or plant. Here its use designates "prescription" or "medicine," which is often a vegetative substance whose function as a charm serves a symbolic purpose. Such knowledge about powerful substances is the providence of elders, like Chibale, who may not reveal everything about its intended purposes, importance, or characterisitics.
10. This ritual is *kuzilika mpanga*—"to close or pacify the bush." *Mutungwa mbabala* (unidentified) is a very thorny shrub. Chibale also used its branches to close the opening to the space where we slept.
11. With muzzle-loading guns, hunters typically never armed the weapon, by placing the firing cap on the nipple, until they had a target in sight and were ready to fire. For carrying the weapon and while searching for game in the bush, hunters embedded the firing cap in bee's wax on the trigger guard.
12. This is Hapi's version of what had happened. *"Mpyayi yaiza pamalo yatampa kunyanta muyuma nyanji. Natampa kwita Lowlenti. 'Nati impyayi*

ikonyanta. Nabuka. Naikala. Kutampa kubwizya baChibale. 'Ifyo nalota tevimene?' BaChibale ati, 'cabipa pantu mpyayi cibanda cili ne minga yabalala. Elyo capwa.' Twaima kuzange inkalamo zikete cholwa.' Elyo BaChibale ati, 'ciloto wangulota eechi, chamoneka pantu cholwa viapalama ne mpyayi kubalala. Ee ciloto cobe capwa.'" My translation: [The porcupine came to our camping place and stepped upon my back. I woke up Lowlenti saying that a porcupine has stepped on me. I sat up and remained awake and then asked Chibale if what I had dreamed was not a bad omen? He replied that the dream seemed bad as the porcupine is a "bad spirit" as alone it has striped quills. When we left and found the lions that killed a zebra, Chibale said that this is the dream that you had because the zebra resembles the porcupine with its stripes. That's (the meaning of) your dream.]

13. Although the terms sound similar to the untutored ear, *mboo ya mumpanga* (literally "buffalo in the bush") is not the same mammal as one described as *mboo yakupanga'* ("buffalo crafted by a witch").
14. My field notes have several accounts of hunters' refusing to stalk eland or failing to butcher a "found" eland carcass because of widespread fear attributed to the retribution of its awesome spirit. Eland are also a "protected species" under Zambian law. Elizabeth Colson (2006: 97–98) notes similar rituals of respect for eland among the Tonga.
15. These characteristics are summary statements gathered by a schoolteacher, Elvis Kampamba, at Chaya School (Nabwalya) in answer to my request in 2001. He sent me interviews conducted with six individuals (twenty-four to forty-seven years of age). For earlier characterizations of the elephant and python, see Marks (1976: 93).
16. *Mundu walila, mwebana banama cenjeleni*—literally "The lion has roared, young animals take care."
17. From my notes and interviews by Elvis Kampamba at Nabwalya in 2003. In June 1988, a trapper showed and gave me an example of this hunting medicine, *citumwa ciankalamo,* that he had taken from a large male lion he had snared. It consisted of a large ball of hair, which the lion reputedly strengthens with fresh blood from its kills.
18. Excerpts from interviews recorded by Elvis Kampamba in 2003.
19. Elvis Kampamba, 2003, report on a local study done at my request. My notes indicate that women, who are accompanied by dogs when they go far in the bush to collect firewood, occasionally return to the village with small or immature mammals caught, located, or scavenged by the dogs. Women also collect bushmeat from predator kills if these carcasses are close to their villages.
20. Event occurred during 1973 and was recorded in my notes as well as an entry in a local diary.

21. "Subsistence use" was emphasized by colonials and early expatriate wildlife administrators because they paired it with the category of "commercial use" engaged in by Africans as "poachers" within the extensive bushmeat trades to feed the developing colonial cities. These officials looked more favorably at subsistence use, although they hoped it would eventually become restricted to household use. Commercial uses legally depended upon state licenses, regulations, and taxes to benefit the developing centralized state. As the large bushmeat trade operated in the unregulated informal economic sector, officials hoped that their efforts to suppress "poaching" would eventually help African "subjects" learn the difference between legally enforced categories, support state efforts and its economy, and save "their game."
22. Elders were successful in controlling the number of local hunters, at least around Nabwalya. My records at Nabwalya show consistency in these figures: one-third of adult men in residence over four decades (1960s–1990s) were hunters. Wildlife was plentiful during the first two decades. During the tumultuous and uncertain economic times beginning in the mid-1970s, the number of hunters and wildlife kills increased slightly and gradually decreased under the new state wildlife policies that offered local employment and enforcement (Marks 2005: 244–45). At least around Nabwalya, the colonial and game department stereotype that every village man was a hunter ("poacher") was useful in promoting their urgency to centralize wildlife management.
23. Both interviews with author, 2006.
24. Twenty-one-year-old woman, interview with author, 21 June 2006.
25. "Ngulube" is a pseudonym.

Chapter 5

Caused to Hunt

Life Histories of Three Generations (1903–2003)

Hunters perish by their prey as farmers succumb to their labors.
Bisa song (1967)[1]

I have known three generations of Valley Bisa hunters.[2] In this chapter, I select two hunters to represent their respective generation. Each story tells something different about the hunt, about hunters in general, and about their relationships to each other and to wildlife. In the rapidly changing world of the twentieth century, money surpassed barter and labor, wildlife supplanted agriculture as a prime commodity, guns displaced bows, and men's economic power distorted gender relations. When state control over weapons and wildlife became repressive, the silent nooses of snares and earlier technologies replaced the decisive bangs of firearms. Each generation learned its craft somewhat differently as changes in wildlife policies, the national education system, and an economy based upon migrant labor, wages, and markets pushed and pulled, influencing local cultural envelopes in different ways. Each ripple of change surfaced new opportunities as well as challenges. Under these external flows and internal influences, some cultural elements have remained resistant, embedded tenaciously in moral and social relationships; other aspects endured temporarily as stubborn eddies before eroding into the comparative flatness of pastimes and memories.

Three Generations of Local Hunters

The role of local hunter typically began with the older generation, within a matrilineal descent group, conspiring in the name of its predecessors (then ancestors) to act as spiritual interpreters and later as a broker for a younger generation of boys in ways to protect lineage lives, properties and provide bushmeat as significant goals. In explaining to me why they became hunters, older men used the causative tense of the verb "to dream" (*kulota*), indicating that the power behind their vision came as a spiritual force. As the actors in these apparitions were unknown, elders identified the causal agent as that of a named ancestral and the dream as a summons and a moral calling to enhance lineage welfare. Through their interpretations, elders channeled and structured youthful skills and behavior in ways that benefited them as patrons while offering other lineage members a means of protection and the prized source of sustenance. Elders provided their hunting clients with the templates of bushcraft, access to agency, structured their productivity, and sanctioned their accomplishments. Youths learned by watching and joining others in hunting as they absorbed its behaviors while listening to its stories and accomplishments. As beginners, later as clients and wannabe patrons, they sought to normalize their ambitions within the welfare of relatives.

The Elder Generation (Born 1900–25)

I first met Chibale Chinsambe soon after I arrived in the central Luangwa Valley in 1966. He was then in his mid-sixties, had been the chief's councilor for sixteen years, and was at the time headman of the area's largest village. Like many men his age, he had worked elsewhere as a migrant laborer (as a foreman with three stays in town covering six years) and was then married to two wives; earlier he had inherited three wives from deceased kin and divorced several others. Chibale had no formal schooling, yet he was a storehouse of information about the past and about others.

> I was born about 1903 when the Europeans still occupied Kanele Boma on the ridge above the Chieftainness' palace. I knew Benny Hall before he shifted the boma east of the Luangwa River to Lundazi. While learning to hunt, I accompanied (my maternal uncles) BaKafupi and BaYumbe[3] both of whom lived in the same village into which I was born. These elders pursued animals together with bows and poisoned arrows. Occasionally these archers, accompanied by their relatives as carriers, left the village for several overnights in the bush. During these years, animals were never plentiful because people killed them with traps, downfalls, game pits, snares and poisoned arrows. This period was shortly after Mulenga affected the animals with a sickness.[4] This sickness was no ordinary disease, for these animals appeared healthy before dying.
>
> BaChibale Bakalamba[5] came in dreams and instructed me to kill animals as he had done earlier. I was perplexed and did not know him or what he told me. When I reported my dreams to my mother and to BaKafupi, they were happy and said that BaChibale, as my guardian, was encouraging me to become a hunter. I made a bow and a bunch of arrows and began to pursue wild animals near the village.
>
> The first animal I killed was a male puku[6] in the tall grasses along Bemba stream. After my arrow hit this animal, I tracked and found it dead. Then I returned to inform my father and uncle. As they had previously warned me not to touch the carcass, this was their first question for me. As I had not touched the quarry, we proceeded to the place of kill. Once there, I watched as my elders took grass and roots from beneath the animal's head and placed these in its mouth, ears, and anus, and on top of its head. Then they hit the carcass with my bow and said "*Tuli nenu mwe nama*" (literally, "We are with you animals"), our lineage slogan. They built a fire, roasted some of the meat "doctored" with the remaining roots,[7] and butchered the rest of the carcass. They presented me with a small portion of this meat served on a knife blade and told me to eat it (*sonda we cibinda*) before they ate. We returned to the village and distributed a share to everyone. After cooking and sharing this meat, everybody danced and sang [ritual of *vizimba*]. I collected all the bones remaining in the relish plates and the next day accompanied my father to bury them in an animal trail.[8]

In telling about his hunting career, Chibale repeated these rituals as he killed his first zebra, impala, warthog, and waterbuck with a bow and arrow. As the largest of the "small game" animals, buffalo required a slightly modified ritual.[9] Chibale was with his uncle Kafupi for this kill near Bovwe Plain. His uncle found the prescription, gave it for Chibale to chew, and then led him, with eyes closed, to the carcass. Chibale spit the masticated medicine in the buffalo's nose, anus, and ears. When he opened his eyes, he hit the carcass with his bow and arrow, and shouted his slogan. Kafupi never revealed to him the roots and leaves used in this prescription.[10] Back at the village, the uncle distributed meat widely; relatives held another *vizimba,* danced, and later Chibale buried the remaining bones in the bush.

Unlike smaller game where only the first animal is ritually treated, many Bisa say that the larger species (*inama ikalamba*: eland, elephant, and lion) possess a more ominous shade that hunters must confront ritually after each kill. Around 1930, Chibale killed his first eland and informs of its nature.

> This eland was pregnant, and when hit by my poisoned arrow, the animal ran a long way before collapsing at Sikapotwe lagoon. BaKafupi was my teacher, for these animals possess a powerful shade (*cibanda cikulu*).[11] Our forefathers learned that this animal's ominous shade was capable of killing the hunter, his wife, and children or make the hunter go progressively mad by appearing in his dreams. BaKafupi used the roots of *kamulebe* to placate (*kuzungula*) separately the spirit of the fetus from that of its mother. This prescription works the same as an injection at a hospital—it forces your head to become peaceful and still (*kutalala*) as it overcomes pain. I chewed the roots, closed my eyes, approached the carcass, spat in its mouth and anus, and hit the carcass with my bow while voicing our lineage motto. I did this as a precaution against bad effects happening. In addition, we severed the eland's head and left it in the bush along with the fetus. BaKafupi and I carried the meat back to the village and we had no *vizimba* [ritual] dancing.

Chibale then relates that by 1930, a maternal uncle had acquired a muzzle-loading gun. "His older brothers" refused to let him borrow it for he was too young and might spoil it.[12] So he borrowed his brother-in-law's gun, as colonial policy had criminalized bows and arrows. Chibale's account reveals another side to the rationale behind this shift in policy. He noted that bows and poisoned arrows were very effective in killing animals, "With guns, a hunter must be very close before firing, [whereas] with a bow, the archer shot from a distance and aimed in the general direction of his prey. Once hit or nicked with the poisoned arrow, the animal soon died." He went on to say that this shift from bows to muzzle-loaders reversed trends for both wildlife and hunters. "In the past," he said, "wildlife was sparse and archers were plentiful, while now [1973] animals are plentiful, but hunters few. Europeans feared that we Africans would kill all the wild animals so there would be none for European children to see or to hunt." He reminded me that a few men still hunted with bows and "that if the laws were relaxed and we took to hunting as we did in the past, wild animals would become scarce or run away (*kubutuka*) as before."[13]

When his uncle and headman died in 1934, Chibale inherited his lineage's muzzle-loader. "My brothers," Chibale related, "were left without anything."[14] With this gun, Chibale acknowledged killing a rhino and three elephants. His classificatory "grandson" Kafupi, who inherited his predecessor's name and built his own village, gave him a prescription to placate the rhino's spirit. The reigning chief provided the prescriptions for his elephants.[15]

To my knowledge, Chibale never used a rifle or shotgun, and he told me that these weapons were constructed using "powerful ingredients"[16] not found in muzzle-loaders. To overcome deficiencies in his inherited weapon, Chibale used a powerful potion [*cilembe*], to deliver a decisive blow when the bullet struck its target, and another prescription to make prey visible (a type of game caller, *ulwito*). He also used a ritual wash once in the 1930s and again in the 1940s to remove a spirit that was preventing his weapon from performing properly. His stepfather gave him the following "secrets" of this wash.

> I use *wolowe,* flexible bark from a white tree made into a shallow bowl. I place roots from *kabankungulu* [literally "something strong and sturdy"] together with *citontoloka* [literally "to cause to consent or to yield"] in this bowl to soak overnight at the household rubbish heap (*cizhala*). Early the next morning, I wash my right arm, my left leg and the gunstock with the bowl's contents. I do not wash my face, but throw the remaining contents onto the rubbish heap.[17] Then I go into the bush to hunt and should find animals soon. I have not used this prescription recently [braggingly], as I might kill so many animals that the game guards would focus their scrutiny on me!

In August 1973, I accompanied Chibale (see his portrait as a patron and guide in figure 4.1, p. 168) and three of his relatives on a camping trip (*malala*) for three days in the bush. Yet by the mid-1970s, Chibale had largely retired from hunting. He taught ten nephews to hunt with his lineage's two muzzle-loading guns. By managing these guns through loans to his nephews, Chibale assured that his clients and kin were protected and received the bushmeat the guns produced.

Chibale was both feared and respected for his knowledge and reputation. People sought his advice on a range of problems and cures. He had liabilities as well—beginning in the 1960s a series of deaths and disputes in his large village caused some to doubt that his knowledge and power were benign. Some accused him of using *bwanga*,[18] asserting that the bushmeat produced by his guns tasted like dirt. They talked behind his back and accused him of witchcraft, suspecting that he was the cause behind their misfortunes, misadventures, even deaths.

I saw Chibale for the last time during a short visit in October 1987. It was a brief conversation, for we were looking forward to my longer stay the following year. Continuing conflicts caused his large village earlier to split and scatter into four smaller factions. Chibale was then in his mid-eighties, rather feeble and didn't move much beyond his own village. He died in December 1987 shortly after returning from a witch diviner and from participating in a séance east of the Luangwa River. A relative challenged him to attend the ritual séance and to prove his innocence from the charge of sorcery. He rose to his accuser's challenge, yet died upon his return to his village.

I met Jackson Katongola before going to Nabwalya in July 1966. He was a district messenger at Mpika and, in that capacity, was appointed by its district secretary to serve as our armed escort and host. A week later, I descended the steep Muchinga escarpment in a small, overloaded Land Rover with him and my wife. For a week, Jackson was our official caretaker and sponsor, introducing us to the chief, his relatives, and numerous others. Even without his blue-and-red-striped uniform, Jackson was an impressive individual and respected by those around the chief's palace. When the week ended, Jackson killed a bull buffalo (for us, he said, yet everyone knew it was for his local clients) and returned to his duties at Mpika.

Jackson was born in 1922, two decades after Chibale, in a neighboring village and belonged to another lineage. At that time, the colonial government still enforced a strict policy against Northern Rhodesians possessing firearms. Jackson's older brothers used bows and arrows to kill their animals, yet, unlike Chibale, Jackson took to hunting only when he got his hands on a muzzle-loading gun.

> The reason my brother [age twenty-five] and I [age twenty-two] went to town was to purchase a gun. What others were doing and saying mattered to us. A man belonging to the Mvula lineage married into our village and mainly talked about his hunting exploits. Since we were only "sons" to his lineage, we received only token pieces of meat when he made a kill. This difference mattered to us as we lived in a small village. So my brother and I worked underground in the copper mines at Luanshya, pooled our low salaries [at thirty shillings per month] to purchase a muzzle-loading gun costing four pounds sterling. This was during the Second World War and my brother was the first to return to the village with our gun. He killed nothing and left to work in the South African mines.
>
> My grandfather was "BaCipampe panama" and my mother his niece. My grandmother and mother told me of his reputation and exploits with bow and arrows as he soon appeared in dreams directing me to hunt. Even today, BaCipampe is the one who tells me where to find animals. Until the heavy rains [1996–97] collapsed my roof and broke everything, I kept a gourd in the rafters in BaCipampe's memory [*lukombo lia BaCipampe*]. Before every hunt, I petition his spirit to guide and protect me from all personal dangers.[19] This blessing also works against game guards for I have never seen them while out hunting!

After BaCipampe summoned him, Jackson learned about muzzle-loading guns by observing Paulo, also of the Mfula lineage.

> There were four of us, all close relatives,[20] in the bush that day. Paulo was the gun owner and senior kinsman. We spotted a female rhino "playing" in a mud hole and the group sent me to stalk it. When my shot hit the rhino, I saw cakes of mud fall from its hide, yet Paulo exclaimed that I had missed. I remained at the spot from which I had fired while Paulo followed the trail and found blood. Then he saw the rhino fall and instructed me to remain where I was.[21] I did not move until Paulo found the appropriate "prescriptions". He cut branches from four short trees and tossed them at the carcass, shouting, *"Vimuti vimudlya"* (literally "these bushes

consume you"). Then we approached the carcass, dug roots of a tree and grass that Paulo placed in the carcass's nose and anus. While placing this prescription, Paulo said: "*Fimuti zakuipaye- tee ine iyoo*" (literally "these bushes have killed you, not me").[22] Then we proceeded to butcher the carcass.

When I asked him if they had held a *vizimba* ceremony in the village, his smile spoke volumes. "Even then," he said, "we knew that this animal was illegal for us to kill. We did not wish it to become known for, should anyone ask what kind of animal it was, we would have been in trouble. That evening, we confessed our success and showed respect by sending some meat as tribute to the chief."

Jackson's next kill was a cow buffalo at Mumbi, which he neutralized in a similar fashion.[23] By then, his lineage had stopped having *vizimba* as the public display might draw the attention of outside authorities. Jackson claimed kills of numerous buffalo, impala, warthog, puku, roan antelope, kudu, waterbuck, eland, and zebra. His local praise name, *Nkombalume* ["slayer of animals"] was reserved for the best of hunters. Jackson [1993] continues:

This was a wonderful time in my life for women flocked to me like bees to honey. Women approached me at gatherings and freely supplied me with beer and food. At this time, I was married to three women belonging to Muti, Lungu, and Ngulube lineages. Even the late chief, when he was celebrating his marriage with a new wife, sent me to stock that household with barter from the animals I killed.

I married six wives because I was killing so much game. Women found me attractive and here's how I got their attention. Once I killed a fat animal, I hung a piece of fat on a nail in a public place. That night women would knock on my door soliciting meat and other things. Then I found it easy to pursue whatever I wished with or from them. Women always like tasty food—whether fat or money. The late Sandford Kazembe, a former elephant control hunter [and contender for the local throne] was married to very many women because of his prowess in the bush. The late chief also had many wives because he always had plenty of meat and money.

In 1949, I followed my elder brother to the mines in South Africa and stayed for two years. When I returned in 1951, the native authority employed me as the chief's constable [*kapaso*]. It was no secret that my employment came because of my reputation, as the chief knew that he could count on me to keep his palace stocked

> in meat. I stayed for four years until a European district commissioner [DC] at Mpika summoned me as a *boma* messenger. I worked and hunted throughout the district and accompanied the DC whenever he visited the valley. When Mr. Liambayi became the district secretary [DS] at Independence, we toured the valley but the game guards provided us with no meat. As we were leaving near Kapili Ndoshi, we saw a herd of impala grazing next to the vehicle track. Mr. Liambayi ordered me to kill them. I killed three, placed them in the vehicle, and carried them with us to Mpika. As the permanent secretary of the Northern Province [from Kasama] was with us, he received one of these impala. Another impala went to the district council and Mr. Liambayi split the third with me.
>
> I worked as a *boma* messenger at Mpika for twelve years. Then I went to Kasama in the same capacity at the provincial level for four years and to Chinsali for another four years. After another three years at Mbala, I retired from civil service in 1982 and returned to live in Paison's village.

Jackson possessed two muzzle-loaders and a shotgun. The shotgun came into his hands through a fortuitous event. Jackson accompanied the last European DC on his final tour of the Luangwa Valley in 1966. The DC killed a buffalo but didn't care for the meat. Jackson killed an impala and gave it to the DC so he would have some meat to take back to the *boma*. For this assistance, the DC gave Jackson some money, which his wife insisted he save and combine with his month's salary. Jackson then purchased the DC's shotgun when that official retired from the colonial service. In 2002, Jackson kept the two muzzle-loaders for his nephews and hoped to counter the customary norms by passing his shotgun, his "prized possession," on to one of his sons.

Because he spent much time working for the government in the Northern Province, Jackson only taught an older son and a brother-in-law to hunt. In 1997, the brother-in-law was caught after killing a buffalo, beaten by wildlife scouts, fined (about $60 US) by the magistrate in Mpika, and sentenced to serve time in jail. During the trial, the accused surrendered the muzzle-loading gun, which, according to Jackson and others, was unconscionable, because giving up the gun denied its further usage by the lineage.

Jackson's retirement and his return home were not easy. Jackson was the designated elder on a *malala* [encampment in the bush] with me in 1989 and elaborated then upon his woes. Since his return, he found the cultural landscape greatly changed. As a headman and elder, he found that younger men paid him little notice, never deference. Even the younger men married to his daughters and nieces showed him no "respect," neither did they reciprocate by delivering bushmeat to his house. According to his account, the land had stagnated (become hot: *ciaalo cakalipa*) with alternating seasons of rainfall shortages and floods bringing prolonged famines. Plagues of rats and birds (*lunzizi*) descended on crops at harvest time during the day. "They [the pests] keep everyone busy chasing them from field to field," Jackson noted, "for at harvest time, there is no rest as the larger mammals [elephant, hippo, buffalo] raid the fields at night having lost their fear of humans." If the wildlife scouts responded at all, they never killed these marauders.

Even the disposition of the game guards, who by 1997 had become wildlife scouts and wildlife police, had changed.

> The colonial officials never gave a case to those of us for killing buffalo because they considered these animals as dangerous. Some administrators even admired us for pursuing them with muzzle-loading guns. Today the wildlife laws are very cruel on us here. Game guards harassed me in 1995 when they found me in the possession of an impala skin, part of a folding chair—a gift previously given me by the chief. I was sleeping in my house when they awoke me. The scouts accused me of lacking ownership certificates. So they handcuffed, severely beat me, and took me to Mpika. I still suffer dizziness from these beatings and mistreatment. The court magistrate only asked me my age. When he learned I was seventy-two years old, he said it was wrong to arrest someone so old, as elders could not work while in prison. My case was acquitted—a total "washout." When released, I went immediately to the hospital at Chilonga and from there sent a message for my sons to come and escort me to the village.

Jackson found no consolation that these wildlife scouts were the ages of his sons and grandsons or that their parents had benefited from his previous largesse. The scouts would not acknowledge such relationships and, from his viewpoint, their contemptuous behavior was unanchored from local social conventions. In addition, Jackson experienced a rupture in his relationship with the new chief stemming from an incident during the harsh famine in 1992. Then Jackson publicly differed with the chief, went against the chief's edict, and accepted outside grain from an external agency seeking a long-term hunting concession in valley land. For this action, the chief wanted Jackson expelled from his territory. Jackson was able to halt his removal by leaning on his former contacts in district and provincial offices, who wrote letters in his support. The incursion in 1995 with the game guards signaled the continuation of this spat, as did the inability of Jackson's sons to obtain employment initially under the ADMADE program.

FIGURE 5.1. Jackson Katongola in good health as a pensioner. (Photograph by the author in 2001)

As inflation continued to climb during the late 1990s, Jackson, as a pensioner, doubted his ability to keep his extended family together. When he went to collect his monthly pension of about ten dollars a month, he found that he had "a very large extended family," who in various ways sought claims to his small check. My visit in 2002 ended on the following note.

> There is no more good in this life. ADMADE has certainly not improved my life; rather, it has made life more difficult. Not only do the scouts earn money for themselves from the animals, they have increased the prices on animals and on meat for all of us. Even the animals have decreased because of the activities of the safari hunters and poachers from the plateau. In the past years when there were no poachers, local hunters killed few animals and the animals remained plenty. Local hunters can't finish these animals, only outsiders.
>
> Yes, I hunt and will continue to do so until death takes me from this world. It is like work, something to do every month.[24]

A Second Generation of Hunters (Born 1929–40)

From my first meeting with them in 1966 until their respective deaths in 2003, Kameko Chisenga and Hapi Luben were for me rich sources of information about the ecological and social changes witnessed by their generation. As younger men upon our introduction, they were reserved, yet our trust grew over the decades. As with Chibale and Jackson, the lives of Kameko and Hapi intertwined in many ways. One of Hapi's grandfathers fathered Chief Nabwalya [Kabuswe Mbuluma] and Hapi often hunted for him. Kameko belonged to the chief's lineage and upon the chief's death was a prominent contender for that title. By the 1960s, both men had obtained whatever formal education was available to them and made several journeys into the migrant work world. Like many others of their generation, they returned to Nabwalya hoping to consolidate their positions there. They had learned their local roles from the previous generation, of which Chibale and Jackson were elders.

Kameko Chisenga was born a member of the Ng'ona lineage about 1929. His mother, Chisenga Kalonda, had four children. She died within a year leaving him in the care of his older sisters and father. Kameko was three when an "elder brother," Kabuswe Mbuluma, became chief in 1932. He finished Standard 3 (grade 5) at Mukombwe School elsewhere and led a comparatively carefree and privileged life.

> We played and danced a lot. When our elders saw us playing and dancing they admonished us to help them cultivate. Then most young people did not drink beer and we didn't have much chance to indulge. Our elders were concerned about our behavior knowing that when we went for bride service (*kubuko*) that, if we were lazy, the in-laws would chase us away and we would never be able to marry and have children. The elders told us to show respect (*mucinzi*), for without it we wouldn't live long.
>
> My brother, London Tembo, joined the Northern Rhodesian Rifles and registered my name in Mpika so that [our mother] could receive some of his salary in the valley. During this time, I became betrothed (*kulumbilila*) and went to join my uncle in Luanshya. I wanted to continue my education, but the school required transfer forms left in the village. So I went to Lusaka where my brother was stationed. When I told him of my desire to continue with education,

> he rejected the idea because I was betrothed. Instead, I worked for the Lusaka Golf Club.

In 1950, Kameko's sister Mirika fetched him from Lusaka as their sister had died and "she didn't want to remain in the village alone." Kameko married, but soon left for Bulawayo, in Southern Rhodesia, to work as an orderly in the hospital. He married a woman in Bulawayo, brought her to Nabwalya, and divorced his first wife. Kameko's Bulawayo wife "liked the valley at first, but then became ill and she didn't like living here much after her sickness." They returned to Southern Rhodesia where he worked respectively for the British South Africa (BSA) Police, then for a dispensary, and finally as a waiter. He transferred to the Nyati Gold Mine and, while playing on its soccer team, injured his ankle and returned to Nabwalya about 1962. Kameko initially found employment as a court messenger, but lost his job, as did many others, just prior to Zambian independence in October 1964. Subsequently, he worked as a Catholic catechist and as a local salesman of clothes, towels, bangles, rings, and pins. As a salesman ("hawker"), he used the capital of his brother, who continued to work in town. Finally, a position opened in the local court that Kameko obtained through the intercession of the reigning chief (his "elder brother"). In 1980, he became the presiding justice upon Mwale's retirement in 2000.

I asked him when and why he started hunting.

> As an orphan and while still a young boy, I accompanied Julius Mwale, my brother-in-law, into the bush. Then Julius Mwale only had a bow with arrows and I was with him when he killed a waterbuck and a puku.[25] When I went into the bush to cut poles for buildings and for collecting wild foods, I became familiar with the land around where I lived. There were not many guns then.[26]
>
> I began hunting in the late 1940s before I married. The chief taught me to hunt with his shotgun and put me to the task of keeping baboons and monkeys from raiding our crops. The first animal I killed was a baboon in the fields. I just threw the carcass away.[27] I also killed a female puku, with a cartridge purchased from the chief, and a number of other smaller game. When I married a woman in 1963,[28] the headman had a shotgun and muzzle-loader that I borrowed.
>
> I did not begin to hunt because of dreams. I just liked to pursue animals, and the chief, together with my father-in-law, encouraged

me by lending their guns. Because we believe that European-made weapons incorporate powerful prescriptions, we do not ritually treat the animals we kill with them. I never use rituals or prescriptions on my animals. I believe that the "spirits"[ghosts] of animals have nothing to do with these guns, yet it is different with muzzle-loading guns.

With muzzle-loading guns, you must find a prescription to protect the gun. Our ancestors told us if we kill a bushbuck or grysbok, its spirit could render our weapons useless and the hunter must do something to prevent this from happening. Other animal spirits also affect the hunter including relatives and kin. These animal spirits are those of zebra, eland, lion, elephant, leopard, rhino, buffalo, roan antelope, and kudu. Of these, the eland spirit is the most dangerous. If an eland makes eye contact with its slayer, makes a noise, or sheds tears (*filamba*), that hunter will die unless that curse is countered. Large males (*bulundwe*) collect prescriptions (*muti*) in the dense hair of their foreheads and dewlaps.[29]

When I married across the river, I used the headman's weapons as he was old. Yet I wanted a gun for myself. In 1963, the chief gave me a hammer and spring from an exploded gun and I purchased a barrel for twenty-two shillings from an in-law. A blacksmith on the plateau assembled my gun for five pounds sterling.[30]

In 1962, I killed my first buffalo on Mbuvwe Plain and many more after that with the chief's rifle. Wounded buffalo are dangerous and I had my close calls—once jumping over a high stream bank to escape a charge. The buffalo stopped at the top of the bank and only looked me over as the sand impeded my movement. I no longer pursue buffalo but these close calls are not the reason. Game guards today [in 1993] closely scrutinize local hunters suspected of killing buffalo. If a resident takes buffalo without having a license, the case may end up in my court.[31]

I sometimes see the guards and scouts when I go into the bush, but they have never seen me with a dead animal or questioned my activities. I purchase a license every year, but we are directed since 1997 to take only small species and leave the buffalo for the chief and outsiders.

When I go out into the bush [from 2002 interview], I always keep moving. When I fire my gun, I run because the scouts may have heard the gunshot. I hunt because I can't remain idle. Who will give me meat these days? The scouts are afraid of me because I belong to the royal lineage and we are the "owners of the animals," but this custom is only a local one. Scouts must enforce national law, as I must do, and for that I fear them. I know something about scout "timing" and about their patrols in the bush. My friends among the scouts supply me information secretly on their planned patrols and locations. When they are on patrol, I only go to Malanda Plain a day or two after the scouts leave their camp. I take a dozen or more impala and a few warthogs each year.

In 1993, Kameko's extended family suffered a major tragedy when a wildlife scout accidentally killed one of his twin sons, who was also a scout serving on the same patrol. He describes how this calamity affected him, his hunting, and his guns.

> In the months since my son was killed in June [1993], I have not been able to kill any animals with my guns. I have fired many times, but have not yet killed anything. Just the day before yesterday [July 1993], I fired at a warthog. It was hit in the lungs; I followed its trail, but I couldn't find it. What has happened recently to my family has disturbed my mind.
>
> I washed my gun because I fired it at the funeral of my late son and I suspect that it is "banned" by some type of spirit. We believe that bad spirits can enter muzzle-loading guns and prevent them from killing animals. Then we must cleanse the weapon to chase away these "bad spirits" [*vibanda,* plural of *cibanda*]. Now that I have washed and warmed my gun, I will go into the bush tomorrow morning to determine if the spirit of my son has prevented me from killing animals.[32]

Kameko continuously supplied his close family with meat. When I last conversed with him at his homestead and at the palace in early October 2002, he was not his typical energetic self. For many months, he had suffered from intestinal pains. Upon my return to the states, residents wrote me of his agonizing death on 24 February 2003. A day later, the community mourned Kameko Chisenga and laid his body to rest in the royal Ng'ona burial grounds at the horseshoe bend of Bemba Stream.

FIGURE 5.2. Kameko Chisenga, dressed for a special occasion and reviewing some older photographs taken of Nabwalya residents. (Photograph taken by author in 2001)

Hapi Luben appeared outside our hut one morning in October 1966 after he ended his seasonal duties as a tracker with Luangwa Safaris. We expected his appearance as his stepbrother, who worked as my assistant, and other residents acknowledged that his hunting achievements exceeded others of his age. Hapi seemed to operate aptly in two cultural worlds and consistently beat the odds both as an exemplar on the local scene and as a tracker and protector for foreign safari clients. Yet his personal costs for such exceptional performance became apparent during subsequent decades.

Hapi Luben was born in 1933 in Luanshya while his father was working at the mines. His mother returned to her village near Nabwalya to raise him. Hapi never attended school; instead, he and others his age set snares and pursued small animals with bows and arrows, fished the river, and used dogs to pursue small warthogs and impalas seasonally. Ancestors of Hapi's maternal and paternal lineages were renowned hunters.

> When I was a young boy, I dreamed of someone very tall, hairy, and with scratches on his face. In my dreams, this individual called me *kaingo* ["small leopard"], shadowed me in the bush, showed me where to aim, and ended with a glimpse of an animal's forelimb. I told these dreams to my mother and elder brother. They told me that there was man in our lineage with those very features. His name was BaMbuluma. Elders dismissed the foreshoulder as no one understood its meaning until I made my first kill.
>
> My father was absent for long intervals during my youth for he worked in town. He taught me how to hunt [*kufunda bucibinda*] with guns for I accompanied him when he visited my mother. On these trips, I saw him kill buffalo, zebra, impala, and warthog. He also killed an elephant in our field.

As expected of someone his age, Hapi followed his father and became a laborer in Luanshya. When he had accumulated some funds, he returned to the valley and married. After another year in Luanshya, he again returned to the valley, married a second wife, and began to assess his chances of becoming a "respected Big Man" [*mukankala*].

> I was born too late to hunt animals with bows and arrows. Others used muzzle-loading guns and, in the early 1950s, I began with one belonging to my elder brother. My elder brother had purchased a muzzle-loading gun a few years earlier and did not use this gun much.

> My first kill was a buffalo. During the stalk, my body shook with apprehension. I would have missed, but instead my bullet broke the buffalo's foreshoulder. My guardian spirit had intervened and I realized then the significance of my earlier dreams. BaMbuluma had demonstrated his strength as a hunting spirit and that I could kill animals.
>
> My grandfather "doctored" this kill. He used two roots [unspecified] both of which he held in his right hand. He gave me a root, instructed me to close my eyes while leading me to the carcass. There he ordered me to spit a potion of the root into its nose and anus. Then I opened my eyes and hit the carcass with the butt of the gun saying, *"Tuli nenu, mwe nama"* (literally "We are with you animals"). I distributed this buffalo meat in Tepeka village where I had gone for marriage. I continued to kill animals, buffalo, warthog, and impala until this wife suddenly died and I left.[33]
>
> The village where my mother lived was very close to Tepeka, but I did not go there to settle. Instead I came to settle in Mbuluma's village because it belonged to my lineage. I soon married another woman of Mvula lineage and spent time [bride service] with her kinsmen. In 1956, I ended my service, paid my in-laws five shillings and brought her to live with me.

About 1962, a district messenger appeared at the chief's palace to recruit seasonal trackers for Luangwa Safaris, which had leased the Munyamadzi Corridor as its hunting concession. Hapi was chosen for this highly sought position that included local employment and bonuses for exceptional service. During dangerous incidents between wild animals and foreign clients, Hapi demonstrated exceptional courage and skill. He rose quickly to become an armed tracker ["real hunter" is how he described it], one whom Peter Hankin [a white professional hunter] trusted when pursuing potentially dangerous game. Foreign safari hunters may have sought to impress him with their wealth and equipment, but Hapi could describe their ineptitude in the face of danger. The operators credited Hapi with firing the killing shot and saving lives on three occasions.

> One day I was working together with my elder brother at the safari camp when a client wounded a lion. Pursued by the professional hunter and client, the lion charged, causing all Africans to flee except my brother, who had an axe, and me. My brother said he couldn't let his "younger brother" die alone. So we remained together with those two white men and faced that lion. Both whites fired their rifles and missed. I fired at point-blank range with a

shotgun right into that lion's face. The lion fell and died at my feet. This client thanked me profusely and paid me handsomely as he had expected death or a mauling.

When not working for Luangwa Safaris, Hapi returned to the village where he hunted for his relatives and for the chief. Throughout the 1960s, his reputation grew, he did little farming, and his conquests became grist for many local stories and embellishments. He married four wives, sired more children, hunted with foreigners, and trained nephews and kin to hunt. When a supportive maternal uncle died in the early 1960s, his father became his patron, backing his material needs and enabling him to continue pursuits in the bush. Gun owners readily lent him weapons so they could benefit from his productivity.

In 1969, Hapi's fate turned grim. After a successful safari season in which three lions, eight elephants, three hippos, six elands, and a variety of lesser game were killed, he returned to the village a very sick person. For eighteen days, his chest ached and he had trouble breathing. Initially he thought this affliction came from his associated safari workers, who were jealous of his close relationship with the European boss. He reflected on the demise of his own grandfather and spirit sponsor, BaMbuluma, who was favored by an early administrator and allegedly bewitched by his kin. One of these purportedly sent an elephant (*nzovu yakutuma*) to trample BaMbuluma. Hapi recovered and returned to safari work, but his troubles were only beginning.

Jealousy [*cisondo/bufuba*] caused me to quit working for that safari company. Peter Hankin, the professional hunter, liked me a great deal, but I feared the witchcraft of others there. Beginning in 1969, I became very sick while still working for that safari company. I was convinced that a kinsman was bewitching me. He was trying to kill me because I had a good job, was doing well, and had goods to distribute locally. The following year at work, I became very sick again. This time, the safari company took me to the dispensary for treatment. When I recovered, I married another woman, Carlosi from Lungu lineage, because I wanted many children.

In 1971, I had no safari work because Peter Hankin was hunting in the Congo.[34] I spent that year moving between my wives' villages and building large gardens. Following the rainy season the next year, I built large granaries. Although I hunted during these times, I didn't kill much. Again in March 1973, I fell very sick. Then I decided I must leave the safari business because my

> relatives were bewitching me. I "died" twice. Other relatives in Lusaka and elsewhere poured funeral beer as they heard rumors of my deaths.
>
> This serious sickness and accusations from within my lineage continued and finally caused me to leave Mbuluma Village in 1977. In 1973 and 1976, crocodiles took two nieces doing household chores at the river. In 1973, my daughter rescued one niece as a crocodile was pulling her under the water. Her father grabbed the second niece when only her hand was showing above the surface. My relatives visited a diviner who designated me as the one controlling these crocodiles.

There were other confrontations as well. In 1976, Hapi's niece experienced a difficult pregnancy—a cultural indicator that her spouse had slept with someone else during that time. After many inquiries, the husband confessed to sleeping with Hapi's elder daughter. To ease the delivery, relatives administered a special prescription (*muti wancila*), but the niece delivered a dead fetus. This niece then accused Hapi of forcing his daughter to sleep with her husband in addition to sending the crocodiles. Under these circumstances, Hapi realized he would find no peace if he remained, as these accusations would continue and his "sisters" would not support him as a headman. But he needed a sponsor and someone to lend him a gun. He found that ally in his father, who had settled in Chief Mwanya's territory east of the Luangwa River. Following his father's overtures, Hapi joined that small settlement in 1977. He took with him his immediate household, reduced by then to a single wife and children, to become his father's client. His father promised to help financially and with mentoring. In 1989, Hapi recounted:

> Settling in my father's small village was not easy. Eventually our standard of living improved as good roads connect to the towns and these roads allow us to get our sunflower, sorghum, maize, and groundnuts to market. A stepbrother employed with the wildlife department helped when we first moved by purchasing elephant licenses and lent me his rifle. So in 1978, I killed him two elephants, sent him the tusks, and we ate the meat. The following year, I also killed another elephant and other game for him. My sons and I continue to kill an occasional buffalo and impala off-license.

FIGURE 5.3. Hapi Luben (left) and his father Luben Kafupi (right) returning to Kafupi's new settlement with a young warthog that they ran down during an early morning hunt. Luben grasps the piglet with his right hand and carries his muzzle-loading gun on his right shoulder while balancing it with his axe over his left shoulder. (Photograph by the author in 1973)

Hapi infrequently recrossed the Luangwa River to visit his sisters in Chibale's village, but he never returned as a resident. Hapi was visiting his mother's sister in April 1984 when Chief Nabwalya died. Hapi's hunting forays then became portentous for that interregnum. When the chief's death became public, Kameko gave Hapi the chief's rifle and three rounds of ammunition and commissioned him to secure a buffalo for the throngs of people gathered at the wake and funeral. The first day out, he wounded two buffalo, which, despite his persistent pursuit, escaped. Out of ammunition, Hapi returned and reported these details to his sponsors.

The next day, Hapi took a Game Department rifle and six rounds. He encountered a buffalo herd and wounded a large bull, which escaped while losing pools of blood. Following this trail, Hapi encountered a monitor lizard (*mbulu*) tracking the same buffalo and consuming the spilt blood. Hapi halted immediately, interpreting the lizard as a very ominous sign, and returned to the funeral. The spirits were preventing him from killing animals and the omen signaled that something untoward had occurred within the chief's lineage.[35] After three days, mourners wrapped the body of the late chief in a zebra skin and buried it with lion claws and teeth behind the metal-roofed palace.

In 1988, Hapi came to hunt again on his old turf. A cousin in town had purchased licenses for two buffalo and sought Hapi and another relative to secure these animals. They used a rifle belonging to the relative's father. Hapi encountered no problem securing the first buffalo, but while pursuing the second one, an itinerant honorary game ranger arrested both him and his accomplice. This European ranger was pursuing a licensed buffalo and had engaged a local scout to help him find a herd. Both groups had pursued the same herd that morning.

Apprehended and charged with hunting without a proper license and using a borrowed weapon, Hapi and the relative were handcuffed and taken to the palace of the interim chief. The chief declined to intervene, reasoning that since Hapi had deserted his chiefdom, he was no longer a subject. Consequently both prisoners were handcuffed to the baggage rack atop the ranger's new Land Rover and taken up the tortuous grind of the escarpment to the Game Warden's office at Mpika. These "poachers" were beaten repeatedly while in jail awaiting trial. At trial, they received a year at hard labor or a fine of ZK 1,000 (then about $125 US). Both men spent a week at hard labor until relatives raised the money for their releases.

In January 1996, a crocodile severely injured Hapi's elder sister while she washed pots at the river. She died a few days later at Chilonga hospital after enduring a painful three-day journey in a hammock. Hapi appeared at the funeral gathering in Chibale's village, and that evening he dreamed of a crocodile biting his legs and arms. He awoke the next morning paralyzed in both limbs. An in-law found him, witnessing black spots and scars on Hapi's limbs that bore the imprints of crocodile teeth.

As rumors spread, people expanded upon Hapi's affliction and condition. Many interpreted the incident as caused by the same crocodile that had killed his sister and had earlier tried to take his nieces. They thought that the crocodile belonged to this brother-in-law, and that he had sent the crocodile to incapacitate

Hapi before he could avenge Hapi's sister's death. Other people suggested that the vice was endemic within Hapi's lineage. These incidents became connected as compelling stories, which, with repeated telling, appeared to vindicate their veracity. One story began with the brother-in-law obtaining *bwanga* (evil substances) for making a crocodile, which he then commanded to attack his detractors and in-laws. As an elder and patron, this in-law had used his victims' labor and power to revitalize his health and to increase his wealth without their knowledge.

Hapi moved from his late sister's residence; four days later, he recovered and recrossed the Luangwa River. Things had not gone well for Hapi in Chief Mwanya's territory either. His father, in failing health for some time and on his deathbed, summoned Hapi and reaffirmed his support.

> There is nobody who can advise you, only me. My son, I tell you, fear people, give them respect, and don't do anything wrong to other people. If you don't follow my advice, you will see that the world is bad for you. I am the only person to give you advice, but now I am too old and about to die at any time.

Since his father's death in April 1991, Hapi lost two married daughters from AIDS, contracted from their husbands who were often on trips into town. In the late 1990s, he separated from his remaining long-term wife and went to Lusaka seeking a cure for ill health. Hapi used his knowledge of "traditional" prescriptions and herbs to support himself as a fertility specialist to older women. While there, he was arrested by a jealous husband, taken to the police, who fined and forced Hapi to register as a traditional healer before releasing him. Living in urban squalor with a "grandson," Hapi continued to decline in health until, back in the village, he was reported to have died of "cancer" in June 2003, revived for a few months to die again on 7 April 2004.

A Third Generation of Hunters (Born 1945–60)

Two individuals stood out to me among this third generation that came to maturity during the 1980s and 1990s.[36] Both were cross-cousins belonging to intermarrying lineages. Both started life in their "grandfather's" village, yet they were rarely co-residents again. Both belonged to renowned hunting lineages and sought that identity for themselves.

Kunda Jacobi, born in 1949, was two years older than Malenga Morgan. Both of their mothers were part of the matrilineal cores of a new village founded by their grandfather. Their mothers supported this headman's initiative: he was comparatively wealthy and, as a hunter, protected and provided sustenance, as their husbands labored as migrants elsewhere. Both Kunda and Malenga remember their early years as prosperous and secure in their grandfather's village. No one suffered from hunger, which was not the case later on. Other residents may have accosted the headman for meat and showed no shame, but the headman had tactics for dealing with this behavior. When he or a client made a kill, he claimed to possess no meat while hiding his share in the bush for several days. Malenga and Kunda remember game pits deployed around fields and snares set by young boys, including themselves.

That some of their uncles spent more time hunting than farming impressed them. These relatives spent their time in the bush, exchanged meat for crops and labor, and had married several women. Their wives managed households in neighboring villages, did most of the field and food chores, and kept their husbands and his friends supplied with local snacks and beer. Malenga and Kunda aspired to this masculine mystique. Later they learned the critical role patrons played in these whims.

Malenga and Kunda left their grandfather's village at different times in the 1960s. In 1963, Malenga and his mother joined his father on the Copperbelt and returned later to Nabwalya for primary schooling. He left again after completing primary school, pursued his formal education for another two years, and worked for several years at odd jobs before returning to Nabwalya to marry in 1980. Kunda left his grandfather's village in 1967 when his mother died suddenly. He went to live with his maternal uncle, another headman of a village some distance away.

This uncle encouraged young Kunda to join him on forays into the bush and to develop stalking skills. Kunda remembers his uncle telling him about the late Mupeta Ng'ombe, a maternal grandfather. This ancestor subsequently appeared as a whirlwind in Kunda's dreams and this occurrence always predicted success.

The uncle reputedly used special medicine to see animals, but Kunda was then too young to learn this prescription.[37] Kunda remembered an overnight encampment when his uncle ritually "closed the forest" by encircling the encampment's periphery with burning prescriptions. He was present when his uncle killed an elephant, rhino, zebra, and buffalo.

Kunda's dream of Mupeta Ng'ombe (guardian shade) impressed his elders, who supported his determination as a hunter. They said that the "spirit should be my model as long as I had a gun of my own." Their awareness of my "calling" "forced them to fetch a gun for me in 1972." Still under the tutelage of his uncle, Kunda killed his first animal (*inama yakuzambilila*), a male impala, with this gun. Unlike earlier generations of hunters, Kunda claimed there was no ritual over the carcass or village ceremony. Instead, he crassly stated that it was "chop and carry," expressing a different ideology and social sensitivity from that of Malenga and the earlier generations recalling first kills.

Kunda killed other species in rapid order: first a warthog, then buffalo, and in 1973 he killed his first elephant in a nearby field. He ritually treated the elephant with a prescription given by an "elder brother." Kunda sold its small tusks for K 30 (then about $40 US) to a stranger from the plateau. His reputation grew; during the heavy poaching of the late 1970s and 1980s, he killed many elephants using assault rifles loaned to him by army personnel and small groups of strangers, who came into the valley to solicit assistance in exchange for money and grain. In addition to benefiting from shares in these elephants, Kunda reputedly killed a black rhino in 1978 and sold its horn for ZK 500 ($600 US).

Despite this success with outside assistance, Kunda was not beyond seeking "traditional" means to enhance and to authenticate his prowess. While still with his uncle, he complained of missing several animals at close range. His uncle gave him some roots that Kunda pounded into powder; he then placed its residue in incisions on his wrists. The generic term for this prescription, *cilembe,* referred to the activating particle to cause massive internal hemorrhaging when a bullet strikes its target. When a wounded buffalo charged Kunda later, his uncle granted him another prescription so that dangerous animals could not sense or locate him. This prescription was a type called *mfenzi.*[38] Kunda attributed this prescription with saving his life from a wounded elephant in 1985.

In 1972, Kunda moved to another village where he married a cross-cousin the following year. In his new locale, Kunda possessed a muzzle-loading gun, but he did not let his in-laws know about this gun until he became secure in his relationships there. As Kunda became more open about his hunting prowess, the aging village headman depended on him for protection and for meat. This headman also provided cover, with plausible alibis, to protect Kunda's nefarious dealings with outsiders or his long absences.

After several stints of employment in town, Malenga returned to Nabwalya in 1980. He immediately found village life "very difficult," for he was unmarried and lacked a proper guardian. His father's sister, who was married to a local court messenger, looked after and fed him. While in town, Malenga was aware of Kunda's growing reputation from mutual acquaintances. For a few months, they resided again in the same village, and Malenga followed Kunda, who always took the lead as the owner of a gun and as the elder. Another cousin and a colleague of Kunda's possessed an unlicensed muzzle-loader that Malenga sometimes borrowed.

As he had returned to marry, Malenga went to live with his prospective wife's relatives. Fresh from town experiences, Malenga found his in-laws conservative and ungracious and recounted that they considered him a "slave," expecting him to perform menial tasks at their whim. Chafing under his depressed status, Malenga had few opportunities to develop his hunting passions, although he made several unsuccessful forays. After nine months, Malenga paid his in-laws cash to compensate for his time remaining in bride service and moved to his maternal uncle's village. His pregnant, youthful wife soon followed and joined him there.

As a young married man living in his lineage's village, Malenga hoped to settle into a hunter's lifestyle—a vision of expectations informed by the stories he heard in his grandfather's village and witnessed in his uncle's life. Farming was not as attractive as hunting, yet to establish his new household Malenga had to spend considerable time clearing the fields allocated to him by the headman, constructing his new household, and renewing relations with those around him. In 1981, he killed his first animal, a male impala, and ritually treated it as instructed and as he had observed with his grandfather. Malenga had access to an uncle's licensed gun as well as to an unlicensed ("control") gun belonging to a late cousin, who had been gored by a buffalo years earlier.

As elephants and the ivory trade became international issues in the 1980s, the Zambian government increased wildlife enforcement and patrols in the Luangwa Valley. These contingencies added to the normal load of social and environmental challenges faced by previous generations of hunters; consequences for disobedience were costly for kin as well as for those who were apprehended. Malenga and Kunda mastered these new conditions in their own, yet different, ways.

In establishing a reputation for himself, Kunda spent his time collaborating with outsiders seeking ivory and meat. In these liaisons, he affiliated with a nefarious game guard, Ngambi, who provided him cover and contacts. While under pressure from his superiors to check on local "poaching," Ngambi became a prime detractor and obstacle to Malenga, who resided close to his game camp.

In 1984, while accompanied by a nephew and stalking a zebra near Bouve Plain, Malenga felt his heart pumping wildly and heard a rifle cocked behind him. Malenga slid his unlicensed gun under some dried grass and instructed his nephew to run. Thus began Malenga's first confrontation with Ngambi, who was just beginning a patrol with his assistant. Initially Malenga denied possessing a gun, suggesting that they had seen the sun's reflection on his axe, and that his companion fled because he feared a beating from the guards. Unsatisfied with these answers, the guards searched and found the hidden weapon. With the gun in hand, they threatened Malenga into giving an oral confession. The guards retained the confiscated weapon, proceeded on patrol, and directed Malenga to return to his home.

When summoned to the game camp, he denied possession of the gun shown by the guards, who then handcuffed and retained Malenga. While undergoing alternating friendly and hostile interrogations, Malenga persisted in this story. He assumed the case was not certain, that the gun was unnumbered and untraceable to him, that he had never been to prison before, and that he would deny the evidence in court. Unable to obtain a written confession, eventually Ngambi dropped the case since he had not followed proper arrest procedures and feared that in court Malenga might reveal Ngambi's involvement in illegal ivory and meat sales. The guards kept the unlicensed gun and ordered Malenga to give them something as a favor, to stop borrowing weapons, and to refrain from illegally killing animals. With the case closed, Malenga sent two chickens along with a token of beer to the guards.

Relatives elsewhere, even strangers from the plateau, arranged with specific local hunters at Nabwalya to secure meat for them by providing these hunters cash to purchase resident game licenses at reduced prices; legally outsiders were required to obtain the expensive national licenses to hunt at Nabwalya. Provided licenses were visible and fees paid, most guards considered these arrangements legal for several years, yet hunters found many ways to bend the formal requirements to fit their needs. One of Malenga's "plateau brothers" purchased licenses for a buffalo, a warthog, and a hartebeest and caught a ride on the Catholic mission's vehicle traveling to Nabwalya. The brother requested Malenga to kill the buffalo immediately so that the mission's vehicle could take the meat upon its return trip. Malenga borrowed a gun from Chibale and used the headman's powder and shot. Even on such short notice, Malenga secured, butchered, and smoked the meat from a large buffalo and loaded it into the vehicle. He remained with the two other licenses and expected to deliver the rest of the bushmeat personally within a few weeks.

Unable to find a hartebeest, Malenga killed a wildebeest and, using the cover of the license, killed at least six warthogs. Malenga instructed his younger brother, who was to accompany him on the walk to Mpika, to place four warthog shoulders into a carrying bag. Instead this younger brother loaded eight warthog legs. As their path to the plateau went through Kanele Wildlife Camp, Ngambi ordered them to halt so he could determine if all the animals were on license. Ngambi cleared the wildebeest meat, but he asked if warthogs had eight legs. Malenga responded that he found an additional hog killed by a predator, which accounted for the extra legs. Ngambi rebuffed this answer and charged Malenga with theft of state property.

Ngambi allowed the younger brother to proceed with the wildebeest meat to Mpika. Upon his arrival there, he contacted the older relative who immediately proceeded to Nabwalya. In the meantime, the case turned grim as Malenga faced an additional charge—that he did not own the gun with which he had killed the animals. When the older brother arrived from the plateau, he found Malenga and Chibale in detention at the wildlife camp. After a prolonged discussion, Ngambi released both Chibale and his gun with the admonition that "although you have disobeyed the law, how can we arrest and prosecute such an old man?" Thus Ngambi showed respect for a local elder even as Malenga held to his story of not using the weapon to kill the second warthog. Sentenced to clear Ngambi's field, Malenga completed that task in two days with the help of relatives and friends. A reflective Malenga realized his best option was to take rather than detest this punishment.

Kunda's stories from the 1980s involve more intrigue and illicit activities. In April 1985, he joined a team of seven others, including a Tanzanian possessing a NATO assault rifle, to hunt elephants along the Luangwa River. The group killed three; then they decided to pursue additional elephants they had heard one evening. After wounding one of these elephants in tall grass, Kunda ran toward it. The wounded elephant saw him and charged. When Kunda's rifle jammed, he turned to run but fell instead. The elephant's tusks punctured Kunda's chest and, while he lapsed into unconsciousness, his companions threw their boots and clay at the assailant to distract it. When the elephant left, the companions revived Kunda, placed him on a litter of poles, and carried him to a nearby village. The group nursed Kunda for two days, then carried him to a clinic by a circuitous route. Their official story was that, while hunting locally with his registered firearm, Kunda was gored in the chest by a buffalo. The "strangers" (the team) were in the vicinity and had carried him to the clinic. Kunda sent the assault rifle to Ngambi to hide from the police and paid him for his compliance and silence.

Kunda spent four months recovering at the clinic and received several visits from police investigating the disappearance of a military rifle from an army unit encamped at Mpika. Kunda repeatedly told about his wounding by a buffalo. Ngambi reported to the police how his guards had repossessed the rifle. They found a "poacher" who had recently killed a buffalo. When the scouts fired two warning shots, this "poacher" fled, abandoning the rifle. The police accepted the consistencies of time and story particularly as the rifle had (re)appeared in their hands.

The expansion of the ADMADE program and local enforcement activities in the early 1990s were particularly problematic to both Malenga and Kunda. Each pressed a strategy that diverged noticeably from the other's as the decade progressed. Malenga built his socially oriented and local strategy upon past precedents steeped in long-term commitment and maintenance of relatives and loyal supporters. Kunda focused on short-term, outside goals, individually circumscribed and mainly financial. Malenga's predicament seemed more precarious than Kunda's situation because his residence was close to both the palace and the scout camp. Moreover, his linage was notoriously proud of its hunting predecessors and was often at odds with the reigning chiefs. In the middle of the decade, he married a second wife and moved to her distant village for a few years, thereby increasing his options. He later resettled her as a co-wife when he was selected as headman back in his former village. Kunda, in contrast, resided some distance from the palace, kept good relations with the chief, and disappeared for long periods, staying out of sight from scouts and other residents.

Malenga's plotting as a dependable leader and provider worked well during the early part of the 1990s. He expanded his fields on the fertile soils north of the Munyamadzi River by recruiting labor, who worked for him based on his reputation as a hunter. His discreet killing and distribution of four buffalo and other game in 1992 clinched his budding reputation and paid back the expectations of his laborers, who were also his kin. Yet his material gains one year were offset the next by drought and severe floods. Moreover, his surreptitious hunting eventually raised the suspicion of authorities.

In October 1993, village scouts arrested Malenga based on information given by a woman married to a member of the ADMADE Sub-Authority. He had killed an impala, and the woman observed him carrying bushmeat during daylight hours. On the basis of this observation, this woman informed the scouts that Malenga had killed another buffalo. The scouts went to Malenga's home, confronted his wife with the charge, and, gaining entry into the house, found a cache of impala and warthog meat. The scouts charged Malenga with a felony, handcuffed him, and took him into custody.

Malenga's story was that he had taken the warthog on his license and that his son and a scout's son had taken another young warthog with dogs. He claimed his father had killed the impala on license and given him a share. When his father produced the license, they released Malenga, including his manhandled weapon, which the scouts had repaired. Later, the wildlife unit leader summoned Malenga to relate that the "informer" complained about her safety as the scouts had compromised her identity in the confrontation. Malenga asserted that she had nothing to fear, yet within a week, a runaway bush fire burned her house to the ground, causing her to go elsewhere to live.

Somewhat later the chief summoned both Malenga and his father to his court on charges of disloyalty. The charge against Malenga was that someone accused him of defaming the chief's name in casual conversation by suggesting that the chief had misappropriated bags of famine-relief maize. The court discussed Malenga's conversations and activities as if he were a dissident and suggested that he should consider leaving the chiefdom. Like Jackson before him, Malenga rejected the court's verdict and decided to remain, and as a headman. The size of his village grew in number as residents of his second wife's village fled their elephant-ravaged fields that scouts had refused to protect. These new settlers sought the amenities of the school, clinic, protection, and prospects for employment that residents closer to the chief's palace had benefitted from.

Rituals, Retentions, and Revisions: Spirits Communicate through Dreams and Elders

Both Chibale and Jackson were clients and later patrons, became productive huntsmen and headmen, spent time outside of the valley as labor migrants, specialized in rituals, and "gathered people." Their lives encompass much of the story of what this book is about, a provincial system of ideas and practices related to wildlife and hunting, lineage husbandry. As young men, whose birth separated them by two decades, each faced tumultuous times as imperial imperatives affected both local livelihoods and their identities. They and the next two generations of hunters achieved a measure of success and respect as each in their own ways embellished or reconstructed roles and goals in keeping with their experiences and changing governmental policies. Only Jackson was to live sufficiently long to witness the unraveling of the resource regime to which both he and Chibale had contributed so much.

These life histories and stories reveal changing local cultural and environmental tapestries emphasizing varied threads in the processes of learning and sharing its knowledge. "Traditional" knowledge may seem limited in scale but it is not necessarily ancient, despite the way it may be discussed in that light. It incorporates new materials as well as exemplifies shared experiences, meanings, and values. It is unlike other knowledge in its disposition and intents (Posey 1999: 4). Hunting skills and associated rituals are forms of such knowledge, which individuals develop through initiation and in associations with others as they create and maintain identities. Yet this confidential knowledge is itself open to adjustment, subject to experience and to the processes of absorption and reconstruction (Bender 1993b).

A person's matrilineage (mother's lineage) was the most important grouping, as this membership largely determined the individual's identity, inheritance, and eventual status. Ancestral spirits played a significant part off-stage among this "community of actors" as they became visible through elders' interpretations of events, implied connections, and performance of rituals. Active spirits "caused" and communicated through dreams while elders' interpretations prompted some men to assume the collective needs for sustenance and for protection. What this generation heard and experienced was passed on to the next through demonstrations and in stories.

Hunting for a few became the local symbolic masculine counterpart to that of women in agriculture and in sustenance preparation. Yet older women as grandmothers and mothers played important roles encouraging their "sons" to undertake social responsibilities. Elders managed younger men by lending them capital equipment (guns, money), defining events (dreams, interpreting contexts), distributing meat, cash, and labor (wealth), and conferring legitimacy (status, prescriptions) during performances. Elders also managed a youth's access to women through enabling initial marriages. Young women, like most young men, were often pawns for their elders' objectives, particularly those of older men.

Internalizing and participating in lineage rituals, sayings, and daily behaviors promoted identities and fostered local loyalties within communities. Yet these boundaries were never impermeable, for cross-cousin marriages between alternating generations brought new information continuously into these mediums. That the slogan Hapi uttered at his initial kill was the same as that voiced by Chibale a generation earlier was no coincidence. Hapi was the latter's "grandson," yet he belonged to a different lineage, never possessing the disposition, relevant "sisters," or wealth to become eventually a headman.

A fundamental premise in local hunting was the sharing of its proceeds according to implicit guidelines. The expected distribution of meat among a wide assemblage of relatives carried political and social weight that in turn influenced hunters' selection of the larger mammals, such as buffalo and elephant, rather than the smaller ones, such as bushbuck or grysbok. Besides the elephant, the former group included the eland, lion, and sometimes rhino, mammals believed to possess a particular potency. Lions were strong symbols of power, killed for protection and when needed for the burials of chiefs. Their carcasses were approached with "respect" as their spirits were "fierce" (*ubukali*) signifying that a complex of ideas, behavior and ritual procedures were warranted if pursued. These species were "marked" in deference to the "owners of the land", a chief's way of monitoring their stocks, as well as later a licensed revenue source for the state.[39] Each species possessed ominous qualities, suggested in how Chibale and Kameko informed about the gaze, tears, voice, and charms carried by bull eland. Hunters in 2006 noted these same attributes as reasons they likewise avoided eland.

Some smaller species encrypted additional cultural significance on the local scene. The grysbok and bushbuck were said to have the capacities to "tie up" (*kuikate*) muzzle-loaders. The bushbuck belonged to the *vizemba* group because of its stripes

and blotches, while I suspect the symbolic clout of the grysbok, the smallest antelope, was its diminutive size, appropriate only for an individual meal or two (ie not for communal sharing). Hapi's account of the monitor lizard, encountered while pursuing his wounded buffalo, indicates that other small animals were predictors of personal quests as well as the transitory traumas surrounding a chief's death. With mammals on either end of their prey spectrum classified as either dangerous for kin or a liability for weapon, most hunters pursued the abundant, frequently encountered and medium sized mammals, the impala, warthog, and buffalo (*inama zha buBisa*), as the "optimal" set given their weapons and in terms of their "costs and benefits" for kin.

As with most aspects of power and authority, a person's knowledge of rituals, prescriptions, and symbols was double-edged and could thwart personal ambitions. As energies fall and resources fail, elderly men, such as Chibale and Jackson, became suspects in others' (particularly kin's) misfortunes and quests for assets. Chibale used his authority as an important headman together with his connections to the late chief as he managed accusations and guarded his person as he aged. Jackson adroitly used his competent reputation to secure employment, initially locally and then regionally. His expanding regional experiences, unlike those of Hapi and of his father with safari operatives, enabled him to use his network to counter the unilateral actions of a disgruntled chief.

Jackson continuously remembered his guardian shade by consecrating a gourd and placing it in the rafters of his home. This type of hunter, called *bacibinda bankombo,* solicited ancestral assistance through offerings and prayers, with the gourd (*nkombo*) serving as its most tangible element. Chibale belonged to another hunting tradition, called *bacibinda bamiti,* which emphasized the potency of plants (*miti*) and other substances for enhancing, if not ensuring, success. In addition to anticipating success, these prescriptions allegedly rendered the prey vulnerable to the hunter's will, protected the person against misfortunes in the bush and at home, and neutralized injurious effects from the world of spirits. Kameko insisted that he belonged to a third class of "natural" hunters, called *bacibinda bacifyalilwa,* as these tasks came spontaneously, or from within as an infliction of an ancestral spirit. Despite this claimed distinction, his sons told me that he also depended on ancestral guidance, knew about hunting prescriptions, washed his guns (see figure 8.2), and used his own rituals and private hunting prescriptions (chapter 8). All hunters felt vulnerable to their detractors, whom they imagined as a fourth type of stalker—wizards (*bacibinda bwabwanga*). These individuals supposedly used ominous and powerful sources (mostly of human origin) as their covert killing force, and they were just as likely to use these substances against kin as against animals. I found no such self-acknowledged wizards at Nabwalya, yet some contemporaries alleged that they might exist camouflaged amongst them.

None of these hunter groups was an ideal type as all used a variety of prescriptions and medicines. The possession of hunting prescriptions linked individuals to significant others, whether relatives or outsiders. If the individual giving the prescription was renowned and a nonrelative, he usually expected cash for revealing the secret formulations; relatives of a hunter provided them gratis as part of lineage lore. Activating components (*vizhimba*) [those substances, usually of animal or human origin, that gave the other substances, usually plants, their potency] were of four generic categories. *Cilembe* augmented the killing power behind a projectile (arrow or bullet) as it struck its target to cause massive internal hemorrhaging preventing the prey's escape. *Ulwito* called game, countered a prey's natural tendency to run away, and made prey visible and vulnerable to hunters. *Mfenzi* provided camouflage or presented a "void" [blackness] within which the hunter had to remain still and invisible. *Ntezi* repelled prey seeking to do harm. An individual's accumulated repertoire of contacts, rituals, and medicines, like degrees and institutions among academics, represented a litany of strivings and affiliations as well as a mastery of local lore and skills. Yet their local moorings and cultural rationales appeared to fade quickly in the hunting voices of those struggling to get ahead in the twenty-first century under the new burdens of economic expediency, inflationary market values, and identity checks.

Cycles of Resourcefulness and Response: Staying with or in the Game

The memories of Chibale and Jackson and other men of comparable age were the basis for my description of Bisa hunting culture in the initial third of the twentieth century. Bows, poisoned arrows, spears, downfalls, snares, and various indigenous traps were then the main means for taking wild animals. Muzzle-loaders were not unknown, as Chikunda ivory hunters, warlords, and other travellers possessed them, including a few prominent valley residents. By the 1930s and 1940s, many valley men had obtained these guns as Jackson had through purchase and Chibale had through inheritance. Once (re)acquired these weapons became the targeted material for local hunters' creative imaginations until the gun's culpabilities reduced its utility below that of earlier and indigenous conceptions (particularly snaring).

Although Jackson was older than I, our friendship was closer and more comfortable than that which I had with Chibale. Jackson was a mentor and escort during my initial week at Nabwalya in 1966. We met intermittently during his outside employments and more frequently after he retired to become a local resident again in 1980s. His hunting exploits had added flesh and ritual trimmings to the skeletal arrangements of the lineage husbandry designed by the young Chief Nabwalya [Kabuswe Mbuluma] and Jackson's slightly older contemporaries. Jackson's disdain for heavy agricultural work embellished the proverbial youthful fantasy for "beer drinking and hunting" (*kunwa ubwalwa na bucibinda*) that distinguished young men's work most profoundly from that of women. Jackson's prowess was also recognized by government employment later as a messenger and armed escorts for administrative safaris into the hinterland districts in Northern Province. Both Kameko and Hapi extended Jackson's contributions through their new opportunities and added new stories of venery. When Hapi claimed large fields and granaries during the late 1960s, his words were more figurative than literal, for he paid in bushmeat for others' labor to accomplish his goal. An impressed nephew, Malenga would add similar feats in the next generation. Establishing a reputation as a successful hunter took time, entailed runs of successes and evasions, articulating entertaining stories as well as the support of a generous father (Hapi), the backing of a material uncle (Kameko) or that of powerful outsiders (Kunda).

Birth situates some men more strategically for enduring life's inevitable struggles than it does for others. Both Kameko and Hapi recalled the carefree nature of their youth and the admonitions of elders. As a member of the dominant lineage and extended family, Kameko had access to lineage assets (for schooling, for information, and employment) and prominent positions that enabled him to become a locally influential person, despite his presentation as "an orphan." The accounts of his initial hunts showed little of the subordination implicit in other accounts. Yet in the end, his closeness to the source of local power did not assure him the title to which he aspired (the chieftainship). Neither was Hapi able to secure the lineage "respect" for which he struggled. Upon the death of his maternal uncle in 1960, Hapi largely depended on his displaced father's diminishing power, wealth, and connections. Initially, safari employment buoyed his prospects and reputation. Yet when that job ended, the circumstances of his resettlement together with his father's deteriorating health and exile worked against Hapi's prospects and longevity.

Malenga and Kunda were at the height of their hunting careers during the 1980s and 1990s. Both acquired muzzle-loaders and employed local knowledge that each continually filtered through the lenses of changing circumstances and new uncertainties. Like Kunda, many men deflected the scrutiny of wildlife scouts by becoming clients to powerful inside and outside patrons, by narrowing their immediate social relations while expanding networks regionally, and by participating in the informal economy while engaged with extralegal activities. Others, such as Malenga, survived through persistence, connivance, adaptive skills, earlier roles, and supportive "spirits" as they sought local "respect" and transient status.

Muzzle-loading guns retained much of their significance in lineage inheritance and protection despite their increasing liabilities as hunting instruments. As outsiders assumed that every man was a "poacher," residents responded by becoming more circumspect in their activities and resorted to earlier technologies, including game pits, poisons, extended encampments, dogs, and snares. This reversion to "traditional technologies" and their random selection of species ran against the grain of what conservation planners expected and could rapidly bring down wildlife numbers. Wildlife officials were stuck in comfortable offices implementing their arcane tasks; outside minds worked slower than those on the ground dependent on sustaining themselves.

Wildlife policies and the "community-based" wildlife program impacted local society and its hunting culture in many ways. A military hierarchy of locally recruited and employed youths, responsive to foreign and state interests, enforced the new laws and imposed an institution on GMA residents. The new regulations replaced residential local use-values with exchange (economic or numerical) standards based in a "quota" system and the purchases of expensive hunting licenses. The denial of local wildlife entitlements existed in Hapi's and Komeko's generation, yet law enforcement was erratic, and less costly options still existed in license fee reductions for GMA residents. Local elites, wildlife scouts, outside residents, and safari concessions monopolized the market quotas for buffalo, which together with elephants, became abundant pests around fields and villages. To obtain inexpensive local licenses, residents had to walk to Mpika to find that more privileged purchasers already had purchased their permits. As legal access to wildlife became more expensive and contested, Kunda and Malenga's cohort of local entrepreneurs had to find separate ways to cope with their dilemma.

During brief visits to Nabwalya in the late 1990s and early 2000s, I sensed profound changes in local behaviors and the welfare of its residents, yet I remained unable to comprehend their profundity. Residents were suspicious of outside inquiries, knew the "acceptable answers" to the questions outsiders typically asked, and were openly cagey about the practical aspects of daily living. I knew understanding would demand commitment and more time for me to grapple with. During a brief visit in 2002, I sought the wisdom of a Nabwalya resident known earlier and asked him to describe the current hunting scene and to speculate on its future.

> In some ways there has been a return to the past. Most people now use wire snares rather than guns. Although no longer found around the palace, game pits are seen in places with fewer people and less interference. Some hunters still profess the use of hunting magic; others do not. Obtaining these medicines [prescriptions]

from outside one's lineage has become more difficult [a gamble] because the owners do not share their real ingredients. If you purchase these medicines with money, then there is no obligation; yet one's relatives may accuse you of using power in the wrong way and against them. So you lose both ways. Now everybody has a limited number of trusted friends. We are afraid that "informers" will squeal and wildlife scouts will appear, beat, and arrest us.

In the past, hunters had the "privilege and freedom" of moving in the bush and of killing animals. They did not farm much if they bartered meat for mealie meal [ground maize] and for farm labor. Now most people kill animals secretly. Many of our hunting medicines will be forgotten as few adults know how to shoot with muzzle-loading guns. Today those who snare don't claim interests in animals the same ways as we once did. Now, they are interested mainly in money.

The current hunters are the game guards, who kill a lot of animals, and the chief, who gets money from ADMADE as well as meat. Those in the royal lineage always kill animals and ADMADE employs the "renowned poachers." Scouts recently arrested such a poacher several times until the chief assigned him as a carrier for scouts. He was a "hard case and a repeat offender" with snares and guns. Since he has not been paid with community funds for several months now, do you think his interests in wild animals have changed?

Later that year, this "hard case and repeat offender" got lucky, finding employment as a valuable tracker by a safari hunting firm operating within the Munyamadzi GMA. He has retained that employment, which provides a salary, benefits, and bonuses contingent upon the success and generosity of sporadic international hunters. I became fortunate as well, obtaining additional funding and time to assess changes and welfare throughout the GMA (Marks and Mipashi Associates 2008; Marks 2014) and could complete this volume.

Notes

1. *Cibinda wanama afwe kunama, cidime wa video afwe kuluse.*
2. From my many interviews with Valley Bisa hunters, I chose six for whom my life histories were particularly "thick," and who resided in the villages near the chief's palace. Their villages were near and I could continue to find and talk with them during each visit. Their knowledge of the surrounding habitats and wildlife contributed to my understanding of local hunting processes. Their sharing in details of their lives and experiences resonated within other interviews I had with their peers. Each interview, engagement in informal conversation, and bush foray confirmed or added new insights into local hunting practices and gains. Some wrote letters about local happenings between my visits or kept personal diaries of time spent in the bush and animals observed and noted community events. I use this collated material to give "voice" to local hunters as they describe their rituals, experiences, and changes over time. I selected the life histories of two representative individuals for each of the three generations within this space, to illustrate the ranges of hunting behaviors, rituals, and perceptions, as well as to catalog their responses to their changing worlds. More of the ways in which these individuals interacted with others within this space is provided later in chapter 9.
3. The "Ba" prefix on a name is an honorific plural convention to show respect when remembering elders or ancestors.
4. Bisa attribute this epizootic rinderpest, which devastated domestic stock and wild animals toward the end of the nineteenth century in East and Southern Africa, to Mulenga, a spiritual essence associated with widespread afflictions such as famine, drought, and sickness.
5. This Chibale (Chibale the Great) was an ancestor in the lineage to which Chibale belonged and assumed his name upon his ascension to headman of this large village.
6. A medium-sized antelope that inhabits the riverine savannas and open lands along the rivers. Males are territorial during the breeding season and encountered often as single animals during that time. The initial kills by Bisa hunters, whether with bows and arrows or muzzle-loading guns, were mainly young males. Noting the place of kill was an important attribute of first kills.
7. Chibale did not give me the names of the grasses and roots that his patrons used. All "prescriptions" are referenced by the term *muti* (a generic term for tree or shrub as well as a mixture in a customary

"prescription" to cause an effect). Its application allegedly made the meat "neutral" (defused all intended harm).

8. *Cibanda cia nama cilondoloke bwino*—"to achieve a state by which the animal's shade became clear or recovered"—was the phrase given for burying these bones within an actively used animal path.
9. Apparently Chibale categorized buffalo as "small game" as it was not considered dangerous or aggressive when found in large herds. Buffalo herds then were growing in numbers and when massed in herds were vulnerable to poisoned arrows arched to fall where it was or was heading.
10. Closing one's eyes affected the dead animal's shade (spirit) with a similar disability. The content of some local prescriptions was privileged information. Patrons did not reveal necessarily their prescriptions to younger clients as its secrets made them dependents.
11. *Cibanda cikulu* (literally a large, powerful shade or haunting spirit) includes lion, elephant, eland and sometimes a few other species. Term is contrasted with other mammals with "lesser spirits." *Kamulebe* is a small, thorny shrub (*Ximenia americana*) bearing bitter-tasting fruits. The fruits (as food) and thorns (impediment) are important attributes in its ritual use. According to some hunters, pregnant female mammals are more difficult to kill than males because the fetus must expire before its mother. This premise is implicit in Chibale's mention of the distance this eland ran before succumbing.
12. "Brothers" as used includes older cousins rather than mother's sister's sons, otherwise Chibale was denied temporarily the use of a muzzle-loading gun that he would later inherit.
13. Chibale suggested that I consult with the chief about bow hunting and earlier hunting methods. In 1973, I took his advice and discussed the possibility with Chief Nabwalya. The chief introduced me to five bow hunters, who also willingly helped.
14. This example illustrates a key principle in matrilineal inheritance. Chibale's "brothers" were those of their father, who had married a wife, who were sisters of Chibale's mother.
15. As a peer, Kafupi took dried branches of *kamulebe* and a nest constructed by a sparrow weaver and burnt these materials on the back of the carcass as Chibale kept his eyes closed. Symbolically this nest represents a "household" whose members were "cleansed" and beyond capture by the rhino's shade. The rhino is known as a lunatic (*lizilu*) or mad animal, implicated among other behaviors through its scattering of its dung over a wide area. Although Chibale did

not mention it, his lineage was related to that of the chief through multiple past alliances.

16. Chibale used the terms *muti* (plant "prescription" as in magic) and *amaka* (force, power) to differentiate between the elements used in constructing these two weapon systems.
17. Rubbish heaps occur on the boundary of the village and the bush. These heaps are where people discard or, using the causative tense, purposefully reject and throw away "things" that were once a part of household life.
18. *Bwanga* is the possession and destructive deployment of particular sinister ingredients, usually of human (a sister's aborted fetus, female menstrual blood) or animal (cobra, hyena nose or brain) origin. Such adverse orientations were either in-born or acquired through purchase.
19. *Tangeleni pantanzhi. Nokofwaya munani. Vizwango vionze ladikeni* ("Go before me. We wish meat [to be fed]. Calm all dangerous animals wishing me harm while I'm in the bush.")
20. This group included a brother-in-law (*mulamu),* a classificatory "brother" in the same lineage, and a "father" of Jackson's lineage (a cross-cousin through marriage).
21. This sympathetic magic requires one to remain in place to restrict the prey's movements.
22. Hunters know that rhinos are browsers and grazers, an observation mirrored here in the uses of both leaves and grass. This ritual was intended to confuse the animal's shade from knowing what killed it—*kuzidika inama yalube aipaye.*
23. The verb *kuzungula* means to prevent the effects of a curse by the performance of a special ritual. Jackson described this kill site as across the Munyamadzi River.
24. I saw Jackson again in 2006. His son brought him on the handlebars of a bicycle to the chief's compound where I was staying. He was hurting, very sick and bent over in pain. I listened attentively, humbled by what he told of his pain and suffering, as we reminisced earlier visits. As I possessed only pills to temporarily relieve his pain, I also gave him money to get him to a hospital. Later he reciprocated by sending a white chicken before leaving for the plateau. We next saw each other again in 2011, enjoying an hour under the shade of a mango grove. He was thin and humpbacked but still in good spirits with a bright smile. After I talked about my four grandchildren, I recorded the numbers of children and grandchildren he claimed from each of his marriages. The number was forty-two.

25. Julius Mwale was married then to Kameko's sister. Mwale was a hunter, later a village headman, and worked for a long time as a justice in the local court. Mwale obtained a muzzle-loader during the 1940s.
26. Komeko knew about hunting with dogs and a prescription for increasing the dog's ability to locate prey. His prescription involved roots and barks of some unidentified shrubs, as well as the fruit shell of a thorny tree (*kalongo, Capparis tomentosa*), which has a strong smell. The roots were placed inside the shell along with a hot piece of charcoal. A child urinated into the shell, which along with other substances, were forced into the dog's nostrils. This prescription supposedly made the dog aggressive.
27. *Kupoza fye.* Although few Valley Bisa ate baboon flesh (perhaps more in famine times), Kameko's attitude of disposing of this carcass is different from what other hunters expressed (apprehension, submission) when they made their first kills accompanied by their elders.
28. Kameko had six sons (one set of twins) and three daughters. His sons became well-educated (secondary schools and university) and worked at various government jobs throughout Zambia.
29. To my knowledge, Kameko never killed any of the marked species; he knew about hunting prescriptions, despite his initial disclaimers. He possessed and used *cilembe* (to knock down prey) and *izambo* (ritual wash) for use with his muzzle-loaders, as well as other personal rituals (conversations with and letters from his sons).
30. Komeko inherited a muzzle-loading gun used by Goliat, an elder Ng'ona clansman. Close acquaintances of both said that he also had inherited Goliat's persistent hunting spirit.
31. *Milandu* (court cases). As a court justice, Kameko was aware of possible consequences of any game case brought against him. Another acquaintance corroborated Kameko's explanation when he told me in 1993, "It is not his past experiences with buffalo as to why Kameko refuses to take buffalo these days. He always walks alone while keeping his social distance. If he kills a large animal, others are likely to know since it is hard now to keep such [a kill] secret. Today, if the game guards kill a buffalo, all villagers know."
32. Kameko showed me the herbs used to wash his gun, and then we walked over to the rubbish pit on the edge of his homestead where his previously washed gun was "warming" in the afternoon sun (figure 8.2).
33. Hunting was part of the bride service Hapi provided to his wife's relatives in Tepeka Village. During these years, his in-laws "tested"

him for proper behavior. He remained at Tepeka for about five years until the death of his wife. He did not remain to marry any of her sisters.

34. A few years later, a lioness mauled and killed Peter Hankin in the Luangwa Valley. There was more at stake in leaving safari work than Hapi was willing to tell me initially. In Hankin's absence, Hapi worked for another professional hunter. When this hunter shot a large lion after season and off-license, Hapi turned both the professional hunter and a game guard in for this illegal activity. Although Hapi remained on good terms with Hankin, the other professional hunters in this company blacklisted him so that he was unable to find subsequent employment with any safaris. Several white professional hunters (South Africans and Zambians) confirmed their decision after I related to them Hapi's claim.
35. *Mbulu* is the name of this monitor lizard whose behavior gave rise to Hapi's interpretation of it as a powerful and inauspicious omen. His interpretation relates to the following proverb *"Abalye mbulu balapalamana"* (literally "those who eat the monitor draw together"), which connotes proceeding against the customs of others while among them, a reference to Hapi's settling in another chief's territory. There were also other serious associations in this sighting and circumstance. Hapi told me in 1988, *"Mipazhi teile bwino pamfwa yamfumu. Yakana mipazhi. Mbulo idi mipazhi ya cibanda. Ikulezha inama ukufwa, kupuzana pamfwa yamfumu. Palupwa luamfumu pali mungu."* ("The spirits were not proceeding well upon the death of the late chief. They were refusing [to assist]. The monitor lizard forecast a spiritual crisis preventing other animals from dying on the occasion of the late chief's burial. Within the chief's lineage there was *'mungu'* ['bad blood'].") *Mungu* is a special kind of gourd whose use here refers to the chief as the last survivor of his mother's line in this "Ng'ona womb" or branch. These Ng'ona spirits were not in sync with Hapi's attempt to feed the dead chief's descendants as Hapi had done so proficiently in the past for this chief while still a resident. These spirits were indicating now their disapproval by withholding animals. Beyond that, these troubled "spirits" were visible in the raucous and back-biting maneuvers then current among Ng'ona contenders over succession to the chieftainship. Four years passed before the state affirmed the next resident Ng'ona as acting chief and another two before he was confirmed as a chief.
36. Individuals of this generation were barely visible to me during my first two decades of study. Yet they had accompanied their elder hunting kinsmen on forays into the bush as helpers, carriers, and

learners. I knew them then mainly as students in the local primary school. They knew my wife as she had volunteered to teach English at the local school during our stay in 1966–67. These earlier experiences and associations contributed to the trust and rapport I had with them during the 1980s and 1990s. I use the third person to tell their stories; their own expressions are enclosed in quotation marks.

37. Kunda recalled that his uncle soaked (unspecified) roots in water overnight and then washed his face with this mixture before proceeding to hunt. This wash "made animals appear" in the bush during forays.
38. The ingredients given for *mfenzi* were roots of *pozo* (a sisal-like plant with slippery roots whose name is similar to the verb *kupoza*, meaning to neglect, throw away), roots of *muntu kufita* (literally "black person," the name of a small, inconspicuous black-barked deciduous tree), and the feathers or head of a *kambaza* (nightjar, a crepuscular bird whose cryptic plumage renders it invisible on the ground during the daytime). All ingredients are ground together, burned, and their ashes are sewn in a cloth band to wear while in the bush. The use of *mfenzi* confused dangerous animals by making the hunter's position opaque (cryptic) and invisible (black).
39. Most large predators including lions were classified as "vermin" by the colonial state, which meant that they could be taken anytime. Lions, leopards and even hyenas became valuable later as "trophies" for safari hunters, requiring expensive trophy and export licenses, and as significant sights during tourists' jaunts.

Chapter 6

Coping with Process and Uncertainty

Gameful Pursuits in the Bush

Seek ye, seek ye [the spirits] of our land,
those who killed without missing.
These spirits move in the evening time.[1]
Bisa song (recorded 1966)

Hunting is about moving through a landscape of sights, scents, and sounds; of plants known and unknown; of spaces connected by historical events and antecedents; of stories and the richness of figurative associations; of autonomy; and of community. It is about direct encounters with other sentient entities that are aware of their pursuers and capable of evasion, deception, and outright hostility. Individuals exercise and hone their skills at every stage of a process that includes assessing the behavior of their prey as well as determining the chances of trailing the prey through the tactics of searching, hiding, following signs, and stalking, which if successful may lead to butchering, preparing, and sharing bushmeat for others to prepare as food. People learn how to perform these tasks by watching and through performance rather than by exclusively listening to a teacher or reading a manual. In the bush, the many factors a hunter evaluates and upon which he makes his calculations are difficult to measure or to capture as they come with time and experience—the distillation of a glance, an impulse, an intuition of circumstance, place, and occasion.

Every foray into the bush is an open-ended opportunity for learning more about the habitats and habits of creatures pursued while updating evidence on mammalian movements and behavior. An individual's local knowledge of place constantly shifts as he integrates new elements and experiences during each successive passage through the same spaces. Hunters consciously expend their time and energy conflating other chores, such as looking for building materials while wandering in the bush or chatting over beer in the village, into keeping abreast of their "game" and making their skills relevant. Whereas one might attempt to flatten some components of this lore to the written page, significance remains within the immediacy of its practice.

To illustrate this integration of local knowledge, history, and culture through performance and to describe the tactical adjustments required by seasonal changes, I begin this chapter with the sequencing of movements, observations, and conversations as I accompany hunter Kabuswe Musumba into the bush on two separate occasions. The passages demonstrate how a local resident partitions the adjacent landscape and thinks about it through stories recalling past events. They also reveal the characteristics of some mammals and their distributions in time and space, how dreams influence hunters, how hunters adapt to intrusions into their domains by government agents, how the remembrances of ancestors affect worldviews, and how some trees have important uses. Conversations and timed observations are from my field notes at the end of the 1988 dry and the 1988–89 rainy seasons. Direct quotations are Kabuswe's commentaries.

Transcripts of Two Bush Excursions: The Integration of Memory and Practice

"If you kill a warthog or impala now, the game guards will not know."

17 November 1988: 1.5 mm of rain fell during the evening, temperature: 19 degrees C.

0520 hours. We leave just as Kabuswe Musumba arrives at my door with his muzzle-loading gun in hand. After our mutual greetings and inquiries as to how each slept the previous night, I fall in immediately behind his shadowy self as we walk toward the still blackness of the nearby bush. The eastern horizon is pale and beginning to lighten, but the ground under our feet remains dark. In this darkness, we feel our way along a well-worn path as we pass through a neighbor's field, discussing in muffled voices our plans for this morning. I ask him about where we might encounter buffalo on this day. Since Kabuswe had gone into the hills west of Malanda two days ago, he suggests that we explore the middle ground of Chela and Bouvwe Plains before heading west toward Malanda hot springs and grassland. In the bush adjacent to the field, we jump a grysbok whose scramble is barely perceptible against the dark ground. Kabuswe tells me, "Most of these smaller animals are taken by snares and with dogs. Dogs also capture the young warthogs, puku, bushbuck, and impala found this time of year close to the fields and villages."

As the morning light increases, he points out the ginger-brown color of a slender mongoose (*lukote*)[2] and notes its connotation as a bad omen should it cross our path. Here Kabuswe translates its presence as a spiritual one, a prophetic message about the world we are about to enter. We proceed, and the mongoose remains on our right rather than crossing in front of us. Slightly later, Kabuswe notes a coucal's (*mukuta*) booming voice in a nearby bush. "This bird wakes us up in the morning, eats snakes and grasshoppers, has red eyes, a brown body, and black rump. It is a very weak flyer, likes thickets, and, if caught, is mostly feathers."

We cross the Mpika-Nabwalya road and proceed along a path into the grassland and open mopane woodland immediately west of Chela Plain. The coolness and dew signal the uncertain beginning of another rainy season. Kabuswe points in the direction of the various streamlets and slight depressions indicating how the rains will soon reconnect sections of this parched landscape with runoff. The water from Malanda hot spring, grassland, and hills on our right (our destination, not visible) will join Mpungu, Chifukula, and other streambeds (now dry) in the west to flow through Silenti, Kabundi, Churchi, Mutantangwa, across Bouvwe Plain (visible at a distance in front of us) eastward across various stretches of mopane forest to the Luangwa River. All of these spaces connect with local stories about individuals, events, or objects.

In response to my question, he differentiates between the local terms of *chunga* (a slight depression, microhabitat) and *chizanze* (riverine savanna), two types of habitats found close together along the eastern edge of Chela Plain. He points to the trees and tall grass of *chizanze* and compares that to the relative openness of *chunga* with its shrubs. *Chunga* has claylike soils and is slightly depressed in comparison to the surrounding terrain. Its main cover is a heavy stemmed grass (*cikongonzi*) with prickly hardened leaves and spines at the nodes. Several small groups of zebras are grazing on the open grasses along the eastern edge of Chela.

Kabuswe says that if we come across some animals today that we should kill one. He knows that will be his decision, as I have no weapon. He informs me of what he has learned from his local network. His wife told him that a "cousin," the head game guard, had purchased a license for a hippo and was on his way to the safari camp to ask the professional hunter for cartridges. For killing this hippo, he tells me that the guard will use a government-issued weapon. "If you kill a warthog or impala, the game guards will not know, but if you kill a buffalo or a large animal, they will find out because the meat will go to many people and cooking pots. So, if some people kill a larger animal now they only take a small portion and leave the rest to rot. For me, leaving meat in the bush is 'no good,'" indicating that he hunts for his lineage as his borrowed weapon belongs to them.

Along the northern edge of Bouvwe plain, we find numerous warthog signs from previous days. These hogs have dug grass roots, scraped the soil, and excavated deeper in places. Their tracks indicate that they have proceeded into the mopane woodlands to our west. Kabuswe notes that *zanda,* a tender-leafed, white-sapped wild plant grows on this plain. Many people collect its leaves as a vegetable to eat with their meals.

From Bouvwe, we head west. He relates that someone coming from the south saw a large buffalo herd in the hills southwest of Malanda. These hills are named after Mupeta, a late person who hunted there. We head towards a depression (*vikwaba*) between two hills expecting that the fresh grass sprouts there might hold the attention of a small buffalo herd. If there are buffalo in this vicinity, Kabuswe expects that they have proceeded to Bouvwe and later towards evening may be found in the abandoned fields around Nabwalya.

0656 hrs. We pause at the northwestern part of Bouvwe Plain where the rainwater from Malanda will flow out across it. Here, the streambed delivering runoff rain loses definition as it spreads out over the level sandy clays before entering the trenches in the deep undulating clays of the central plain. Its eventual outflow on the far side is more structured and confined as Mwebe Stream, named for a village formerly along its banks. Further east, this stream becomes a deeper channel, Muvuzia, as the seasonal water flows across a nearly level landscape before reaching the Luangwa River several kilometers to the east. During the early rains, catfish from the Luangwa concentrate in these deep channels waiting for the smaller tributaries to fill with runoff so they can continue to swim westward. Along the edge of Bouvwe Plain, the fresh prints of wildebeest suggest their presence ahead in the mopane forest.

0710 hrs. A large group of warthogs flushes from a patch of stunted mopane; we only catch glimpses of their scurrying legs among the dense bunched stems near the ground. We detour to check their tracks and count thirteen (two adults, three subadults, and eight young piglets). "Warthogs are very afraid these times of [domesticated] dogs, owls [*mancici*], and eagles [*lukozi*]. When grazing in the open, adults frequently leave their smaller ones [*ntunta*] in the shade where they can't be seen. Young warthogs are very stupid—if they are born five in a litter, in a few weeks only two are left."

We check animal prints around a small mud hole and note what animals used it yesterday. "If you want to shoot an animal here, you must build a hide or blind (*citutu*), then come back about 1100 hours and stay quietly until about 1500 hours, when you should be able to shoot something."

0719 hrs. Kabuswe surmises that the buffalo must be at a small stream ahead or at Malanda. We cross a narrow arm of grassland that extends from Bouvwe. The grass is short, there are no buffalo tracks leading north, so he changes his mind and suggests that the buffalo are to our west, where the upland grasses "are a bit tall." When grass ends, we find again the fresh footprints of wildebeest and, in the adjacent woodland, the tracks of several eland.

0731 hrs. In the nearby stunted mopane woodland, we come upon a herd of more than twenty-five wildebeest, including many sub-adults. We hear their moments before observing them, but the cover makes an exact count impossible. "Animals really like this place for they can see any approaching danger at a distance. Wildebeests are not very brave animals, but they have very good eyes and are always afraid."

0745 hrs. He shows me where an impala ram used its horns last evening to rub and remove leaves from a mopane sapling. Kabuswe picks up several leaves. "This impala is only playing [*kwangala*] or exercising, for if it had met another adult male [*kakonje*] they would have fought and the ground would have been disturbed."

0751 hrs. We hear a tickbird (*nyanvi*) flying overhead and observe that it is flying northeast. "The bird is looking for buffalo, which are now coming out [of the hills] in our direction searching for a place to sleep and chew. Since we have seen no warthogs here, they have gone into the hills where the soil is a bit brown and dry."

0809 hrs. A band of impala in the distant woodland gives Kabuswe the opportunity to instruct me how to distinguish between rams and ewes. Rams hold their ears to the side of their heads because of their lyrate horns on top of their skulls; ewes can hold their ears straight above their heads.

Passing through an open stretch of mopane woodland, he points to a large tree with a thicket to one side. In October,

Kabuswe saw a large warthog sleeping in this thicket and he demonstrates how he approached across the open ground. "It was a very hot day about 1300 hours. As the wind was blowing from the warthog directly at me, I walked right up to it within four paces and shot it. I could have killed it with my axe by breaking its backbone, but there were too many branches in the way. Warthogs sleep very soundly on hot days. They are very brave animals and like moving in thick places."

The idea of sound sleep reminded him of a story that took place much earlier. As a youngster, he had accompanied his grandfather and uncle in the bush where they came upon a sleeping rhino lying on its side, and presumed that it was dead. His grandfather immediately exclaimed the extraordinary generosity shown by his ancestral spirits (*ficolwe*) as he claimed the carcass by placing his muzzle-loading gun against the immobile beast. The rhino awoke, rose to its feet, the weapon crashed to the ground, everyone took cover, and the startled mammal ran away.

As we approach a small streambed, he relates another story that took place there. His grandfather, accompanied by his two sons, came upon several lions feeding on a recently killed buffalo carcass. As they approached, the lions became aggressive and roared, causing the younger son, known for his cowardice, to plead that they leave and go elsewhere. This spineless behavior so annoyed the others that they belligerently advanced, shouting and pelting the lions with sticks, causing the lions to abandon the carcass. The trio butchered what remained and took the meat home.

The mention of Kabuswe's grandfather's younger son stimulates a story of my own. I remember that this son possessed political savvy, yet lacked bush skills. He was the chairman of the local branch of the United National Independence Party (UNIP) when my wife and I first arrived at Nabwalya in July 1966. Jackson Katongola, the district messenger from the area, escorted us from Mpika to Nabwalya and spent ten days with us, making sure that we were well received. Early one morning, Jackson went into the bush and killed a bull buffalo to consolidate good will and facilitate our welcome. When he returned to the villages, he suggested that we accompany him to see the carcass. That late afternoon, accompanied by Jackson and the UNIP ward chairman, my wife and I went as far as we could in our Land Rover before proceeding on foot for more than a kilometer to the kill site. Jackson had covered the carcass with branches so vultures could not detect it from the air and tied a prescription (*muti wecimbwi*) to the front and hind limbs so no predators would find it on the ground. After telling us of his pursuit and stalk that morning, Jackson stayed behind to butcher the carcass.

The chairman, my wife, and I began our walk back to the Land Rover. Despite Jackson's implicit directions, the chairman became desperately disoriented. Realizing that dusk was rapidly approaching and that he was lost and unarmed, he panicked, climbed a nearby tree and began shouting at the top of his voice, "BaJacksoneeaye, twalubaee!" ("Respectfully Jackson, we are lost!"). Jackson was butchering when he heard this disturbing summons. He arrived on the run, shirtless, out of breath, blood on his arms, clutching his butcher knife in his hand—and very relieved to find us only lost. After locating our vehicle, Jackson returned to the butchery where he was soon joined by his relatives and clients.[3]

0828 hrs. We arrive at the southeastern edge of Malanda grassland and see two men crossing the plain on its western side. Kabuswe notes that they are following buffalo tracks, carrying one muzzle-loading gun between them, and proceeding from the hills. Although he knows them, he does not disclose their identities, neither do I ask. Their presence is the reason we saw no animals on the grasslands for, "when an animal senses a person or catches his scent, it goes far away. Had it not been for these people, we would have seen plenty of animals. The buffalo must be southwest of Malanda for, if they were north of us earlier, we would have heard them as we passed Chela."

0830–0930 hrs. We sit in the shade of a tree bordering grassland, watching for the appearance of wildlife, discussing local events, listening, and enjoying stories. Kabuswe informs me that the acting chief has made a rule for people burning sedges around Ngala hot spring to burn and use the ash as traditional salt. Now people must ask the chief's permission before burning sedges.

My companion speculates that since the buffalo herd was at Malanda and Kawele for the past few weeks, they must "feel somewhat harassed and now have gone south to the Mupamadzi River." He informs me that in 1986, while across the river, he came upon an elephant feeding on a thicket and shot it in the head. In answer to my questions about what he did next, he says that he had rested (*kuzungula*) the elephant's spirit as his grandfather had instructed him. Otherwise, his children would have become sick, gone mad, or died. When I ask him about what he used, he replies that he used *lenje* roots—a grasslike reed that grows near the river—mixed it with the bark of the marula plum (*muzebe*) taken from the east and western sides. He was alone and, given the distance to the village, took only a small portion of the meat, as he did not want people to know about this kill as they might inform on him. He removed the small tusks, buried them, and later retrieved them to sell to someone from Mpika for ZK 120 (about $170 US). "The problem with something stolen is that if you try and sell it for too high a price, people learn that you have that something and then you will be caught. So you let it go for a small amount of what they are really worth." Thus began a string of stories about elephants, various people, and changing circumstances and roles.

Kabuswe tells me that he and a relative found a dead elephant south of Malanda near the site where we were resting. They returned and told the senior game guard, who instructed them to bring the tusks because there was an amnesty and rewards for people who surrendered ivories. They never received payment, so he surmised that the guard claimed ("ate" or "pocketed") the reward for himself.

He then describes the game guards' strategy when they hear that someone has killed an elephant. They proceed to the suspects' village, assemble all gun owners, and take their guns to their camp, promising to return them only when the elephant killer confesses. He notes that the guards have learned this scheme from Chief Nabwalya [Kabuswe Mbuluma] and admonishes that "rumors can sometimes get one into trouble." Years ago, Lemson Kalemba killed an elephant north of the Munyamadzi River, distributed some of its meat, and then buried the tusks near the river crossing. The chief heard about this elephant kill, yet could obtain little substance about its details from the area's headmen. When the headmen's stories became that the elephant had died and that someone had found the carcass, the chief changed tactics. As a found elephant, the chief insisted that he should have received some meat as customary tribute. So if this elephant had died as the headman claimed, where were its tusks? The late chief refused to accept the headmen's answer that they had thrown the tusks into the river. Instead, the chief sent his constable to collect all guns within their villages and declared that if they did not recover the tusks soon, he would send their guns to Mpika. Kabuswe's grandfather was implicated along with the other headmen; he pleaded that he was too old to go into the bush, that he was suffering because of another person's misbehavior, and that he was keeping his gun only for younger men to use. Headman Kalemba sent his son, who allegedly had slain the elephant, to confess that the shooting was in self-defense. The chief tolerated that reason, yet advised that if the son had shared the meat with the chief, everything would have been acceptable. Hearing that the tusks were disposed in the river, the chief demanded his constable bring a whip. Lemson evaded the whipping by bringing the tusks immediately to the chief. The chief sent both the tusks and Lemson to Mpika where the culprit served three months at hard labor.

Our conversation shifts to where and how we should proceed. The wind persists from the east and we had seen people come from the southwest. Kabuswe notes that the heat was beginning and animals would begin looking for a safe resting place.

0930 hrs. We proceed east into the wind.

0935 hrs. We observe two adult warthogs digging roots in a slight wallow of grass (*kazenze*) within a stretch of woodland. They are flighty and alert from the earlier passage of hunters. The warthogs soon detect our presence and flee.

Near an active game trail in this woodland, Kabuswe points to a stack of dead limbs leaning against a trunk and indicates it is a hide (*citutu*). "This hide is well placed, for the hunter can spot at a distance any wildlife using the trail in either direction. When coming along this trail, the animals don't pay attention to anything. As they get closer to water, they circle, becoming suspicious. Out here in the open woodland, they trot along and are not very alert. A *citeba* [an elevated hide in a tree] is sometimes better because animals are not suspicious of anything above them and [because of wind drift] they must be far away to detect a person's smell. Even then, they may get confused and forget about what they may have sensed. *Citutu* is only good if the wind is blowing favorably.

0958 hrs. A small group of sixty buffalo appears in the woodland ahead. They are moving from cover in Kawele thicket toward Chela and Bouvwe plains, exactly the places that the tickbird we had noted earlier was headed. "These buffalo are headed to rest and to ruminate on Chela Plain, where earlier we saw that group of zebra." The woodland before us is so "open" (*kuabuta,* "very white") that the buffalo see our approach.

Moving cautiously toward the buffalo, Kabuswe spots a sow warthog with three piglets. They are busy rooting in a small depression; he crawls behind an intervening tree trunk toward the hogs. He is within thirty-five paces before the sow spots him and runs. Kabuswe notes that we could have run the piglets down if we had wished. "Small warthogs have two speeds—fast and straight or fast and zigzagging, which tires them quickly. Young zebras have three runs, which is why we can't catch them with dogs."

1006 hrs. Kabuswe resumes stalking the buffalo, closing on them as they amble away. This small group, mainly cows and calves, is nervous from a previous encounters and moving rapidly. They break into a run as we approach within 100 paces. Analyzing this situation, Kabuswe admits that the only successful tactic was to get as close to the herd as the wind allowed and then reduce the distance further by running and firing on the run. He remarks that the buffalo had noticed my blue shirt and fled, yet I suspect that he was teasing me as he was only testing how close he could get, because if he had fired, the gunshot would have been audible to those in the villages. He speculates that the large herd that his friend noticed yesterday must still be behind us and expects it will visit Malanda later in the day. The two men observed earlier at Malanda had flushed this small group from Kawele thicket.

Walking back to the village, he tells me that buffalo ingest burnt grass ashes for its salt content; the internal lining of their stomachs subsequently become black from eating so much soot.

1127 hrs. We arrive at the villages.

We "criminals" prefer to hunt in March through July, when the grass cover is tall and it's impossible to see very far.

10 May 1989. After several weeks of no rain, this clear, crisp morning became a hot day between 16 and 29 degrees Celsius, with few clouds appearing after 1230–1330 hours.

0550 hours. As Kabuswe Musumba appears with a muzzle-loading gun, we leave and head toward the bush. I ask if he had dreamed last evening. He responds that he dreamed of celebrating with roasted meat at a beer party. As he grouped this dream in the propitious category, he wonders if he will realize it today. He is suspicious of the unanticipated game guard reinforcements from Mpika, who had arrived almost a week ago. His wife mentioned that most scouts had crossed the Munyamadzi River yesterday and proceeded north. Earlier in the week, he saw no scout tracks and would check for their signs today.

While crossing Kapume depression at the edge of the cultivated fields, Kabuswe points to fresh buffalo tracks. The prints indicate that these bulls (*kakuli*) were staying in the nearby thickets and had been grazing the previous evening in the nearby fields. As we pass a thicket, he tells of a buffalo kill he made there in 1984. Before daylight, he found four bulls in this field and followed them as they left for dense cover close to the late chief's palace. His first shot entered a bull's neck, paralyzing it standing. As the other bulls fled, Kabuswe saw late chief walking in his compound, wrapped in a blanket. When he shot again to bring this buffalo down, he heard the chief exclaim, "Someone is trying to shoot me!" Kabuswe left quickly and returned his borrowed gun. His father instructed Kabuswe to tell the chief "everything" and to follow his directions. The chief was grateful for this serendipitous appearance of meat, claiming that his larder was bare. The chief instructed Kabuswe to keep this kill confidential and to share the meat with few others.

0600 hrs. Crossing the Mpika road, Kabuswe says, "I think zebras have gone to graze in *cikwaba* [low-lying area] where *kampusu* grass [with a spike like a cat's tail] grows on the rocky slopes. Warthogs have begun to graze along Chela Plain, but it may be difficult to find them in the tall grass. Other animals are now grazing on Chela as well as Mutantangwe and Bouvwe. Elephants are the most likely animals found now in Kawele."

He notices that "the wind is blowing from the Mupamadzi [south] so there are not many animals here [at Chela] now. When the wind shifts to blow from the Munyamadzi side [north], then animals will be closer to the villages."

Walking along the path he picks up wind-blown seeds of *cilembelembe,* a climbing plant shown me elsewhere weeks earlier. Seeds of this plant have plumes for dispersal through the air and are ground into paste for poisoning arrows.

0633 hrs. Three adult zebras are grazing in the mopane woodland; two minutes later we see two adult impala. I ask about small, scattered habitats, and he answers that all the major habitats have smaller areas of grasslands, sandy soils, and depressions within them. He knows where these areas are located for they frequently hold game.

Kabuswe points to hyena prints in the path and says that he heard them howling last evening in Kawele thicket. "They may have found a snared animal there and killed and consumed it." We pass a downed mopane tree from which a person has collected bark. This observation prompts a comment about uses for the two varieties of mopane. "*Mopani waluzizi* has long sections of bark that cling to the trunk. Now it's possible to strip the bark from the tree immediately when you chop it down. Later on during the dry season, you have to wait for several days before stripping this bark. The other type, *kapani mamba,* has very short bark sections and is not useful for tying materials."

0650 hrs. Reaching Malanda stream, we stay on its northern side.

0702 hrs. Six adult impala are standing in shrub mopane. In the sandy soil, Kabuswe points to footprints with a crooked line separating each set of prints, signs of a monitor lizard dragging its tail. Several hides of stacked wood occur along well-traveled trails leading to Malanda grassland and stream. "This year will not be a good one for stealing animals. Better you buy a license for them." And then to let me know what's on his mind, he continues, "We won't hear any gunshots until people learn that the game scouts from Mpika who came expecting to find poachers here have returned to the plateau."

0711 hrs. For two minutes, Kabuswe stalks a group of five (one subadult) zebra grazing along the edge of Malanda grassland. He uses stunted mopane on the plain's edge as cover until the zebra run off. A baboon troop, which may have alerted the zebra, barks and dashes away. "Zebras have small numbers now because food is becoming less. When we saw them in December and January they were in big herds, food was very plenty. Even now, the buffalo herds are beginning to break into smaller groups."

0718 hrs. We cross Malanda stream and proceed south into another stretch of mopane woodland. We find a herd of nineteen alert impala, mostly females.

0733 hrs. A male impala stands next to cover along a streambed. A tall grass (*mape*) particularly favored by buffalo grows along its banks, and Kabuswe shows me some recently cropped stems. In the soft soil at the stream's edge, he points to a hunter's footprints from two days earlier.

0745 hrs. Approaching the Mupete hills from the east, a slight rise in topography yields a rocky surface where the grass (*nzoche*) is noticeably different from that in the mopane woodland. "In 1985, I wounded a zebra on lower ground and it came here and died about 0800 hours. I covered the carcass with grass and left for home. When I returned later with relatives, we found a large lion resting on the zebra's buttocks. The lion had not touched the carcass, but if we had come later, we would have found it eaten. We chased this lion away for it wasn't very brave." We turn and go southwest into the hills.

0805 hrs. Nyavi stream rises in these foothills and separates Mupeta from the Chalukila Hills. Chalukila Village was located here in the 1930s.

0808 hrs. A southwesterly flight of tickbirds causes us to shift directions slightly.

0813 hrs. We pass near Bokoboko Spring, which is a vital watering point during the dry season for game in this mid-space between the two major rivers. A baboon barks as we are leaving this spring.

0830 hrs. Ahead of us, an adult warthog digs *tukwe* roots in a depression. Kabuswe relates how warthogs change their habits with the seasons, as their diggings leave behind a specific configuration of signs. "Earlier in the year, warthogs graze [*kupunga*] wild grass seeds [*lupunga*] and do not dig much for roots. Now they search for roots and tubers in waterlogged depressions. They blow air out their snouts [*ukuzubula*] to scatter the water and mud from the roots. Later in the dry season, warthogs dig [*kuzinda*] for roots, and their excavations are not so extensive."

Scattered tracks and dung indicate that a buffalo herd rested here a few days ago. With the wind blowing on our backs, we are not surprised that the warthogs caught our scent and that we observe few animals. A female knob-billed goose flies overhead, signaling her goslings to remain motionless. Kabuswe checks the reeds along a stream's margin, but fails to locate the brood.

0847 hrs. An impala ram stands ahead of us. Kabuswe comments on recent animal signs: "There were plenty of animals here yesterday and this place smells of fish, so there must be plenty of fish now in nearby Nyanvi stream." The wind shifts; now it blows from the south. We change direction and head toward the Mupamadzi River, which is discernable in the distance by an occasional tree along its banks.

Further on, eight vultures bank and circle as we watch them. The vultures keep rising rather than pitching, so we resume walking. In the sandy bottom of a streambed are buffalo prints from the previous night. Kabuswe says he encountered a herd here two days ago, and, as if to underscore the accuracy of his claim, he steps on a pile of buffalo dung. The outside of the dung is dark and dry from exposure to the day's sun, yet under pressure of his foot, the core oozes out a vivid green paste.

0900 hrs. We pass cautiously through some heavy cover while still watching the circling vultures overhead. "Buffalos were here yesterday"; Kabuswe comments on more droppings and points to a wilted small purplish flower (*kantona*), noting that it was dropped from a buffalo's mouth. These herbs occur in people's fields so this herd must be the same one he followed from Bouvwe Plain two days ago.

0926 hrs. Our scent flushes a warthog from a small depression; its sudden flight is unexpected and startling. We examine where it was digging *tukwe* rhizomes. My companion relates that elephants and rhinos created these wallows (*chizambo)* and that they retain water in the hinterland after the rains end. In the past, the Bisa used these water sources when hiding from the Ngoni and the slave traders.

0935 hrs. We spot three alert warthogs (two adults, one subadult) in a depression ahead and watch for five minutes hoping they might settle or come our way. As they do not, we resume walking along Nyanvi Stream. On its banks we find where the buffalo grazed the previous evening as well as a bleached wildebeest cranium. We cross the stream, making a detour to avoid a deep bank and the water immediately in front.

0955 hrs. Fifteen impala (fourteen females, one male) and fifteen zebra (fourteen adults and two subadults) graze and browse along Nyanvi Stream.

1009 hrs. Kabuswe stalks for five minutes toward three impala rams that are grazing along a margin of grassland. The impala eventually see him, as does another band of five rams in the adjacent mopane forest. We pass along the edge of an open area, devoid of trees (*citemaleza*) and inundated and slippery to walk through in the rains. The shrubs change (to *mubambangoma*) noticeably as we approach the Mupamadzi River.

1025 hrs. Three female waterbucks and a ram impala graze in a small patch of savanna woodland. A flock of green pigeons takes flight as we pass under a large fig tree. As we pass through a rather large grassy area, I ask Kabuswe why we have encountered no mammals in the site through which we were passing. He tells me that the main grass is *nzoche* interspersed with smaller patches of preferred grasses. In this place, the animals must work harder to find their favorite grasses, which are more plentiful elsewhere.

1053 hrs. Still walking along a well-defined game trail in tall grasses along a deep tributary, we come upon a lagoon that was once the riverbed of the Mupamadzi River. Kabuswe relates what he has heard about an earlier chief's attempt to change the course of the Mupamadzi River and is uncertain about the date (perhaps in the 1930s). The chief hoped to reap the benefits of "plenty of fish and some animals" when this lagoon filled again with river water. This chief assigned each of his subjects a week of tribute work to remove the sandbank that separated the lagoon from the river. This diversion of water lasted only a year, for the river reverted and the chief gained only "a bunch of turtles" for all this effort. I inquire if this project was the chief's response to the demarcation of the boundary with the former South Luangwa Game Reserve in the late 1930s and that if shifting the river channel southward was what had left some Bisa villages isolated from their sources of water. Instead, Kabuswe suggests that the assigned tasks were to punish subjects refusing to relocate to the chief's palace, which would have enhanced his client base along the Munyamadzi River. Since the event occurred before he was born, Kabuswe suggests I ask his elders.[4]

1118 hrs. Along the south side of Kafupi Lagoon, we come suddenly upon four puku ewes, which quickly run away. After we go a bit farther, Kabuswe hails an unidentified man on the far side and asks him to wait. As we backtrack to cross the lagoon, we aren't surprised that the man has disappeared. We laugh, suspecting that this man's stories upon reaching his village will be about "a white game ranger accompanied by a scout" who tried to outsmart and arrest him. That rumor will pass to neighboring villages, and reports about the recent wildlife scout reinforcements from Mpika will bolster these suspicions surrounding us.

1128–1155 hrs. We rest along Kafupi Lagoon in the shade as Kabuswe talks about his great-grandfather, BaKafupi, and how this area (adjacent thicket, woodland, lagoon) now bears his name. As a resident during the 1910s and 1920s, BaKafupi came here with a bow and arrows, built a tree platform above the lagoon, and killed many warthogs. "BaKafupi accompanied Fox-Pitt[5] during the 1930s when he visited the valley and was with him when he killed a very large elephant near here. Fox-Pitt departed the next day leaving BaKafupi to butcher the carcass, but BaKafupi lost his way back to this elephant. Others found the carcass the following day by listening to the hyena calls and by following the descending vultures. I occasionally come here when short of "relish" [meat] at home. I start out from my village at 0400 hours, arrive here about 0700 hours, and hope to intercept impalas and warthogs coming to drink from this lagoon. When I kill one, I smoke and dry the meat, leave late in the afternoon carrying the smoked meat to the village, and arrive at night. Already, I have killed two warthogs and two impalas around the lagoon this year."

1155 hrs. We proceed northeast through several open-type habitats, find an impala skeleton, and along the way encounter small groups of ram impala and zebra. The zebra group has four adults and a young foal. "Young ones make adult zebras very clever."

1254 hrs. A large group (thirty-five adults, five subadults) of wildebeest rests in the mopane woodlands near Chipuma Stream. Some are lying down but others have seen us for some time. As we approach, they run away, then come ambling back and pass at close range.

Shortly after the wildebeest run by, Kabuswe briefly stalks a group of seven zebra together with twenty-nine ewe impala, which are looking in our direction even when we first observe them. We resume walking and detect several other smaller groups of impala. A large baobab tree is missing bark from its higher limbs. Kabuswe admits his son-in-law removed this bark to make into mats.

1342–1357 hrs. We come upon five (two adults, three subadults) warthogs rooting and wallowing in a wet section of Kalale Stream. As Kabuswe begins to stalk, the warthogs leave the stream and move into the adjacent open woodlands. The adult male smells us from forty paces. Apparently thinking of his recent excursions, Kabuswe exasperatedly exclaims, "Until now we have heard no gunshots, as if there were no people here! Is this the way the game guards want this place to become?"

1403 hrs. An extensive mopane woodland between Kalale Stream and Bouvwe Plain is named after the late BaBedi, who frequently came here to hunt and to collect honey. We find the crania of a zebra and an impala together in the same place and ponder why. We proceed past Bouvwe plain on our right.

1432–1435 hrs. Kabuswe notices a grysbok (*katili*), the smallest antelope, resting in the shade of a mopane tree and quickly moves behind a small shrub. From there, he crawls within twelve paces of the grysbok, removes the firing cap from the wax on the trigger guard, places the cap on the firing nipple, cocks his weapon, and cracks his knees together. The grysbok responds by standing, whereupon Kabuswe immediately fires his gun. The grysbok careens out from under the black gun smoke and runs into the woodland. I am amazed at how this small mammal can run at all since the shot has broken both back legs. Kabuswe runs after this small antelope and gains possession. As I approach behind him, he is walking away with his back towards the kill site, holding his gun and axe on his left shoulder with the grysbok held in his left hand. (figure 6.1) He keeps the distance as I follow him to a clump of bushes in leaf. He washes his hands with fresh leaves and cut some branches and ties the grysbok in a bundle of leaves. When I join him, he tells me that he had overloaded the gun with powder and shot a heavy load for such a small mammal. He had loaded the gun several days ago expecting to find and kill a buffalo. "The capacity for powder was too much. It [by inference, his guardian spirit] wanted my dream to come true. This animal is *inama yakapashi* ("a gift of the spirits") for the spirits covered the animal's eyes, so it couldn't see me approaching in the open. This animal was ["professed/ ordained"] by the spirits in my dream." I notice Kabuswe holds the grysbok's body in his left hand with its head facing backward. He looks only forward. As I am following him, he talks and we head toward Malanda stream. At the stream, he washes his hands and drinks water. He strips fiber from a downed mopane tree and ties the grysbok's legs together. Kabuswe tells me that he cracked his knees so that the grysbok would stand and present a larger target. He would have done the same had it been a buffalo lying down.

1519 hrs. In this mopane stretch on the way back to the villages, we notice a small band of six impala. When we reach the split in the footpath east of Kawele thicket (1521 hours), Kabuswe and I go our separate ways (1539 hours). Kabuswe proceeds toward the Ngala with the grysbok while I continue on and reach my house in a field near the school at 1546 hours.

FIGURE 6.1. After killing a grysbok (described as "a gift of the spirits"), Kabuswe Musumba turns his back on its recovery site and walks directly away holding the small antelope in his left hand. His muzzle-loading gun is slung over his left shoulder along with his axe. (Photograph by the author in 1989)

Context for the Two Bush Excursions

Both forays occurred in the year that the new "community-based" wildlife program, ADMADE, became operationalized at Nabwalya. As reflected in Kabuswe's comments and actions, local hunters were keenly aware of recent changes in wildlife policies, particularly the shift toward employing local young men and favored others as wildlife scouts, and understood the dangers that this deployment posed for their practices. Rather than needing bush skills, wildlife scouts required a primary education and appropriate connections. These jobs also required young or eager minds accepting outside directives rather than those schooled in the wisdom of place and processes vetted through cultivating local livelihoods. ADMADE expected their scouts to obey officers' orders rather than show loyalty to kinfolk.

ADMADE presented new hazards to local hunters and placed social objectives in jeopardy. Local residents were not consulted about these new wildlife laws, learning about them instead as they, their kin, or their acquaintances were apprehended and prosecuted for their violations. Arrests for game-law breaches entailed beatings, physical isolation, harsh language, and the confiscation of weapons and meat; a conviction in a court resulted in expensive fines or long imprisonment at hard labor. Individuals felt crippled personally by these indictments, as families and lineages experienced the serious loss of a relative, including his labor and their security. More significantly, their experiences taught them that the new scouts, state wildlife regulations, and the imposed local Wildlife Sub-Authority sought to silence their histories, including their local knowledge, practices, and identities.

Local hunters were aware that scouts used gunshots as indicators for their activities, that the they had to monitor scout movements to know when and where they might hunt, and that constant vigilance and movement in the bush helped them avoid detection by scouts. Hunters also knew that persons who had relatives employed as scouts were worth cultivating as friends. Adjustments to prior hunting routines were apparent, including the flaying and smoking of meat in the far bush, its portage to a homestead under the cover of night, and masking conversations in code, and strategies for distributing meat were also altered.

Whereas many recruits of the ADMADE program were young boys with some formal education, few had bush skills. The greater threat to local hunters came from the few "notorious poachers" who gained employment as scouts. The employment of these men was the program's strategy to control their clandestine activities (supposedly to "reform" them) and to deploy their knowledge to capture additional recalcitrant individuals. As a consequence, the greatest threats to hunting enterprise shifted from the pursued prey in the bush to the village—to village jealousies and gossip among kin and neighbors.

The wildlife observed even within a short walk from the village was remarkable in its diversity and abundance. In mid-November 1988, Kabuswe and I spent just over six hours in the bush, one of which we spent at Malanda talking and watching animals come to graze or drink. For our time in the bush, we saw 113 zebra (on seven different occasions), eighty-two impala (eleven occasions), nineteen warthogs (three occasions), some sixty buffalo, a herd of twenty-six wildebeest, one grysbok, and one mongoose for a total of 302 mammals. This number was slightly less than one animal per minute we spent in the bush. There was little cover, and most species were aware of their vulnerabilities and stayed beyond the "flight distance" of local hunters using muzzle-loading guns. At the end of a harsh dry season, most herbivores' movements become limited by diminished forage and water sources, and they seek to conserve their activities. Anticipating the return of the rains and lush transformations in habitats, some females (impala) drop their young early and pregnant, or nursing females are secretive or flighty.

At the end of the rains (often in April), dense vegetation and tall grasses grow in some areas and scattered shallow pools of water dot much of the hinterland. Game is dispersed and less visible, demanding longer hunting forays. In May 1989, we were in the bush for almost ten hours, covered some forty kilometers and saw 122 impala (nineteen encounters), thirty-nine zebra (six), ten warthog (four), thirty-five wildebeest, one grysbok, three waterbuck, and four puku. Kabuswe stalked two groups each of impala, zebra, and warthog and killed a female grysbok.[6]

These notes show a hunter remembering his past while constantly working new experiences and associations into his strategies. Kabuswe reconstructed and updated his political, economic and cultural worlds while walking across this landscape, actively engaging his identity within the larger political-economic world in which he had become designated a "criminal." Passing through the physicality of local landscapes, individuals constantly interweave the threads ("habitus") of actions, beliefs, encounters, and engagements that generate personal subjectivities, collective identities, and changes (Bordieu 1977; Bender 1993b).

Local stories of ancestors, events, and even indigenous place names were significant stakes in linking past entitlements with the present, as well as in reinforcing their rural identities. For Kabuswe Musumba and for many others, having ancestors who hunted and remembering them by recounting their adventures while passing over the same terrain bespoke of inheritable rights and identities invested in place and assets. To legitimize local hunting practitioners, knowledge of their genealogies as well as their ancestors' histories, residences, and activities were essential claims to identities and their rights to exploit assets within place (Harwell 2011).

Kabuswe's behavior upon his encounter with the grysbok reveals its meaning when cross-referenced with his dream and the role of this animal within lineage lore.[7] He acted to vindicate this dream, and I found his explanation compelling that the grysbok's unexpected appearance was a "gift of the spirits." Even that small antelope was positioned in reference to its hunter, as the blast of Kabuswe's gun was focused away from the villages rather than toward them, as was the buffalo during the dry-season excursion. Later I would find that many of the hunting practices and beliefs recorded during earlier years would diminish, certain accounts and organizing ideas would be forgotten, and some accounts reputed as esoteric facts would become judiciously guarded and reluctantly shared during the rapid changing political-economy of the new wildlife regime of ADMADE. The answers depended on the types of questions asked, the vocabulary used, and who asked whom as materials solicited appeared reliable only once long-term association, experience and trust had been established. Answers to formal questions asked from strangers became shallow and lacked substance compared to the depth gained through participation and observation among practitioners performing their craft.

Cultural amnesia rapidly closes upon lapsed practices. Memory alone remains surface in contrast to the depths of watching, learning and interacting with artisans in action.

When the Chips Are Down: Local Knowledge and the Pursuits of Buffalo

I began accompanying local hunters during my initial studies in 1966 and continued this practice in subsequent visits until 1988. These forays, as well as the overnight encampments (*malala*), provided opportunities to know individuals (and for them to know me), to observe and to ask questions, to learn the landscape and the animals. These occasions allowed me to inquire about motivations and life events and to accumulate knowledge about encountered prey. Most local hunters directed their forays toward securing cape buffalo or remaining current with their movements and dispositions, while other pursuits happened as vulnerable prey presented themselves, as the grysbok did. My earlier excursions into the bush around Nabwalya served as local tutorials in buffalo ecology, behavior, and movement, as well as times to reflect on scholarship (Sinclair 1977; Mloszewski 1983). These practical forays were also explorations in listening and learning how to ask appropriate questions and attach cultural meanings to happenings (Marks 1982, 1996).[8] Local materials about buffalo were always changing, reflecting the skill and experience of those who pursued them. In addition, shifts in environmental conditions and disturbances changed buffalo movements and numbers within the range of local hunters just as surveillance by authorities and individual motivations influenced hunting strategies and careers. My challenge was to systematize and translate this local knowledge as well as to make connections with other cultural processes. In this section, I summarize materials from these exercises about how hunters' pursuits reflected important components of local social lives.

As buffalo recovered from the rinderpest epizootic at the end of the nineteenth century, they became the largest mammal unfettered by colonial regulations and local prohibitions still available to local hunters. With local subsistence subsidized through the

FIGURE 6.2. Julius Mwale sits upon a buffalo bull he has killed in Kawele thicket earlier that morning. That afternoon he returned with two cousins and the court messenger prior to butchering the carcass into portable slabs of bushmeat, which they along with several women who arrived later would carry back to their households for further distribution. (Photograph taken by the author in 1966)

migrant laborer economy, buffalo meat became the local currency sustaining residential solidarity and protection, which elders and headmen used to promote, accumulate, and enhance the numbers of their dependents within the local lineage system. Patrons and their hunting clients embellished their reputations as "slayers of buffalo" (as well as other dangerous and menacing beasts) to mark their separation from other men who, despite their pretensions, were without the means to sustain such broad alliances.

Only a few local hunters knew a great deal about buffalo and how to pursue them, as this knowledge was not commonly known. Consequently, my understanding stems from long-term association with ten hunters, who were frequent residents in the villages around Nabwalya's palace and whom I accompanied on bush forays or listened intermittently to their discussions and interpretations of events. Between 1966 and 1993, these individuals were in their prime as hunting clients or as headmen. They provided specific information on buffalo movements and behavior, as well as on the cultural attributes ascribed to the species. Much of this lore and knowledge has become history, as its practices and ideas changed quickly under the prosecutions and pretenses of the imposed "community-based" wildlife program beginning in 1988. My use of the present tense within this section is to provide a summary of my accumulated understanding of these processes learned between 1966 and 1989 and to stylistically communicate with the reader, not to validate the anthropological "ethnographic present tense," especially its past associations rendering its practices static.

Local Knowledge of Buffalo Ecology and Behavior

At Nabwalya, hunters recognize three classes of buffalo. Beginning with the smallest unit (based in numbers), these groups are single (*cizunkulu*) or small bands (*kakuli*) of bachelor bulls, small herds of either sex containing up to around fifty or sixty buffalo (*kazakaombe*), and large herds (*ibumba*) that may number in the hundreds. Except when severely wounded by predators or hunters, cows are encountered in herds. Local hunting practices and off-takes acknowledge variability in behavior, ecology, and vulnerability for each of these groupings.

Older bulls retired or forced from the herds make up the first class of buffalo. These bulls often restrict their ranges to areas where they can find cover, grazing, and water in close proximity. They forage in cultivated fields and small patches of grassland close to rivers during evenings and early in the morning, retreating to nearby cover during the day. In the dry seasons, bulls stay close to villages and are often disturbed by residents' daily routines in the nearby bush. These interactions enable hunters to know the general locations of specific individuals. When wounded or assaulted with firearms repeatedly, these bulls and other wounded animals may take refuge in the tall grass and thickets near villages where they become a menace and add to the dangers of village life.

During the leanness of the dry season, small clusters of buffalo break from the larger herds and temporarily establish their own grazing, resting, and watering routines while intermittently remaining in contact with the herd. This is the second class of buffalo. With fewer animals, these groups do not travel as widely or as far to find enough food or water, as larger herds do. Their smaller numbers and sporadic movements make them more difficult for hunters to keep track of than the final class.

Large herds, the third class of buffalo, move daily, sometimes over extensive distances, during the late dry season. With most of its cows pregnant, the herd constantly and deliberately searches for sufficient grass and water. Herds often drink late at night especially in the environs of villages, graze in the late afternoons and early mornings, and spend most of the daytime resting and ruminating. In the dry season of 1973, the home range of the three herds around Nabwalya overlapped. During the rains, the herds ranged more widely between the Munyamadzi and Mupamadzi Rivers, sometimes crossing these rivers in their searches for suitable pastures. Depending on disturbances from hunters and predators around Nabwalya, these herds would spend several days or weeks near each major river. Malanda and the scattering of springs along the base of the foothills were intermediary sources of water. Large herds often rest in open habitats as these sites offer little cover for approaching predators.

A large herd moving at night is a noisy procession of bellows, plodding of hooves, and a cacophony of coughs as the procession proceeds past the villages and fields on its way to the Munyamadzi River. Buffalo often pause to graze on the green sorghum sprouts in the harvested fields in the late dry season. In contrast, the crackling sounds of dry stalks under hooves in the nearby fields are the only signals of a small herd's presence or passage at night. If there is moonlight, hunters sometimes attempt to ambush these herds on their way to the river. If a hunter is successful in securing a carcass close by, his tactic reduces his pursuit time, yet it could become more difficult to manage meat distribution between the claims of kin and the demands of others.

Bulls typically space themselves along the margins of herds. As herds feel threatened, bulls position themselves between the danger and the cows and calves. Small bands of "commando" [local idiom] bulls allegedly are the vanguard in searches for fresh pasture. These groups leave the herds and return at intervals to inform their leader(s) about what they found and where they found it. Elder cows serve as guides (*cilongozi*) in front of moving herds. The last buffalo, the "defender in the rear" (*cizendanzela*), is a bull.

Buffalo utter a range of sounds that local hunters interpret. *Kukamba* are the mooing sounds made while grazing or resting. Hunters use these sounds to locate a herd before they see it, and from these sounds they can assess wither the herd is dispersed and undisturbed. Stragglers left behind a moving herd utter deep sounds (*ukwita*), both to locate the herd and to monitor their process of catching up. Their behavior (*kukonka mucikuba)* tends to make them "act stupid without paying attention to wind or things happening around them," and their vocalizations alert hunters to their positions. Wounded or charging buffalo utter an ominous grunting sound (*kungonto*); another prolonged utterance (*mupano*) implies a targeted buffalo's imminent demise.

Estimating Local Hunter Harvests of Buffalo (1973)

When large herds are present around Nabwalya, hunters find them readily as they are noisy and their movements leave copious signs on the ground. Hunters base their predictions about buffalo activities on the animals' leavings and on past experience with a particular herd. While in the bush, hunters note fresh buffalo sign (prints, dung, sounds) and share this information with other hunters. By examining the fluidity, consistency, and color at the center of exposed dung, hunters say that they can calculate the time since its disposition. Early morning and subsequent flights of certain birds indicate where hunters might find buffalo. Oxpeckers pluck ticks from buffalo while egrets graze alongside the herds to catch insects disturbed by their movements. By noting the locations of bush fires and the sequence of burns, hunters' narrow their searches to plausible grazing sites.

From their own experiences, hunters may "mentally calculate" the vulnerabilities of each sex and age class of buffalo. Cows are allegedly more tenacious of life than bulls, as hunters say that the fetus within a pregnant cow must expire before its mother succumbs. Cows are also more aggressive and protective of their calves. Like young humans, young buffalo (*mapete*) are hardy, unpredictable, and energetic. With muzzle-loading guns in the 1960s and 1970s, hunters predominantly targeted buffalo bulls, mainly from herds and occasionally from the bachelor bull category. The behavior of herd bulls in positioning themselves between the cows and calves contribute to their vulnerability, as do the known small home ranges and sedentary locations of the older bulls near villages. Such older bulls are tough to kill and sometimes survive repeated attempts by hunters. Even lions reputedly fear older bulls, preferring instead to prey upon the herds and cows. Cows are more readily available to outside hunters with rifles. By ambushing or dividing and forcing herd members to run selectively, predators selectively choose cows and younger buffalo rather than mature bulls. This complementary selection by local hunters and large predators was a documented feature at Nabwalya during the first decades of my studies (Marks 1973b; 1977b).

To estimate the buffalo harvests of the "Nabwalya community of hunters" in 1973, I sequentially mapped the known appearance and movements of small and large buffalo herds on the Nabwalya

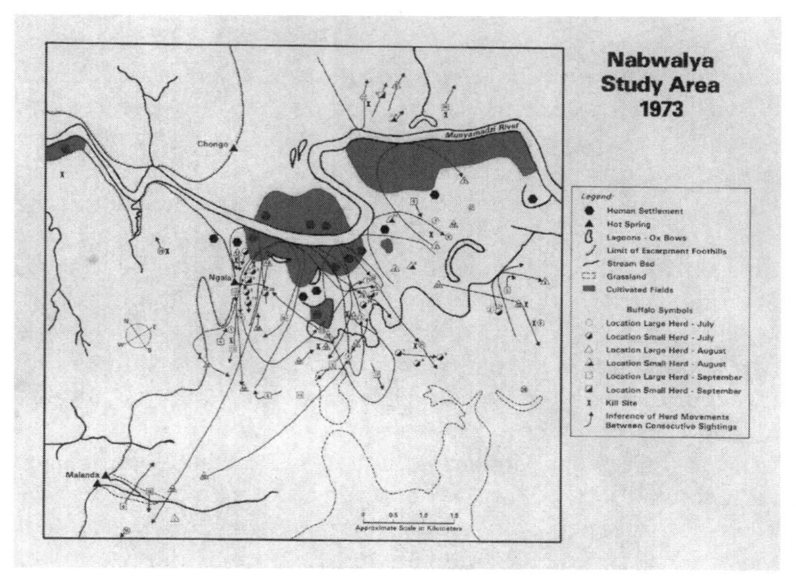

FIGURE 6.3. Buffalo movements, encounters by local hunters, and kill sites, Nabwalya Study Area, July–September 1973. (Map modified after Marks 1977a; 1982)

Study Area during a seventy-nine-day period. As much of this information was obtained with the collaboration of local hunters, I presumed that, if a herd were present, that one or more of them were following its movements and perhaps stalking it as well. If a herd wasn't present when a hunter needed to kill a buffalo, he could always find a bachelor bull. These observations on herd availability over four months that dry season became the first calculation in a four-part conditional probability statement summarizing these hunters' relational success in killing buffalo that year. Other data linked to these observations provided the remaining three statements in this conditional probability equation:

Probability (P) of killing a buffalo on a given day =
P (a buffalo class is present)
× P (hunters find buffalo when present)
× P (hunters stalk buffalo when they find them)
× P (hunters kill a buffalo when stalked)

If a class of buffalo were in range on any given day, the hunter had to locate and encounter it in space (given prior information from other hunters or by following indicator signs such as tracks, dung conditions, flight directions of tickbirds, chance encounters). The information on herd availability came from observations by villagers as well as hunters. Since I knew that most villagers kept track of the location and ranges of bachelor bulls, their presence and availability was much higher than that for herds. Beyond the actual presence of a bull or a herd within range on any given day, hunters still had to find the buffalo in space, an exercise that could never be certain. Once encountered, a hunter must decide to stalk (conditions of cover, wind, or buffalo and other species permitting), or not, and if so hopefully make a kill. However, a herd is more likely to perceive a hunter's presence and flee, the weapon may misfire or explode, or the targeted buffalo might escape after being wounded. Based on outcomes of forty stalks observed that year, the probability a hunter killing a buffalo was eight chances in a hundred.

Using my probability statement, my calculations indicated that hunters have a higher probability of finding and killing a bachelor bull on any given day (five chances out of a hundred) than they do killing a buffalo from a large herd (three chances out of a

hundred). The main contributing factor to this difference relates to the first variable: whether the buffalo class was accessible.[9] Hunters can find bachelor bulls nearly every day, but these bulls are aggressive and difficult to kill. Hunters seem more willing to take their chances with the bulls within large herds—these bulls are less belligerent and, if wounded, usually flee with the rest of the herd. The sporadic movements of the smaller herds are more difficult to determine, monitor, and track over a sequence of days.

Since hunters' pursuits of buffalo are treated separately for each buffalo class, the sums for all buffalo encounters suggest that Nabwalya hunters have nine chances in one hundred encounters of killing a buffalo on any given day. Therefore the prediction (0.09 chances per day x 365 days) predicted a harvest of thirty-three buffalo in 1973. When I left in mid-September 1973, hunters claimed to have killed twenty-one buffalo. During the remainder of that year (111 days), I projected them to kill an additional ten buffalo for a total of thirty-one. By tracking these relationships and interactive processes over time, I hoped to translate their cultural logics into meaningful terms given its local circumstances.

As with all theoretical calculations, this simplified formula needs an examination of its assumptions and limitations from its field application. Hunters may not have given a full accounting of all buffalo they killed during the earlier part of 1973 prior to my arrival in June. If so, their contribution was probably only three to four additional buffalo, as most adults were engaged in agricultural chores, including crop protection, during the rainy season. I did not enquire about buffalo meat acquired from predator kills. These occur throughout the year, but are more noticeable in the open terrain of the dry seasons. I could not obtain the subsequent number of buffalo killed after my departure, so the number of ten remains an estimate as the local demands of buffalo meat for Christmas feasts were particularly strong. Prior to my departure, hunters complained of limited supplies of saltpeter (sodium nitrate), which was mixed with local charcoal and used to manufacture gunpowder. The scarcity of this component indicates that they made outside connections and purchases to enable their local wildlife harvests. As traffic and communication increased with those on the plateau, a few hunters seized these opportunities to turn their expertise and enterprise into making money as well as producing meat for local consumers. Money was also useful in supplying lineage needs. Yet for this dry season, vehicular traffic remained sporadic as petrol remained expensive and patchy during the global energy crisis.

In 1966–67, I recorded that this same pod of hunters took thirty-five buffalo over a twelve-month period. The close agreement of the 1973 exercise with the earlier extended period of observation indicates some validity for the probability calculations. Most of the buffalo taken in both periods were bulls. This selection could imply that these hunters were acting prudently to conserve the stock, but I never heard this from them (Marks 1977b).

The economy was just beginning to inflate in 1973, creating hardships throughout rural Zambia. The wholesale slaughters and traffic in valley wildlife during the mid-1970s and 1980s by outside commercial gangs went largely unchecked, as the under-funded wildlife department came to depend increasingly upon donor support and NGO advice. In the late 1980s, the massive transfer of donor funding along with NGO expertise produced new wildlife policies and regulatory institutions to stem the widespread "poaching problem"; yet there was little if any socio-economic research on what was happening within rural villages.

Buffalo as Cultural Creatures

Buffalo share a number of ascribed characteristics with their human pursuers. I learned these metaphysical aspects as hunters responded to my inquiries as to why buffalo behaved in certain ways. Yet local answers depended on respondents' ages, experience, and whether they had given the behavior much thought. Here I attempt to summarize responses through 1989. The local orientations were compatible with models in which the management of resources and welfare resolutions are conceptually linked and prioritized through human relationships (past and present). They were different from bureaucratic wildlife management models in which the value of each wildlife species is defined specifically as a state property, whose wealth is calculated as prospective revenue, while the larger environmental and human relationships are discounted as externalities.

Buffalo herds belong in lineages similar to humans. Under the influence of "benevolent spirits" and conditions, herds expand in numbers, while under less favorable times and "spirits," their numbers decline. Thus buffalo increases (between 1967 and 1973) and their decreases (1978) were attributed largely to inclinations of buffalo spirits (*fikolwe*) rather than to human actions. Sometimes, local buffalo herds came together and comingled, an event witnessed and described to me by a hunter in 1973. According to one hunter source during these times, such large groups were considered "unassailable," protected by their "spirits" and the leader's charms (*canga*), and watched over by other "spiritual" (white) animals (Marks 1976: 96–7).

Like humans, individual buffalo possess "essential characteristics" (states) variously described as demonic, protective, or aggressive, compelling them to engage in normal or unexpected ways. Certain individual, dangerous animals, catalogued in the *cizwango* group, could allegedly be "concocted and controlled" by witches. I recorded, sometimes witnessed, specific instances of this alleged buffalo type. Other types of "spirits" protect buffalo when the animals appear more powerful or cunning than their assailants, either predators or hunters. Buffalo "leaders," described as substantially larger than the other animals observed with it, succumb only after prolonged, aggressive fights. Hunters know that they've killed such a leader (*cilongozi*) if the remaining group despondently mills around the carcass and has to be driven away. Both mystical and other animals associating with herds presumably provide a protective "covering." Mystical protectors (*kakoba*) described as small, white animals with long tails act as "priests" to large aggregated herds (*cilangano ciamboo*). Eland, the largest African antelope, has been known to lend its protective presence to buffalo found accompanying it. Bull eland supposedly carry powerful charms in their tangled forehead hair and in their dewlaps.

Buffalo mauled by lions or wounded by hunters become *bukali* (wicked, ferocious) and are found often close to villages, where in the dry seasons they take refuge in the tall grass formerly protected and used as public open latrines. Ascribed as a public menace (*cizwango*), particular buffalo becomes a problem that demands collective action directed by a chief or headman to dispatch it. Since 1990, the disposition of such "problem animals" falls within the duties of the wildlife scouts.

The most nefarious buffalo, or other dangerous species, are those "concocted" by an unknown but suspected sorcerer. The following incident begins the day after a buffalo attacked a young man in 1989, and later upon his return to his mother's village within another valley chiefdom.

> In 1988 a young Kunda [stranger] man, Banda came seeking his fortune by netting fish in the numerous ponds and lagoons of Chief Nabwalya's territory. He married a Valley Bisa woman in a village along the Mupamadzi River—a marriage that gave him access to the neighborhood fishery. In late April 1989, his mother-in-law brought prepared beer to the field hut, where Banda's task was chasing birds and monkeys from the ripening sorghum in her field. She found him already drunk. She instructed him that upon returning to the village towards evening that he use the regular path, rather than the shortcut through tall grass and thickets. Residents knew the thicket were the haunts of a "dangerous" buffalo. Proceeding against her wishes, Banda took the shortcut where the "concocted" buffalo [*imboo yakupanga*] attacked and gored him. This buffalo broke several of Banda's ribs and gorged his buttocks leaving him temporarily crippled. A funeral the next day in his in-laws' village delayed his portage to the clinic across the Luangwa River and enhanced his infection. As a concocted and "sent buffalo," the local consensus was for the headman to approach an African diviner/doctor (*nganga*) to determine the buffalo maker's identity and to explain how the community could rid itself of the menace. The village also talked of holding a game oracle (*lutembo*) to settle the identity of the guilty party, a determination based upon the sex of the animal killed by a designated [commissioned] hunter. We met Banda the day following his incident and interviewed him and others in his mother-in-law's village.
>
> Later that same year, I found Banda in Chief Nsefu's area [east of the Luangwa River in Eastern Province] as he had returned to his village of origin. He was healed and showed few signs of the buffalo attack. He told us that he would not return to his former wife's village, that he was convinced the "witch" that commanded the buffalo was not among his relatives, and that his in-laws never held the game test.[10]

Spirits are also known through what they leave behind. *Mulenga wampanga* is a "spirit" sometimes linked with rivers or muddy depressions and recognized by the carnage extracted in such sites from buffalo herds. These simultaneous deaths of

healthy buffalo while drinking or crossing rivers seem to occur frequently, at least around Nabwalya, beginning in the late 1990s. These deaths are likely linked to the decline in local hunting intensity because of local law enforcement and recent increases in buffalo numbers there.

Buffalo Pursuits as Markers among Men

The sheer size of a buffalo, the qualities and quantities of meat obtainable from a carcass, a powerful sounding and lethal firearm, the devotion of time to keep track of a herd's movements were just a few particulars and tasks to bring this prized beast to the ground for distribution. Yet, more was a stake than these repetitive qualities of a quest and the demonstrated abilities that some hunters voiced and demonstrated to mark their standing beyond those who took "lesser game." The hunting of buffalo with muzzle-loaders was a local cultural vision and personal strategy developed (1935–1960) by the first generation of local hunters (Chibale and Jackson), mastered briefly (1960–1985) by the second (Komeko and Hapi) and radically transformed (1990–2000) by the last (Kunda and Mulenga) under the structures of a new wildlife program and the national traumas of disastrous political-economic policies. Even as the currency of local patronage shifted from meat and protection to money and local employment under the ADMADE wildlife regime, the pursuit of buffalo retained its former lure as a status marker. A few privileged and salaried residents monopolized the meager, annual quota of buffalo residents' licenses to sustain this symbolic fulcrum from the past. I briefly summarize some main components of this unfolding "buffalo mystique," which will be choreographed later in chapter 9. The aura of buffalo was not enigmatic for those living with them in place, but only to those who thought they knew better values.[11]

Initially, this faunal centerpiece was the creation of a young Valley Bisa chief, Kabuswe Mbuluma, who was appointed to bring stability to a chieftaincy, which, from the colonial administrative view, had floundered under a sequence of prior drunken, disorganized, and short-lived elderly appointments. He and his supporters imaginatively shaped versions of indigenous ideas

FIGURE 6.4. A local hunter cautiously stalks a serially wounded bachelor buffalo that has grazed over several days in sorghum fields near Luba village. (Photograph taken by author in August 1973)

and practices on the anvil of colonial policies that produced a working format for wildlife husbandry that lasted for most of the six decades of his reign. This attentive and polygynous chief kept abreast of the inevitable rivalries, conflicts, and changes within this lineage husbandry through his marriages with significant lineages throughout his chieftaincy. The antagonistic relationships within the local scheme's inconsistencies and limitations built over time and surfaced during the national political-economic crises and international responses of the 1980s.

Secondly, this system of wildlife husbandry required a large mammal, in sufficient stock, palatable to consumers and whose taking remained beyond the scrutiny of outside regulators. As other large mammalian species, elephant, rhino, and hippo, were already captured by the state and regulations, the increasing number of cape buffalo satisfied these conditions. Not just a dangerous beast and one that frequently destroyed local fields, buffalo was known to transmit diseases to cattle, an industry that colonials sought to promote and were not likely critical of Africans killing buffalo. Jackson told us that much in the last chapter.

Another large licensed species, hippo was restricted in distribution within the valley to rivers and lagoons, was also a destroyer of gardens, yet protected locally by a prohibition on consuming its flesh. In this local husbandry management, hippo became a reserved, "surplus" meat supply, killed only during droughts and famines and, under the chief's authority, for exchange with villagers on the plateau for their "surplus" grains. These stocks of this wild game, buffalo on the hoof, with hippos held in reserve within the "trenches" for exchange use in times of famine, provided the wherewithal to store, sustain, and even transform a small, dispersed society for decades.

Thirdly, hunting and defense were cultural roles, through which some men sought provincial identities as protectors, providers of bushmeat, and procreators with women, were engaged in competitive rivalries with other matrilineal men, who as migrant workers accumulated other valuable skills and products. The killing of wild animals depended upon guns, owned or loaned, and by other methods; yet, a slain buffalo carcass was more than just a statement about a hunter's prowess, it was a social strategy. For these men, hunting lasted a lifetime, promulgating their means and reputations as prominent adults. For them, each kill was invested with ambition, each distribution screened and monitored by significant others, who were eventually the ones to accord an individual's recognition. In the crazy quilt of village histories, transcendent standing depended progressively on demonstrating "economies of scale" and convertibility of products, which cash made possible, as well as the political skills to distribute judiciously needed products rather than the limited skills exclusively honed in the neighboring bush and villages.

Fourthly, the possession of capital and guns, even as custodians, remained in the domain and hands of elderly men. Muzzle-loaders represented important capital, even if it were a "placemaker's weapon" left by an absent worker. The combined roles of hunter and distributor of bushmeat may have fitted together for men returning from migrant work to their villages armed with a new muzzle-loader and flush with material goods during the earlier decades (1930s–1960s). Given time, the small number of men remaining in home villages, and the discrepancies in their ages, created a social space and boundary between sponsored youthful hunters, who had the time to keep up with the game as producers of bushmeat, and their elders, who remained as youth sponsors, as custodians of weapons, and the micromanagers of this wildlife enterprise. Elders also promoted the ancestral cult to substantiate their higher role, a function that became contested by those with wider and different exposures through education and work. The clashes between these two roles led to the youthful assertions and wannabe antics expressed during the latter decades, including the rapid infusion of hidden, unregistered muzzle-loaders in the 1980s.

In the next chapter, I examine how local hunters have interacted with and selected their prey over time as well as the various factors influencing their success within the bush together with how changing conditions affected and restricted meat distributions within villages. Both types of relationships with the fauna and between residents have changed rendering local wildlife a national and international "resource" as the Zambian economy inflated.

Notes

1. *Konkobela, konkobela, muno mwezu, mumayamba kulizia mianye, Mizimu yenda ciungulo.*
2. This mongoose's name is associated with *kukota*—to become old rapidly, incapable of accomplishment.
3. "Jacksoneeaye, twalubaee!" subsequently became an in-joke and rallying cry during my stay at Nabwalya in 1988–89. Among us, the expression became a trope for the deteriorating Zambian economic and political conditions that were accumulating under the long-term governance of the United National Independence Party (UNIP) party and the one-party state. The Movement for Multi-party Democracy (MMD) party replaced UNIP in the 1991 election.
4. I followed Kabuswe's suggestion but got no further than the information he had already given me. Chief Kabuswe Mbuluma's motives remain unclear. It might have been a local initiative to reconstruct the environment to make it more productive, or to assure access to water when the Mupamadzi River changed course and the colonial government (particularly the wildlife department) refused to grant passage to water (see chapter 3).

5. A well-known district native commissioner stationed at Mpika during this period.
6. As the smallest antelope, the grysbok allegedly possesses a spirit that affects muzzle-loading guns should the hunter not follow protocol. Its effect may occur even months after a kill. The year following this incident, the muzzle-loading gun Kabuswe used to make this kill malfunctioned, and he consulted with his elders and a diviner to determine the cause. He washed the gun to restore its potency.
7. As grysbok may affect a muzzle-loader, see endnote 6 and reference in chapter 5 (Komeko). Rather than guns, snares were the appropriate technology for normally taking this small species.
8. Part of what I was exploring with local hunters was how they remained current with buffalo movement and, if on the day I was accompanying them, they could show me buffalo. Despite coy attempts to conceal prowess, local hunters were cooperative in sharing their knowledge and kills of buffalo. In any case, their kills were indicated by other hunters or had appeared as fresh buffalo meat in relish dishes.
9. In 1989, I asked the hunter I was accompanying on six occasions if he could show me buffalo on that day. On five of these days, we observed buffalo. During the remaining occasion, the buffalo herd had been disturbed for several days by hunters and we saw only recent signs of its presence.
10. I was a member of the chief's party during his week's walking tour of villages along the Luangwa River in April 1989 where we found a recently wounded Banda. I listened to the discussions of the various parties describing the circumstances surrounding the event and how site residents should proceed with this case. A member of the chief's party gave Banda a handful of penicillin tablets, which probably saved his life. That September, I was doing a survey of hunters in Kunda villages in Eastern Province when I met Banda again and asked him about the events subsequent to our initial meeting.
11. I extend respectful deference to Jean and John Comaroff (1992, 2012) and to James Ferguson (1994) for their insightful analyses of domesticated bovine cultural systems in southern Africa. Chapter 9 contains my glimpses into how this "buffalo mystique" worked chronologically over the decades in the lives of Valley Bisa hunters. See Robert Cancel (2013: Chapter 4 "Bisa Storytelling: The Politics of Hunting, Beer Drinks, and Elvis [Bisa StoryTelling 2 by Laudon Ndalazi] about a local puzzle about buffalo and of a video taken during its telling at Nabwalya in 1989.

Chapter 7

Changes in Scope and Scale

Lineage Provisioning through Hunting

I consume as lightening [God]
Lions drag their prey to the ground.[1]
Bisa hunting song (recorded 1967)

In philosopher Albert Borgmann's terms, game meat for the Valley Bisa is a "focal thing," a cultural placeholder that engages both the imagination (mind) and behavior (body) of those who procure and consume it. Through interweaving activities and meanings honed through practice, focal things become the stuff of everyday lives. "Focal things demand patience, endurance, skill, and the resoluteness of regular practice—a focal practice," David Strong and Eric Higgs wrote, "even a certain character developed in order to become a match for the thing." In these terms, game meat is a focal thing that engages the minds, energizes the bodies, and centers the lives of those seeking to obtain it (Strong and Higgs 2000: 23; Borgmann 1992: 119–20).

Like an elephant within its complicated web of ecological relationships dispersed over a range of supportive habitats, a focal thing is not an isolated entity, but is the material center of an intricate network of human relationships set within a cultural landscape. In this context, bushmeat becomes a "total social phenomenon"—a common thread given meaning within the cultural fabric connecting people with others and to the land (Mauss 1954). Local hunting practices therefore are a culturally eloquent and practical means that provide an important subsistence element within an uncertain environment.[2]

Naturally occurring "goods" produced in fields and gathered from forests are shared and consumed in ways that bind people together in a material and cultural sense that makes Valley Bisa society cohere and work. On a daily basis, the reiterative rituals of acquiring, sharing, preparing, and eating food are the crucibles forging local identities, making possible enduring connections between neighbors based upon learned patterns of relatedness. As a tangible part of the social and political glue of lineages, game meat is a rich source of sentiment and symbol, a key for understanding local values and welfare. Its production, controlled by few men, is complementary to the provisions provided by most women and features interpersonal rivalries and competitions mediated under the canopy of a corporate ancestry. I examine in this chapter some of the cultural attributes of meat and hunting that have made them focal concerns in Valley Bisa life. This cultural centrality, focused on survival within a chancy world, becomes visible as the older ways eroded under the surges of globalization, commoditization, state regulation, and the intensified hardship of frequent famines.

As a focal objective, hunters have to find their game, bring it to ground, and butcher it before they have a valuable product to share. Even before the hunt, they have ideas about the appropriate mammals to pursue and, once encountered, how to approach their prey under the circumstances that it is vulnerable. I begin with a comparison of hunters' stated preferences, including observations on their interactions with different species over several decades. These comparisons allow an assessment of whether their words match their observed selections in the bush. Once prey is killed, its size determines its disposition, its utility, and social functions.

The distribution of meat from small game killed in 1989 and 1990 shows its duration within the household as food during this period and its continued relevance within the diminishing bonds of earlier, more extensive kinship. In subsequent years, prolonged famines and intense scrutiny of wildlife scouts affected these integral distributional social threads, rendering them more restrictive, circumspect, and tentative. With the ongoing commercialization of game, some hunters sold their skills and its produce as individuals rather than depended on the reciprocal relationships of earlier years, adversely affecting the bonds of local cohesion and social capital.

Changing Patterns of Wildlife Use and Distribution

Speaking of Selecting Mammals

Hunters generally answered questions about their preferences by listing various combinations of predominant species—buffalo, warthog, and impala (table 7.1). Buffalo usually headed this list of preferred species, yet there were exceptions. These exceptions took time and a broader exposure for me to grasp, as respondents' choices reflected their standing within their social world. In a telling way, these subtler aspects voiced in their selections embodied their identities, both personal and social. In 1966, when an elder epitomized his choice relish dish as consisting of warthog, female impala, or buffalo meat (grouped together as *inama zha buBisa* [meat or meal of Bisa-ness]) combined with porridge (*nzima*) made from an indigenous wild rice (*lupunga*), he was telling me more than I initially realized. Embedded in his answer was a rich texture of history and symbolism, as this combination of wild species (animals and plants) had enabled the Valley Bisa to survive through lean times of scourges and droughts. Moreover, these mammals grow fat and remain healthy feeding on the extensive beds of wild rice during the late months of the rainy season. In their own seasons, men gathered the game while their women reaped the wild rice and prepared the meals celebrating the richness of valley life; history and social structure reviewed and renewed through extended household consumption.

TABLE 7.1. Preferences and ranking of major wildlife species stated by the main lineage hunters on the Nabwalya Study Area in 1966, 1973, and 1989.

Lineage	Birth date	1	2	3	4	5	6
		Ranked species preferences of hunters interviewed in 1966					
Ng'ona	1900	Elephant	Buffalo	Eland			
Nkazi	1903	Buffalo	Warthog	Impala	Zebra		
Nzovu	1904	Buffalo	Puku	Warthog	Bushpig	Impala	
Mbulo	1914	Buffalo	Warthog	Impala			
Mvula	1918	Warthog	Female Impala	Buffalo			
		Ranked species preferences of hunters interviewed in 1973					
Muti	1912	Buffalo	Elephant	Warthog			
Muti	1928	Warthog	Impala	Reedbuck	Buffalo	Elephant	
Nzovu	1928	Buffalo	Impala	Puku			
Muti	1933	Buffalo	Warthog	Rhino	Female Impala	Zebra	Bushpig
Mvula	1937	Buffalo	Impala	Warthog	Elephant		
		Ranked species preferences of hunters interviewed in 1989					
Nzovu	1922	Buffalo	Elephant	Warthog	Impala	Zebra	
Ng'ona	1929	Impala	Warthog	Zebra			
Muti	1947	Buffalo	Zebra	Warthog	Impala	Bushpig	
Nkazi	1949	Buffalo	Warthog	Waterbuck	Puku	Bushbuck	
Muti	1952	Buffalo	Impala	Warthog	Zebra	Bushbuck	Porcupine

Source: Personal interviews during various years (Modified from Marks 2014: 231)

Chief Nabwalya [Kabuswe Mbuluma] (Ng'ona lineage, born 1900) was one of two residents who owned a high-powered rifle. This possession was a significant status symbol and distinct from historical precedent, when the government would loan antiquated rifles that often malfunctioned, along with a restricted number and uncertain supply of ammunition. The chief purchased this rifle from a departing European administrator on the eve of Zambian independence. This chief frequently used his rifle to reinforce his patronage, as when he killed elephants raiding fields during visits to his many wives' households scattered throughout the chiefdom. At other times, he loaned the rifle to hunting clients, often his own sons, when the palace needed meat. Unlike other local hunting prescriptions, those implemented over the carcasses of killed elephant, eland, and lion were homogenous, indicating the chief's stamp of long tenure and hegemony of these mammals.[3] His answer to my query about his preferences (elephant, buffalo, and eland) carried his claims of status, largesse, and authority.

The Nzovu [lineage] headman, born in 1904, struggled to keep his village together in the face of frequent and prolonged absences. His objective, like the concerns of his other Mbulo [lineage] (birthdate 1914) and Nkazi (1903) peers, was to encourage those who borrowed weapons to kill buffalo, thereby meeting his own need for generosity as a provider of bushmeat to clients. An exception was Mvula (1918), an elderly chief's counselor, who lived in a detached settlement with few kin. In the past, he had built his reputation upon carpentry and hunting, but by 1966, many residents feared him as a magician. His occasional kills of smaller game were sufficient for his small settlement of dependents.

The individuals listed for 1973 were then in the prime of their careers. Muti (born 1928) was headman of a village along the Mupamadzi River who, as a chief's counselor, made periodic trips to Nabwalya where I interviewed him. His listing of reedbuck as a preferred species indicates a slightly different ecology, as this species occurred sparsely on the extensive plains near his home village along the Mupamadzi River. Both Muti (born 1933) and Mvula (1937) were younger men pursuing their hunting dreams through service to patrons. Muti hunted for a number of patrons, local and outsider, while Mvula, a son of the chief, fulfilled his father's demands and assisted nonresidents when they came in search of game. As a few rhinos were legal targets on game licenses in 1973 and were primarily sought by foreign safari hunters, Muti included this species to distinguish himself as a safari tracker.

The lists of the two older hunters in 1989 contrasted reputation with that of role. The listings of Nzovu (born 1922) reflected his past as the chief's hunter and associations with authorities that gave him access to elephants, a species he hunted earlier in his life. As the state no longer licensed elephants since 1980, as an elderly man, he now hunted smaller game. The Ng'ona clansman (born 1929) had killed several buffalo in the past, but he did not list this species in 1989. As a member of the royal lineage and as the presiding local court justice, his targets were mainly impala and warthogs together with an occasional zebra.

Buffalo remained the obvious preference when most hunters elaborated their choices. As described by local hunters, buffalo were usually approachable ("easy to stalk as they get absorbed in grazing and never mind about other things"), its flesh bore no stigma ("everybody eats buffalo as no one fears an illness"), and its meat was palatable (contained "fantastic fats and a good taste"). Everyone acknowledged the dangerous nature of wounded buffalo and recounted how these mammals tended to hang around in the tall grass next to the rivers and villages. Despite this reputation, most people relished its flesh. As "traditional" fare in local recognition, buffalo remains a favorite meat even after the possession of its flesh became increasingly vulnerable to scouts' scrutiny under the ADMADE program. Buffalo were susceptible to snares as well as to guns.

Among the smaller species, impala were abundant and "eaten by everyone," yet their "big eyes" and alertness (smell and hearing) remained challenges for stalkers. During the breeding season, single rams without accompanying ewes were more vulnerable to stalkers. Residents associate the flesh of two other common species, warthog and zebra, with several illnesses (including *bwakakazhi* and *ubukozhi*), a reason that some adults and children gave for their refusal of this bushmeat. Some people preferred warthog because its fat deposits enable the meat to remain moist and pliable for days after curing. Others prepare crackle by charring its skin. Zebra meat is suspect in physical afflictions of those who consume it, as it is a prominent member of a marked class (*vizemba*). Those relishing zebra meat referred to its "yellow fats," "density of the meat," and "sweet taste."[4] Another hunter conditioned the taste of zebra meat upon the quality of the grass it was eating at the time it was killed.

Residents used similar qualities of texture and taste to describe the flesh of most species. One claimed that waterbuck flesh, a stringy meat, "appropriately cured by hanging for several days, tasted like that of buffalo." Another person described porcupine as "the best tasting animal" as its roasted back skin (quills removed) was like roasted maize.[5]

TABLE 7.2. Recorded wild meat prohibitions from a survey among residents of Nabwalya Study Area by gender and age, May 1989.

Species	Ever Eaten?	Men* (Age Categories)					Women* (Age Categories)				
		<20	21–35	36–50	51–65	66+	<20	21–35	36–50	51–65	66+
Hippopotamus	No	2	12	14	11	8	15	47	37	26	2
	Yes	0	18	7	4	1	3	11	1	1	0
Baboon	No	2	23	15	14	8	18	57	36	26	2
	Yes	0	7	6	1	1	0	1	2	1	0
Zebra	No	0	13	3	2	3	5	22	10	8	0
	Yes	2	17	18	13	6	12	35	28	19	2
	NOP**						1	1			
Eland	No	0	5	3	2	1	4	15	12	6	1
	Yes	1	20	17	12	8	5	30	21	19	1
	NOP**	1	5	1	1	0	9	13	5	2	0
Bushbuck	No	0	6	2	2	0	4	18	10	5	0
	Yes	2	24	19	13	9	12	38	27	22	2
	NOP**						2	2	1		
Warthog	No	0	4	0	0	0	1	7	3	3	0
	Yes	2	24	19	13	9	12	38	27	22	2
	NOP**						1				
Elephant	No	0	0	1	0	0	0	2	0	2	0
	Yes	2	30	20	15	9	16	54	38	25	2
	NOP**						2	2			
Rhinoceros	No	1	0	2	1	0	0	5	0	1	0
	Yes	1	15	14	12	8	2	20	26	23	2
	NOP**	0	15	5	2	1	16	33	12	3	0

* Men (n = 77 respondents); Women (n = 143 respondents).

** Respondent replied s/he never had the opportunity to taste this mammal's flesh.

Severe and extensive famines beginning in the 1980s eroded for many an earlier (1966–67) prominent prohibition on consuming hippo flesh (Marks 1976: 78–80, 99–100). This stigma then [1960s–70s] signified local identity and largesse compared to the stinginess and stranger status of government residents who openly relished its meat in those days. In the 1990s (table 7.2), more local households were less averse to hippo flesh and claimed to have consumed it at one time or another. Hippo flesh was a commodity that the ADMADE Sub-Authority and some local hunters sold or exchanged to outsiders for grain and for cash during the droughts.[6]

As a valued attribute to wildlife, fatness varies seasonally. Sow warthogs and ewe impala generally acquire fat prior to parturition before the early rains, but lose this quality quickly upon birthing and nursing their young. Bachelor buffalo bulls remain fat throughout most of the year, while those in herds become "healthy and fat" after feeding for months on the abundant lush grasses of the late rainy season. At other times, buffalo herds move constantly to find pasture and appear undernourished in the late dry season. Eland are renowned for the deposits of fat on their bodies, yet local hunters were reluctant to shoot this infrequently observed visitor around Nabwalya.[7]

Dreams and omens sometimes influence hunters' choices while in the bush, as exemplified by this hunter's choice of the game he selected and the circumstances in 1989:

> Dreams are very important in hunting for they foretell what is going to happen. Whenever I go into the bush, I always ask my wife what she has dreamed. If she tells me that she dreamt about having fresh meat at home, sure I always kill; or even if I dream I am in the bush looking for animals and kill one, it's true I am going to kill.
>
> Yesterday, my wife told me that she dreamed that my cousin's wife appeared at our doorstep carrying a dish of meat to exchange for sorghum. But my wife did not have sorghum and gave ground maize instead. When I reached the house of my hunting friend, I told him my wife's dream. He told me that last evening he had dreamed of white men standing across the river, all had guns, and were pointing at some unknown object. From then, I was sure we were going to make a kill.
>
> So in the bush we looked for buffalo. That afternoon, we saw them resting in the middle of an open plain where there was no cover for us to approach. We waited for hours hoping that the buffalo might move and begin grazing. As darkness was approaching, I became anxious and crawled slowly toward them. I rested in a small bit of cover some 125 meters from this herd. Within five minutes, a large, male warthog approached and came twelve paces [meters] of where I was resting. I changed my mind and imme-

> diately shot the warthog. It fell dead on the spot. It was such a big animal that, when we butchered it, we had to leave portions hanging suspended from a tree. When I took the meat home, my wife said, "Sure, I dreamed meat last night."[8]

This warthog was another example of what hunters refer to as a "sent animal" that they attribute its unexpected behavior to the beneficence of their guardian spirit. Despite the initial dichotomous labeling of dreams as either "good" (*mupazhi*) or "bad" (*cibanda*), these visions sometimes produced counterintuitive behaviors and expectations. One hunter said that even when he had a bad dream, such as an elephant or buffalo chasing him, he still went into the bush anyway; otherwise, he might lose his courage (Marks 1977a: 12).

Besides dreams, the sightings of various omens sometimes may influence a hunter's behavior, curtail a hunt, or even shift its direction. Since a sighting or behavior is usually unexpected, the witness assumes that the "spirits" must have sent it and there is meaning behind its appearance. Remember Kabuswe's watching carefully the movements of a slender mongoose in chapter 6 (17 November 1988) before making his decision to proceed? Or Hapi's dream of the porcupine in chapter 4 and asking his grandfather for an interpretation of its meaning? Like dreams, people sort omens into good or bad expectations based on the connotation of a name, color, or associated attribute, assuming that its occurrence predicts a shift in personal fortune. Two species of mongoose (banded, *munzulu*, and slender, *lukote*) suggest opposite expectations. The appearance of the gregarious banded mongoose advises that a hunter's route to game or a carcass will be direct and soon. In contrast, the appearance of the solitary slender mongoose implies that the hunter will become old quickly (should he continue his direction of search).[9] Despite the ubiquity of this animal lore, its significance varies, as do the sightings and inferences made from them. From my experiences, some hunters might immediately assign meaning to a sighting, or they may do it later, when asked, upon their return to the village. Other hunters appeared oblivious to and unaffected by such sightings. Much of this lore circulates as proverbial statements or as local expressions, yet its persistence argues against dismissing it unilaterally. Whether accurate or not, these connections, even when used in hindsight, offer some order in a world where death and disaster can strike without warning and good fortune may appear out of nowhere.

Observing Hunter-Prey Interactions

Individual preferences about prey and decisions to stalk a species vary seasonally and depend upon a hunter's assessment of the circumstances during each encounter. I sought answers to three questions of observed hunter-prey interactions: I wondered if stalks and retrievals matched hunters' verbal statements and if these choices changed with time, and I was also curious about how hunters' stalks of a species varied in outcome.

My observations on hunter-prey interactions support hunters' verbal statements on their preferences (table 7.3). If "preference" for a species is measured in terms of stalk attempts per encounter, buffalo remained the major target of forays through 1990. During 1966–67, hunters stalked buffalo on 84 percent of the eighty-seven occasions they found this species when I accompanied them. In these encounters, hunters killed nineteen buffalo while wounding thirty-four. In subsequent years, my recordings, including accounts volunteered by hunters returning from a foray, show that the percentage of stalks per encounter dropped progressively. In 1990, hunters stalked buffalo 54 percent of the time they observed this species, but killed only one while wounding ten. That hunters wounded more buffalo than they killed outright later seemed realistic as hunters exhibited a precautionary tactic on firing their weapons. If the animal did not immediately appear mortally wounded, they rapidly moved away from that site to avoid detection by wildlife scouts.[10]

In terms of stalks per encounter, warthogs were the second most sought species. During the 1960s, hunters came across warthogs infrequently, suggesting that their numbers were at comparatively low levels. In subsequent years, warthog numbers increased and the decreased again in the 1990s—a result of alternating heavy rainfall and drought conditions together with increased snaring by residents and the taking of smaller (particularly juvenile) game with dogs. Residents came upon warthogs as singles or in small groups (usually a sow and offspring) and, with favorable wind conditions, were able to approach them quite closely.

With the exception of single, territorial rams and small groups, hunters found impala more difficult than either warthogs or buffalo to stalk. In the late 1980s, some hunters killed impala while hiding along animal paths leading to water. Few local hunters took zebras, whose numbers fluctuated from year to year. During the 1990s, trappers with snares and owners of dogs became the main takers of smaller game, including an occasional zebra and buffalo often snared near their villages, settlements, or fields.

TABLE 7.3. Stalk outcomes of Valley Bisa hunters of four abundant mammalian species, Nabwalya Study Area, Zambia, 1966–90.

Year (months)	Species	Number of Encounters	Stalk Attempts	Kills Retrieves	Wound, Escape
1966–1967 (12 months)	Buffalo	87	73	19	34
	Impala	197	72	2	4
	Zebra	122	11	1	1
	Warthog	57	38	7	3
	Source: Marks 1976, Appendix C				
1973 (July–Sept.)	Buffalo	55	40	3	15
	Impala	167	25	2	4
	Zebra	103	23	0	2
	Warthog	86	34	10	4
	Source: Marks 1977a, Table X				
1988 (Oct.–Dec.)	Buffalo	34	22	6	4
	Impala	314	18	2	3
	Zebra	115	1	1	0
	Warthog	191	30	5	3
	Source: Unpublished Observations; Hunters' notes				
1989 (12 months)	Buffalo	102	61	19	10
	Impala	2661	326	32	38
	Zebra	904	91	4	15
	Warthog	1134	180	23	31
	Source: Marks 1994; Hunters' notes				
1990 (12 months)	Buffalo	37	20	1	10
	Impala	945	181	17	23
	Zebra	265	57	1	7
	Warthog	374	74	9	14
	Source: Marks 1996; Hunters' notes				

Outcomes: "Wind and Honey Bird Makes Animals Very Clever"

A stalk is a good indicator of a hunter's attempt to kill an animal (Marks 1977a).[11] Once the hunter commits to stalking a particular mammal or group, the eventual outcome depends less on the hunter's skill in reaching a desired distance prior to firing his weapon than on the awareness and reactions of his intended prey (table 7.4). During 1988–90, recorded stalks ended overwhelmingly (74 percent) with the prey moving away from the hunter. All prey depend on scent to detect danger as most wild animals generally face upwind when foraging and traveling. During stalks, shifting wind was the main (59 percent) variable that alerted game of impending danger. Detecting potential harm by sight is a less acute sense for most species (27 percent); however, sight seemed important for zebras.

Wild mammals pay particular attention to the behavior of other nearby species, particularly birds, and respond to their alarms. The constant chatter of honey guides, starlings, and sparrow-weavers are particularly vexing for hunters stalking impala and warthogs. At times, honey guides persistently shadow hunters as nearby game responds instructively to its signals of danger.

During 1988–90, hunters killed or retrieved prey (45 percent) almost as often as they wounded it (47 percent). A weapon misfired or missed its target on twenty-six occasions (8.5 percent). In terms of successful stalks (kills or retrievals), buffalo provided the highest return for effort. When shot at, more impala and warthogs got away wounded than were killed or immobilized. Given the inaccuracy of smoothbore muzzle-loaders, this outcome may reflect the difficulty of hitting a small target in a critical area. In 1990, a serious drought and apprehension over the new ADMADE enforcement program contributed to the comparatively low level of hunting effort. That rainy season, government agencies employed most adult men to ferry bags of relief grain from the plateau into the valley for distribution as famine relief.

TABLE 7.4. Summaries of hunters' stalks by species and their outcomes, Nabwalya Study Area, 1988–90.

Year (months)	Species	Hunters (numbers)	Stalks	Hunter Options				Prey Options		
				Kills	Wound	Miss/Misfire	Acquire Predator kill	Wind	Sight	Alerted by Birds
1988 (10–12)	Impala	7	18	1	3	0	1	10	2	1
	Zebra	1	1				1			
	Warthog	5	30	4	2	1	1	19	2	1
	Buffalo	5	22	5	4		1	11	1	
	Wildebeest	3	3					1	2	
	Puku/ Waterbuck	4	5	1				1	3	
	Other species	3	5	2	1				2	
1989 (1–12)	Impala	9	326	32	36	2		152	59	45
	Zebra	6	91	4	12	3		33	28	11
	Warthog	9	180	23	26	5		72	31	23
	Buffalo	9	61	13	6	4	6	23	9	
	Wildebeest	8	32		1	4	2	21	2	2
	Puku	1	5					2	3	
	Other Species	6	24	8	1	1		7	4	3
1990 (1–12)	Impala	3	181	17	20	3		89	29	23
	Zebra	3	57	1	7			23	24	2
	Warthog	3	74	9	11	3		28	14	9
	Buffalo	2	20	1	10			7	2	
	Wildebeest	3	31	1	2			10	14	4
	Other Species	2	9	3	1			3	2	
	Totals		**1175**	**125**	**143**	**26**	**12**	**512**	**233**	**124**

Source: Individual hunter's notes, reports, and personal observations; only the stalks for which there is complete information on hunter and prey options are included in this table.

What about Those "Spirits"?

Hunters are mindful of a host of intangible influences affecting their activities, yet these pressures are rarely visible. Signs observed by hunters and interpreted as sent by "spirits" sometimes become an integral part of their awareness within a conceptually hostile bush environment. This awareness, a search for meaning in the behavior of other animals, suggests that human fortunes are connected profoundly with other forms of life. Such associations come readily to a rural people who regularly encounter wild animals and must contend with forces that could turn unsuspectingly upon them.

One hunter recalled that dreams, prompted by his guardian shade the night before a hunt, successfully foretold the outcome of his bush forays, at least within a day or two. For him, protection together with spiritual engagement depended on his kin conforming to normative behavior; he attributed earlier failures, as well as feelings of increased vulnerability in the bush, to moments when his lineage is bickering. Similar associations show in the following account, experienced by a hunter who lacked paternal support while an understudy. Restitution of lineage support and his will returned when a diviner clarified ancestral demands.

> During my initial hunts in 1972, I moved with Mbola, my friend. I saw how he stalked, shot animals, and watched them die. He explained that once the weapon was on my shoulder with the barrel pointed at the animal, I should stop breathing until I discharged the weapon. I was "troubled" at the thought of firing at an animal. I decided one day to hunt by myself with his borrowed gun, but I missed the animal with several shots at close range. Then Mbola instructed me how "to become sufficiently brave."
>
> The following day, I stuck to my friend's instructions and killed an impala. I felt happy with my first kill, but my father did nothing to please me even when I shared the meat with those in his village. After I killed a buffalo, I realized that I was no longer able to kill animals. This failure occurred immediately upon the death of my [maternal] uncle Casho,[12] who was the agent responsible for my failures. I discussed this problem with [lineage] elders and together we sought the guidance of a diviner.
>
> The diviner told us that my uncle wanted to "inhabit" me as a hunting spirit. My relatives then prepared a gourd [*nkombo*], put sorghum flour in it, summoned the spirit of Casho, and instructed "him" not to harm me. Instead, he should "make me like he was in the past." Since then I have used this gourd, prayed to Casho before going hunting, and have become successful.[13]

In the early 1990s, some older residents still acknowledged that their ancestral spirits monitored them and affected their welfare, and caused unusual happenings. In March 1991, Mutombo and a (lineage) "brother" went hunting, and by midmorning Mutombo had killed a warthog. They butchered this hog and spent much of that day drying the meat on a platform. Carrying the meat to the village in the late afternoon, Mutombo chanced upon and killed another large warthog. They gutted and suspended this carcass in a tree, then proceeded to carry the meat of the first warthog to the village. Reaching his village at 2000 hours, Mutombo noted,

> My wife told me that my [paternal] grandfather had died. I was very sad and wept. At 0300 hours, I woke up, loaded my gun, and went to wake my brother. We went into the bush where we found the second warthog carcass unmolested, divided its meat, and came back in the village before sunrise. I bathed and then went to the funeral at my sister's house.
>
> At the funeral, I told my relatives how I moved in the bush and killed two animals without difficulty. They told me that my grandfather was a great hunter that his "spirit" was in front of me that day in the bush. That is why I killed two animals.

At other times, spirits may cause a hunter to miss repeatedly or cause game to become unusually flighty. These failures may also signal a relative's death or portend malice among relatives in the village.[14]

Outcomes: Brief Summaries and Comparisons of Hunts, 1989–90

The records kept by local hunters reveal outcomes in their interactions with various species of wildlife. We know that hunters encounter more groups of mammals than they stalk and that few stalks end as bushmeat to distribute. We also expect that the ratio of success varies by age, experience, and prey sought, as well as by year, depending on personal and environmental circumstances.

Figures 7.1 and 7.2 compare the results from two hunters' interactions with prey during 1989.[15] Both used muzzle-loading guns. Hunter A is middle-aged and renowned for his abilities to find and bring home game, particularly buffalo. He is a searcher whose forays take him far into the hinterland as he updates his knowledge of animal movements. Hunter B is an older individual whose major faunal exploits are behind him. He hunts alone or is accompanied by one or more of his sons, and uses hides along

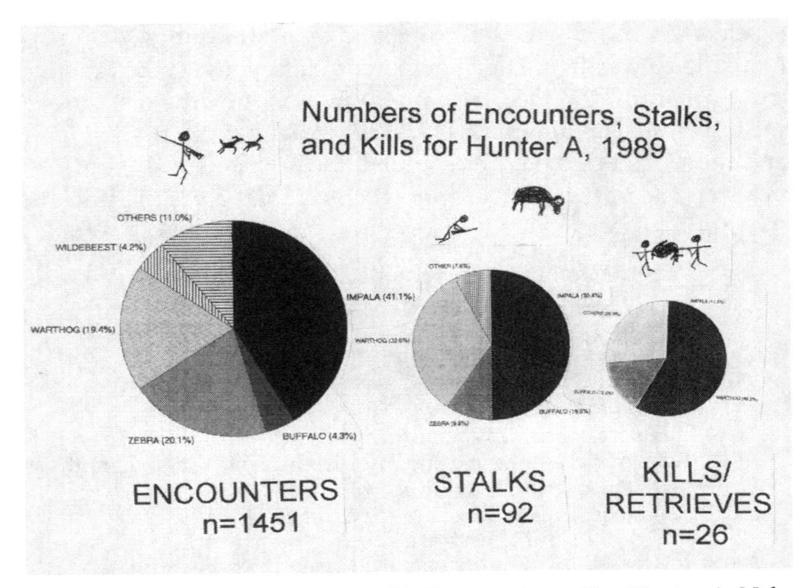

FIGURE 7.1. Encounters, stalks, and kills experienced by Hunter A, Nabwalya Study Area, 1989.

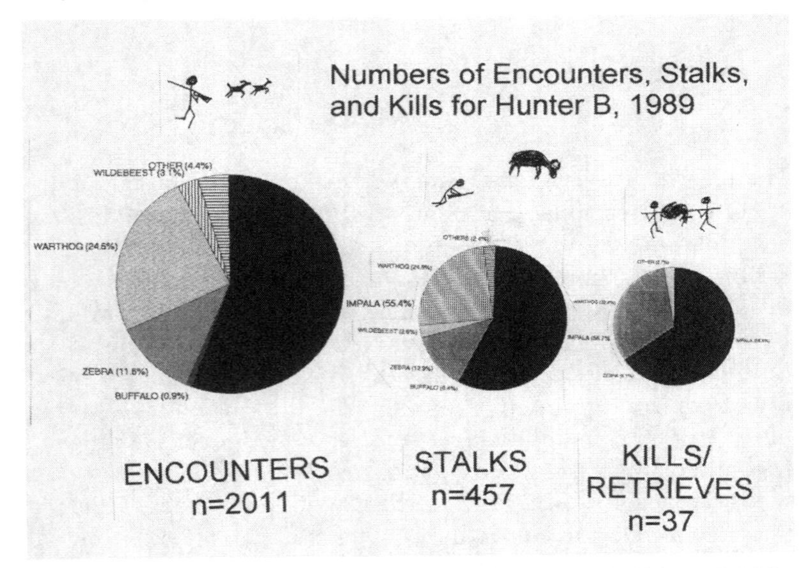

FIGURE 7.2. Encounters, stalks, and kills experienced by Hunter B, Nabwalya Study Area, 1989.

game trails within three kilometers of his homestead to wait for small game to come his way. These profiles reflect two different tactics for securing game.

In 1989, Hunter B encounters slightly more than two thousand groups, mainly impala, warthog, and zebra, and stalks 457 times (figure 7.2). Perhaps because he is instructing his sons, he produces meat for his household only eight times out of one hundred stalks. Hunter A (figure 7.1) experiences fewer encounters, fewer stalks, yet retrieves more and bigger carcasses (twenty-eight of one hundred stalks were successful). This year was a comparatively lush and productive period for both agriculture and for wildlife abundance around Nabwalya where both hunted.

For both people and wildlife, however, 1990 was different and a more difficult period. Heavy rainfall ceased abruptly in March, wilting most cultivated plants. The resumption of heavy rains in April caused massive flooding on the valley floor, destroying gardens and crops on the alluvial soils along the rivers. Thus began several years of alternating droughts and floods. This was also the initial year of intensive wildlife enforcement using locally employed village scouts. Both the inclement weather and enhanced enforcement factor into the depression of hunter-game interactions recorded for 1990 (figures 7.3 and 7.4). Hunter A experienced fewer encounters and stalks while taking less bushmeat. Hunter B showed similar decreases in encounters and stalks, with a slightly decreased ratio between his stalks and kills.

Environmental conditions affect all dimensions of valley life. They influence wildlife reproduction, dispersion, and behavior; they also shape and shift Valley Bisa lives and livelihoods. New wildlife policies transform hunters or, perhaps more tellingly, alter their practices in more subtle and shadowy ways. Figures 7.5 and 7.6 illustrate how the different environments of 1989 and 1990 affected wild meat production from three local Nabwalya hunters. Hunters A and B are the same as those above, whereas C is a younger man who mainly snared game. Whereas off-takes from wild populations were down in 1990, they increased in subsequent years as residents narrowed their focus to smaller prey and reduced their largesse to close kin, adjusting for scout activities once it was clear what they could expect under the new wildlife program.

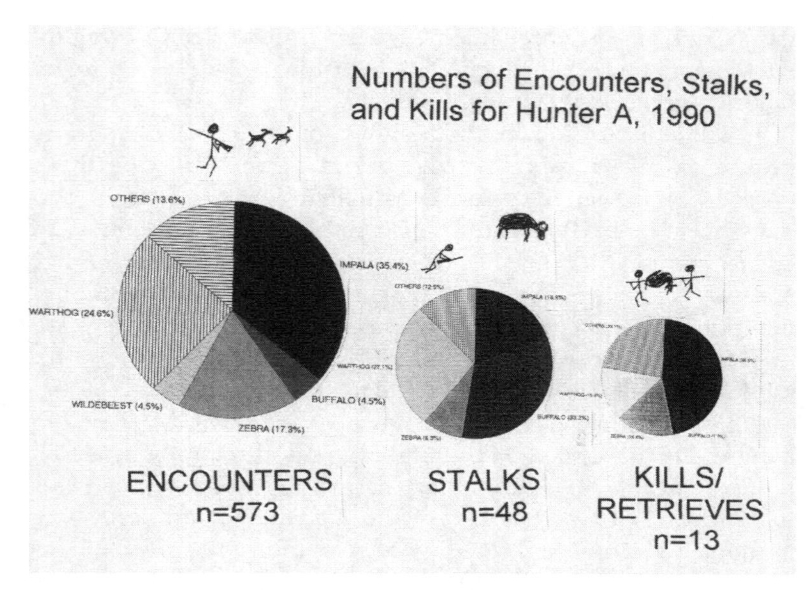

FIGURE 7.3. Encounters, stalks, and kills experienced by Hunter A, Nabwalya Study Area, 1990.

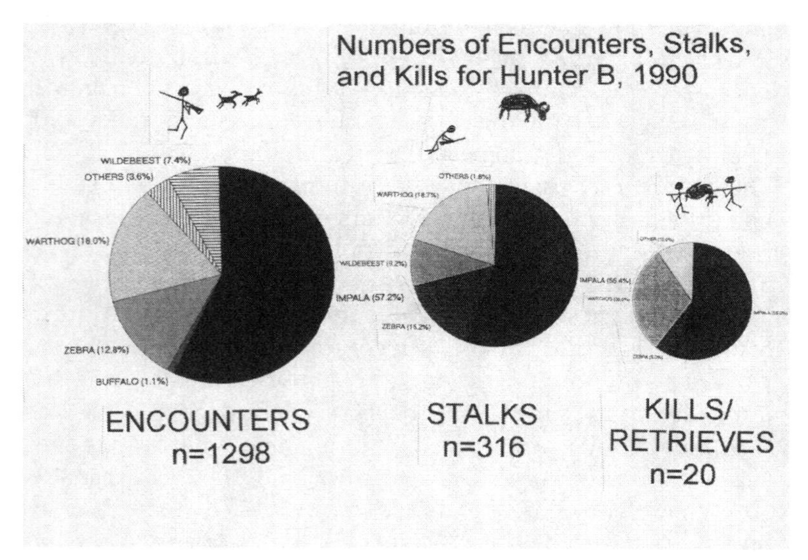

FIGURE 7.4. Encounters, stalks, and kills experienced by Hunter B, Nabwalya Study Area, 1990.

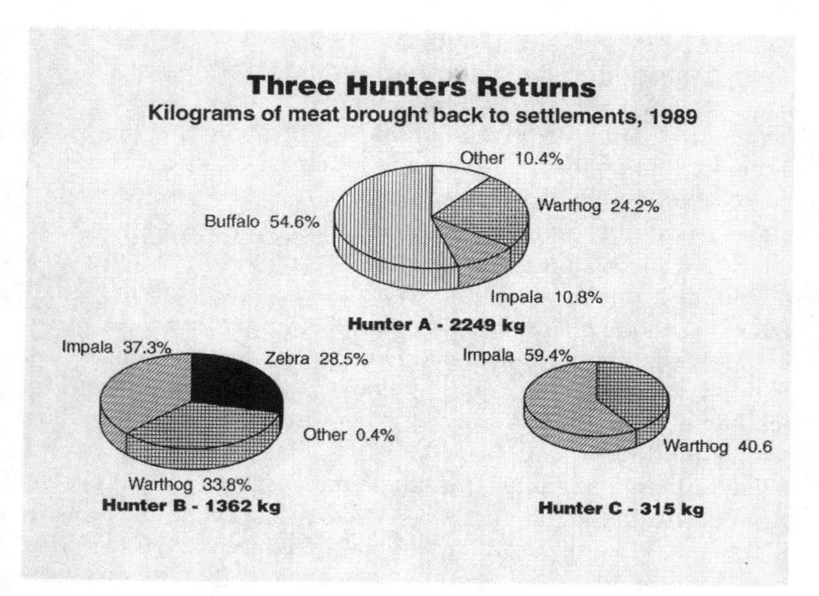

FIGURE 7.5. Three hunters' returns with bushmeat (kg) to their respective settlements, Nabwalya Study Area, 1989

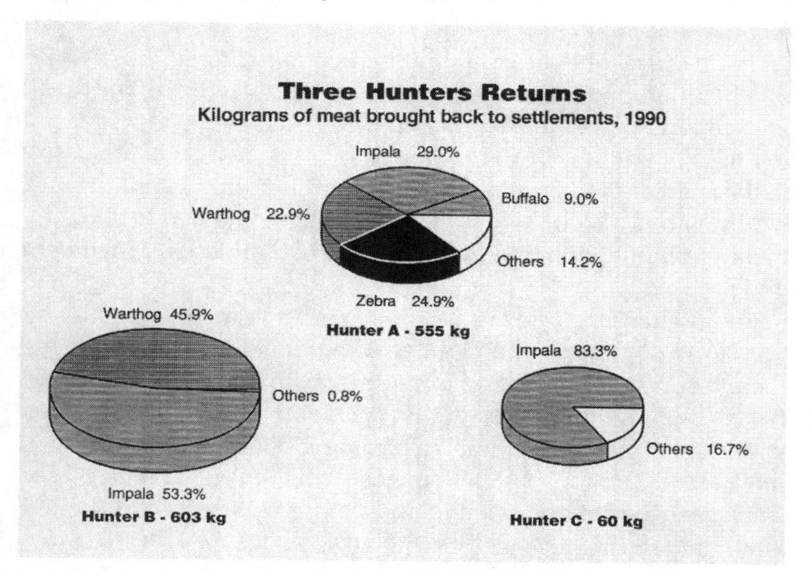

FIGURE 7.6. Three hunters' returns with bushmeat (kg) to their respective settlements, Nabwalya Study Area, 1990

The Limits of Lineage: Distribution of Bushmeat and Social Relations

Game meat is the preferred and most esteemed relish dish consumed with any meal of porridge. Although produced by a few men, women within a household turn this raw product into a dish enjoyed by everyone. In cultural significance, bushmeat trumps agricultural products and is an important signature of community connectedness. Individuals evaluate their links within the village and with the land contingent upon whether or not they recently have tasted bushmeat. Continuous access to wild meat connotes healthy connections with the land as well as with kin. Having ready access to wild meat as well as visible wildlife was a source of Valley Bisa identity, pride and even despair. These attributes made them significantly different from most other Zambians. When residents go elsewhere and are identified as "from Nabwalya," people inevitably recognize these two traits. Strangers as guests in this homeland always want to "see and eat the animals!"[16] Some residents also make the connection that having so much wildlife is the reason they have stayed so poor.

The rules governing meat distribution are different from those of other foods, yet these conventions have changed with the times and circumstances. In the past, the advantage of killing a large mammal, such as a buffalo or elephant, was that its carcass was sufficient to accommodate the needs for most within walking and hearing range. Even then, patrons and hunters aimed to keep the meat distributed to all the cooking pots of their lineage members and clients.

To butcher large mammals, men disarticulate the carcass by removing the skin and limbs. From the limbs, they flay strips of meat from the big bones and lay these chunks on skin or leaves. Butchers hack the rest of the body including the rib cage into sections and remove chunks of meat and bones in portable sizes. In the field, they distribute the meat together with the bony sections to those (mostly women and youths) assembled at the kill site. Upon returning to their respective households, these porters may distribute some of their shares to others who were not at the butchery.

Hunters typically retain some portions from their kills, usually the heart together with portions from the head and intestines. Hunters also save the soft cartilage, accumulation of fat and tender meat of the brisket (*luzao*), to present to a respected elder or associate. On occasion, the chief might receive the brisket, a whole limb, as well as other portions. Since the mid-1990s, licenses for buffalo have escalated in price and become legally restricted within the limited annual quota system to certain privileged safari and outsider groups of hunters. These licenses are available in district and provincial towns, not in rural villages. Even so, wildlife scouts kill elephant and buffalo nearly every year; both of these species supposedly are killed in protection of life and property, but they are taken also informally for rations and personal uses. Safari and other licensed hunters take most of these buffalo, and within the GMAs some of this meat may find its way into local relish dishes. Very few local residents can acquire a buffalo license from the small quota allocated each year for the "community." Once a resident obtains one of the few legal licenses, he has the option to resell them at inflated prices to civil servants, who then employ local hunters (or relatives) to kill the animal (1988–91). Most of this bushmeat ends up on the plateau where it is sold to meet the needs of the residents' school and clinic fees as well as for clothes and other necessities. For civil servants, these meat sales augmented their salaries and allowed them to purchase goods not available in the valley.

Until 1989 and often with the help of a companion, local hunters eviscerated smaller game such as impala and warthogs and carried these whole carcasses back to their homesteads. Others might flay these carcasses into distinct sections (hind leg, ribs, fore shoulders, intestines) and, if alone, carry these sections serially to their households. Although hunters were cautious initially, most volunteered information on the sex and location of kills when asked and were open with me about their quests and motivations. During my residences, I could also cross-check most of their claims against logs kept on village relish dishes and general activities. Yet by 1993, the invasive and aggressive prosecutorial activities of wildlife scouts had changed the willingness with which many local hunters shared information. To decrease the possibility of detection by those not favored in a meat distri-

bution, hunters subsequently butchered and smoked their kills in the bush before ferrying the meat to their houses at night. Darkness covered these activities providing some assurance against these transactions ending up in the wrong hands or ears.

With the agreement of recipients to keep logs on the length of time meat remained in their households and the numbers of people they fed each day, I documented the meat distributed from the carcasses of two small mammals. Meat from a warthog killed on 1 April 1989 found its way into 6 households, constituted an important part of 35 meals over six days and fed 100 adults and 65 children (table 7.5). On 11 April 1989, a hunter killed a male impala and dispersed the meat to eight households. While butchering the carcass in the bush, he was approached by two boys, who

TABLE 7.5. Meat distribution and meals consumed by adults and children from a male warthog* killed, 1 April 1989, Nabwalya Study Area.

Parts	Initial Distribution	Meals/Persons Fed**/Day April 1	2	3	4	5	6
½ brisket ½ intestines ½ lungs	sister	2a3j	2a3j 2a3j	2a3j			
rest of organs front leg backbone ribs	hunter	2a3j	4a3j 2a3j	4a3j 3a4j	2a3j 2a3j	1a2j 2a3j	2a 2a3j
haunch front leg backbone ½ brisket	maternal uncle		5a 5a	5a2j 4a2j	5a2j 4a2j	4a 3a	3a2j
haunch	mother		3a2j	4a3j 2a2j	3a2j 2a2j	3a	
½ head	cousin		2a	3a2j			
½ head	cousin		2a	2a	2a		

* carcass yield about 42kg
** a = adult, j = child, youth eating from the same dish

were given the sacrum to make them allies and to keep silent. An elderly village headman, who was a former patron and the hunter's maternal uncle, received the brisket while the chief got a hindquarter. I shared my portion with my research assistants and others who appeared at mealtimes. Over the subsequent days, meat from this impala appeared in 43 meals (one to three meals a day) and, consumed together with maize porridge, fed 109 adults and 62 children (table 7.6).

TABLE 7.6. Meat distribution and meals consumed by adults and children from a yearling male impala,* killed 11 April 1989, Nabwalya Study Area.

Label	**Initial Distribution (relationship)**	**Meals/Persons Fed**/Day June** 11	12	13	14	15	16
½ sacrum	none#		2a3j	2a			
½ sacrum	none#		3a2j	2a			
right haunch	Chief		4a6j	4a7j	3a	2a	
brisket	Headman "uncle"		5a 4a	4a 5a	1a		
½ stomach ½ liver head, neck front leg	assisting hunter (cousin)	2a3j	2a3j 2a3j	2a3j 2a3j	2a3j 2a	2a	
½ stomach heart, lungs spleen front lrg	Hunter	2a2j	2a3j 2a3j	2a4j 2a2j	3a2j 2a3j	3a4j 3a3j	2a
haunch tenderloin	Machisa none#		3a 4a 3a	2a 3a 4a	2a 2a	1a	1a
½ liver, ribs	2 cousins		2a 2a				

* carcass yield 35 kg
** a = adult; j = child, youth eating from relish dish
# not related, individual learned of kill and claimed a portion

With the breakdown of the earlier patronage system and the splitting of households into scattered settlements, meat distribution came to reflect smaller-scale alliances and fractures within lineages. In 1989 and 1990, the chronology of one hunter's distribution illustrates his frequency of kills and his social strategy (table 7.7). Both his mother and brother lived in adjacent settlements while his sister resided some distance away. The "others" listed in the table are those who borrowed his gun and who retained portions from their kills. He may have given some bushmeat to other people as well that went unrecorded.

Knowing who killed an animal or who possesses fresh meat is tantamount to establishing a claim. Anyone slighted might inform the wildlife scouts, and anyone accepting meat becomes a co-conspirator. Therefore, a hunter or carrier returning to his homestead did so as discreetly as possible. If anyone noticed someone carrying meat, they were readily offered a portion (see reference to nonrelatives shares in Table 7.6). Similar rules about knowledge and claims to bushmeat applied to snared game as well, yet different sanctions enforced these norms. Trappers claimed to smear their snares with a special medicine (*luambu*) that would afflict, with a powerful curse, any person stealing their catch. Therefore, an observer of a snared animal either remained in its vicinity or returned to the village to locate its owner as Joseph Farmwell did in chapter 2. In either case, neighbors generally knew who the trappers were and where they set their snares.

In the following incident, recorded in 1989, a hunter's observation of a snared animal goes unrewarded.

> I came upon a buffalo strangled by a wire at 0800 hours, yet the buffalo was still very much alive. I knew that the snare belonged to a neighbor, but I could not identify him, so I set a fire around the animal as a signal. I waited for its owner to appear so we could discuss "what to do with this animal." Suddenly, the buffalo slipped off the wire from its attachment at the base of a tree and ran off. I feared that the snare had magic to protect against thievery. I would have helped the owner finish the buffalo had he appeared earlier in time.

What were the older rules governing the distribution of meat within a village, and how were these norms applied to meet these expectations? Within the daily face-to-face encounters of village

TABLE 7.7. Chronological tactics of a hunter's initial bushmeat distributions of his kills of small game, Nabwalya Study Area, 1989–90.

Year Date	Kill*	Initial distribution of bushmeat on day of kill					
		Self	Mother	Sister	Brother	Lineage members	Others**
1989							
8/17	1m impala	x	x			x	
8/29	1m impala	x	x	x	x		
9/6	1m impala	x					x
9/23	1m warthog	x	x	x			x
10/21	1m impala	x	x	x	x	xx	
11/10	1m warthog	x	x		x	x	
11/30	1m impala	x	x				x
12/11	2m impala	x	x			x	x
12/28	1m warthog	x	x		x	x	
12/31	1m impala	x	x				
1990							
1/7	1m impala	x	x			x	x
1/18	1m impala	x	x			x	x
1/30	1f impala	x	x		x		
3/4	1f impala	x	x		x		
3/24	2? warthogs	x	x		x	x	x
4/24	1? warthog	x	x				x
5/10	1m impala	x				x	x
5/25	1f impala	x	x				
6/8	1m warthog	x	x		x		x
8/14	1f warthog	x	x				
9/16	1m impala	x	x		x		
10/12	1f impala	x	x			x	
10/23	1m impala	x	x			x	
11/10	1m impala	x	x			x	
12/7	1m warthog	x	x			xx	
12/23	1m impala	x	x		x		

* Sex of kill when listed: m = male, f = female, ? = unknown
** Borrower of owner's weapon or accomplice

living, the distribution of bushmeat often depended on the size of the mammal killed, the status of individual households within the village, the household members on site, and those butchering, carrying, or hiding the flayed meat. The following individual recalled what he witnessed as a youth.

> In the past [1960s through 1970s], government did not consider animals so very important and there weren't many game guards. During my grandfather's and uncles' times, they killed wildlife anytime they wanted. Once they made a kill, hunters called villagers to carry the bushmeat from the kill site to the village publically. Once in the village, they shared meat according to the importance of lineage members. In most cases, the lineage elder, not the hunter, allocated the meat to households. Upon arrival in the village, these other individuals would give elders some meat out of "respect."
>
> At the kill site, villagers made a fire to roast some meat and made a lot of happiness noise in broad daylight, even when carrying the bushmeat. When they reached the village, they made a drying rack (*butabo*) or dug a hole for coals to dry the meat in the open. Even people who came to the village from a distance were given meat, which they carried away without having to hide it. At the village, the hunter's "uncle[s]" and "mother[s]" were given bigger shares as well as the village headman. The individual, who accompanied the hunter, was also given a large share. "Sisters" and "brothers" were given less meat. Unmarried women were given meat, but not unmarried men and boys. Hunters often exchanged meat for grains without problems and this was done openly. This openness began to change rapidly in the 1980s when NPWS got tough and game guard numbers increased. ZAWA has done nothing but brought misery for many people.

When I asked this individual if he remembered instances of conflicts and their resolutions over meat distributions, he provided the following anecdote about an uncle.

> In the village where my uncle stayed there was a different problem, for within that village were three "wombs" of lineage "sisters" and their descent "wombs" had hunters, but the most active was Hapi [remember him for chapter 5 and elsewhere?], a descendant of the youngest ranking "sister." The elder sister expected him to show her and her line of descent conventional respect in terms of bushmeat distributions. As one of her grandsons accompanied Hapi in the bush, she knew of his kills and of his largesse. This grandmother always expected more than she received. When Hapi killed a zebra nearby, the mother of Hapi's companion in the bush refused to accept her allotted share of meat. She told Hapi, "You

were in the bush with my son and killed a zebra; I want more meat. Without being with my son, you can't go hunting alone. So from today, you must move alone in the bush. If you move with my son again, lightning will strike you."

Hapi immediately stopped hunting with the elder grandmother's son and instead chose the younger son of a niece to accompany him. They followed the rules and continued to provide meat to everyone. But in the 1960s, Hapi forgot the earlier "curse" and took his elder mother's grandson to accompany him. They went north of the Munyamadzi River and killed a buffalo. Before they could leave the kill site, thunderclouds gathered and heavy showers fell. The men took shelter under a large tree. While standing under its limbs, a lightning bolt struck the tree, covering the men beneath with falling limbs and embers. Hapi was bruised, escaped severe burns, and remembered his elder grandmother's words. He resisted beating his companion, but upon his return he immediately informed his mother and her older sibling of what had happened. The younger "sisters" accused the elder of being a "bad woman" [witch] and found "leaves" [a bush prescription] for Hapi to prevent future lightning strikes. He never moved again with this particular cousin.[17]

As a locally abundant and focal resource, game meat had values beyond those embedded in the structures and sentiments of the valley's lineages and neighborhoods. As migrant laborers, Valley Bisa men are aware of and exposed to these other assessments as well and increasing costs. Their involvement brought progressive changes in social arrangements and behavior, many of which were not previously tenable.

Taking Stock: Bushmeat as a Regional Resource

Beginning in the late 1970s, the inflationary spirals of a deteriorating Zambian economy escalated the costs of "essential commodities" for most households and made these supplies in rural villages expensive and uncertain. If they could, men and women everywhere expanded their economic activities during these indeterminate times. Informal relationships and transactions between valley residents and outsiders became widespread, personal, and individualized, and in the process transformed the smaller-scale lineage valley economies based upon local patronage and subsistence. The uncertainties and insecurities of larger scales impacted valley homesteads and showed progressively in the smaller circles of expected relatedness through increasing intergenerational hostilities. As an abundant local asset, game meat was a marketable "good" and even increased in relative value as other required products' prices rose. Possessing cash and goods to exchange, outsiders drove this "secondary or shadow economy" in wildlife exploitation and often linked themselves to valley residents and their environmental knowledge to make the most of this new shift in their welfare (MacGaffey 1991: 12–14; de Soto 1989).[18]

Valley residents sold wildlife products to outsiders whenever they could, and some of these activities were sanctioned, as some claimed, by Chief Nabwalya [Kabuswe Mbuluma], who championed local access to wildlife as a safety net in times of famines. Given the low morale within the Department of National Parks and Wildlife Service (DWNP) throughout much of the 1970s and 1980s with this chief's declining health and death in 1984, wildlife became largely an "open access resource." For many years, the prices for legal hunting licenses and permits remained comparatively low in comparison to the revenues derived from selling meat on the plateau. Consequently, even small-scale operators used their connections among valley relatives and friends to make a "killing," in literal and financial terms, while operating seemingly within the law. Contacts and access to money conveyed an initial advantage to outsiders with connections in Mpika, where people procured these licenses;[19] residents at Nabwalya were then employed to do their hunting and curing the meat. License holders manipulated relationships and friendships further to minimize the cost of this bushmeat procurement and of its portage to plateau markets (Marks 2016). The following ethnographic accounts show how two individuals used these relationships and social dynamics to their advantage. They also demonstrate the consequences for local residents.

A Greedy Head Schoolteacher

Mr. Bwalya was caught abusing funds at the plateau school where he taught, so his options included losing his government employment or accepting the head teacher position at Nabwalya, the lowest-ranked school within Zambia's largest district. Bwalya decided to make the most of a bad situation and decided to live among the wild animals and "wild people" in the Luangwa Valley.

As the head teacher, Bwalya became the prime contact for most government officials visiting the valley. For these itinerant officials, the school was the only "civilized" space where they might find a place to relax, food to eat, and a safe haven for the night, as well as contacts for potential deals. Their prejudices against local residents together with their need for a familiar cultural space made these itinerants and Bwalya mutual allies. On his grounds, Bwalya learned quickly to meet his visitors' need for food, quarters, and ease of access to resources they might desire from the valley. In turn, he expected reciprocity when he visited Mpika where he sought easy credit or cash to purchase hunting licenses and "essential commodities," as well as ammunition not available in the valley. In these ways and with these connections, Bwalya lived beyond the means of an ordinary schoolteacher.

As a teacher, Bwalya enjoyed status in the eyes of parents who glimpsed opportunities and prosperity as their children advanced through education, found employment, and gave back to the lineage. The head teacher was amenable to promoting students provided that their parents complied with his wishes and gave gifts. He accumulated many local debts, but he was slow and delinquent with repayments. During school vacations, he used students as carriers to ferry his "goods" to the plateau, paying them vaguely with promises of promotion. As the self-appointed leader of the local Catholic congregation, Bwalya extended his reach into it to assure space in the priest's vehicle whenever he needed transport. He claimed "relatedness" to both the district magistrate and game warden, who presumably gave him immunity from prosecution and coverage for his dealings. His subordinate teachers complained that Bwalya rarely set foot in a classroom. Instead, he wandered up and down in the villages, carousing and drinking local brew, all the while searching for new "business."

In a puzzling conversation I had with Bwalya in 1988, he confided that residents had moved beyond their past relationships with wild animals. "In the past," he declared, "wild animals, were just something to chase down and to eat. Now these same animals bring development and are important resources for 'business.'" Outsiders paid high prices for meat, ivory, or skins, while tourists paid merely to look at the animals. I thought he was paraphrasing the new community-based slogans, but Bwalya's ingenuous comments were merely assertions about opportunities for himself.

Although he owned no weapon, Bwalya always possessed cartridges that facilitated his ability to borrow weapons from residents and visitors alike. Hunting at night with a lamp in the bush and on the green lawns surrounding the school grounds, he reputedly shot several leopards and numerous antelopes. Shots fired in the dark of evening left no evidence the next morning of what had taken place. Skins and meat commanded favorable prices when smuggled to the plateau and these commodities often reached their markets through the courtesy of visitors and officials. Within the school and neighboring community, Bwalya acquired a reputation as someone who could "sweet talk" anybody out of anything. Residents watched intently as Bwalya attempted to put "the squeeze" in his conversations with me for ZK 300 ($50 US) shortly after my arrival in October 1988. This amount was more than half his monthly salary and about five times the local monthly wage for labor. As friends had already warned me of his reputation and of their experiences, I was able to deflect his demand.[20]

In early 1988, the schoolyard learned that someone across the Luangwa River possessed two rhino horns for sale at ZK 3,000 (about $500 US). When he failed to raise money among his coworkers and neighbors, Bwalya secured these funds from friends in Mpika. Accompanied by two young men into Eastern Province, he pursued this offer to make this purchase of rhino horns. Instead he became the victim of a clever ruse with someone he met there. He returned to Nabwalya with his two purchases, cleverly stuffed immature buffalo horn cores that looked somewhat like the real thing. This deception added to Bwalya's debts and to his local reputation. Rather than seeking restitution from those who had hoodwinked him or waiting for his lenders to call in their loans, he came up with another quick scheme to raise cash. Bwalya was apprehensive that his failure to produce genuine rhino horn or seek restitution from the sellers might prompt investigations, as trafficking in rhino products was illegal and taken seriously by the police. Such investigations might bring his other clandestine activities into focus and land him in jail.

In October 1988, it was rumored that Bwalya possessed four licenses to kill buffalo.[21] Sensing no overt challenge to this supposition, which was likely false, Bwalya proceeded to secure this treasure trove of buffalo. He took up residence with a local hunter in a distant village, out of sight and scrutiny of local scouts and of his detractors at the school. The local hunter whom he employed was naïve about Bwalya's reputation and contracted to secure the buffalo at a bargain price.[22] At one point, the rumor was that Bwalya had gone missing for several days in the bush. His wife reported her concerns to the chief, who took no action. No one was surprised when Bwalya reappeared at the school a few days later to buy additional gunpowder and to obtain cash for his hunter. He and the hunter had shifted to another village without informing anyone.

Bwalya's hunter killed his first buffalo late one evening. During the night, lions discovered the carcass and consumed much of the meat. Bwalya and a few helpers salvaged and cured what remained while the hunter killed a second and third buffalo nearby. The hunter shot a wildebeest as the fourth buffalo.

With the meat cured, Bwalya gathered students as carriers and hauled the meat to Mpika. He calculated his arrival in time for the Christmas feasts when game meat nominally fetched the highest price. With some of the meat spoiled or consumed along the way, Bwalya received ZK 2,000 ($330 US) from these sales. He dismissed his students' demands for pay, paid off his most pressing debts, and enjoyed the festivities in drinking binges.

In January, Bwalya began his journey back to Nabwalya already late for the opening of school. He had no money, but he envisioned grander schemes for "business." His expansive style made him new friends, several of whom, enthralled by his stories of easy, early returns on their capital, accompanied him. The group took its time descending the escarpment as Bwalya stopped at villages to make deals and consolidate plans. When his party ran out of salt, mealie-meal [ground maize], and local hospitality, Bwalya sent someone to fetch supplies from his wife. He had not counted on her resentment for being abandoned at Christmas or on her isolation within the community because of his reputation and

behavior. His wife went to the palace to report his whereabouts and plans. The acting chief dispatched a retainer to fetch Bwalya and commanded the wildlife scouts to check on his activities.

When he was brought before the chief's court, Bwalya claimed sickness as the reason for his delays. He was mum about what had happened to the other civil servants' salaries that he collected and had promised to bring with him. The chief scolded him for not teaching and accused him of "moving about killing animals and stealing money from the Ministry of Education." He summarily dismissed Bwalya and sent him and his wife back to the plateau, who set out on foot with only their clothes on their backs. His debtors claimed what they could of his possessions, mainly his flock of chickens, as he possessed no cash.

Another Teacher Uses Inside Connections

Kabi's father was born in the central Luangwa Valley but worked as an adult on the Zambian Copperbelt. Kabi was familiar with both valley and plateau worlds, as he had completed secondary school while in town but worked as a schoolteacher near Mpika. A teacher's salary during the rampant inflation of the 1980s was woefully inadequate, and Kabi wished for capital to further his training to become a businessman. While his wife hawked scones and foodstuffs in town, he spent school vacations and leaves in the valley, where he organized his relatives to generate capital for his transition.

As a nearby resident of Mpika, Kabi purchased game licenses as soon as the quotas became available. Others living further from this administrative center often found the preferred licenses for buffalo or warthog already sold out or spoken for when they arrived at the center. Without owing a weapon, Kabi used the "blue book" (to establish local gun ownership) of a valley relative to procure inexpensive hunting permits and licenses.

In 1986, Kabi purchased a game license for a buffalo and a permit to hunt in the Munyamadzi GMA at the resident's rate (ZK 65 versus ZK 125). He then engaged a relative to hunt this buffalo and paid him ZK 300, a market rate that reflected his lack of personal familiarity with this hunter. Kabi himself oversaw the butchering and smoking of the meat and employed four local carriers to transport this cured meat to the plateau for marketing. Carrier compensation for the four-day foot journey to Mpika required ZK 340 (ZK 85 each). At market, the meat realized ZK 1,916 and returned a profit of almost 300 percent, double Kabi's monthly salary. He had spent twenty-two days overseeing the process, but did not include his time as a cost.

Around Christmas of the following year, Kabi again purchased a license for a buffalo and called upon his relationships in the valley. His expenses were ZK 90 for license and access fees together with wages of ZK 450 for five carriers. Even without paying a close relative to kill this buffalo, he turned a profit of ZK 1,710.

Six months later during the 1988 dry season, Kabi found his entrepreneurial and social skills challenged by new conditions. After securing licenses for two buffalo and an impala, he engaged an "uncle" to hunt for them in the valley. This "uncle" secured the first buffalo and the impala without incident. While pursuing the second buffalo, he and an accompanying relative were arrested by an itinerant wildlife official who charged them with hunting buffalo without a license and using a borrowed weapon. The official took these "poachers" to Mpika and placed them in jail to await trial. While all this was happening, Kabi was only a short distance away, supervising and curing the meat from the earlier kills. When he learned of the arrests, he delegated the remaining tasks of assembling, packaging, and carrying the buffalo meat to another "relative" and set off on the four-day trek to Mpika, carrying the impala meat with another relative. He arrived in time for the trial and witnessed the magistrate sentence the two "poachers" to a year at hard labor or a fine of ZK 1,000. The uncle spent a week at hard labor while Kabi raised the money to pay the fine and obtained his release. His neighbor served the sentence.

Kabi recovered nothing from the first buffalo. The man charged with that task reputedly took it to market, sold it for an unknown amount, and subsequently disappeared. Kabi claimed ZK 200 from the sale of the impala; clearly, this episode was a costly turn of events. In late October 1988, Kabi obtained leave of absence from his school and returned to pursue his second licensed buffalo. The rains had begun so there was little chance of interference from itinerant game rangers. Another relative

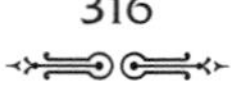

secured a buffalo that, when the meat reached market, fetched ZK 1,000. Consequently, Kabi ended the year with a small profit (ZK 145), but with the expenditure of considerable time and social capital.[23]

Making a Buck: Turning Bushmeat into Capital

These stories suggest some ways that insiders and outsiders exploit kinship and ethnic affiliations to gain access to wildlife. Even within the framework of licenses and permits, there are many ways to cover one's tracks, and the various transactions (reciprocities, barter, gift exchanges, theft) between parties are rarely clear. Yet these transactions in themselves provide opportunities for developing trust, cutting costs, and meeting locals' and others' needs. Calculations are a consummate part of these deals, and one must be wary of con artists, who manipulate personal and legal boundaries and appear at all levels. The real dilemmas often come from those outside of these informal arrangements, such as foreigners and government officials, who are empowered to enforce the more exclusive, costly legal system that sustains them.

The organization of this meat trade depends extensively on personal ties of trust and confidence. Relationships and common cultural background bestow a level of reliability and predictability that is lacking in the official system of permits and licenses. Ties among family and kin serve both as leveling and safety functions, as depicted in Kabi's assistance to his jailed uncle. For most participants, much of the capital raised from meat sales is transitory, for rather than generating additional capital, it meets immediate needs and may cover only some "opportunity costs." Moreover, for residents, the plateau market grossly undervalues these products such as meat, skins, and ivory from the valley in comparison to their prices further along in the chain of sales. This undervaluing results from the perishable nature and illegality of game products as well as the personalized nature of the relationships between purchasers and sellers. Personal networks are not necessarily exploitative, for in hard times, they may become a security and survival net. Were the valley residents not connected to those on the plateau, many would have suffered from lack of food and other important products.

Schoolteachers are not the only participants in these deals, as employment itself is a special type of social capital presenting individuals with the possibilities of rendering service, creating obligations, and receiving payment in cash and kind. Many government officials and administrative postings earn additional income through such "rent-seeking" opportunities. Bribes to facilitate public functions and extralegal activities supplement salaries and are prerequisites for tapping into opportunities. Officials may withhold licenses from the allocated quota for a friend or choose to overlook certain infractions for unspecified later benefits. Such payoffs, in addition to ordinary ruses and deceptions, such as the sale of the fake rhino horn and the desertion of disloyal agents, are part of the high costs experienced by participants in the rural and secondary economies. Within Zambia, the extralegal and informal system of wild meat production extends over vast distances and includes several ranks of marketing intermediaries, from its production in rural areas to its consumption within the cities and towns (Barnett 2000; Brown and Marks 2007). Many participants in these activities claim that they would pay for licensing and legal transactions if the legal processes were equitable and fair.

The state seeks to regulate and restrict access to wildlife primarily through its licensing functions and enforcement. Such measures seem reasonable to those who can afford to work within the restrictive legal system, yet rising costs and a deteriorating economy force many more to take extralegal routes. High costs and repressive enforcement do not stop the taking of wildlife; rather, they encourage those in the secondary economy to find increasingly subtle and ingenious ways to meet demand and to protect their own access. Reverting to widespread snaring within Zambia's GMAs, reducing meat distribution to close and trusted kin, preserving kinship connections with village scouts for access to information, maintaining secrecy, and carefully scripting dialogs with outsiders are just some of the many ways practitioners of these informal economies sustain themselves. All of these ways signal significant transformations and breaches in previously accepted local boundaries and relationships.

Powerful political and economic stakes inside and out of the Luangwa Valley are part of the stiff resistance to current attempts by the state and donors to squelch or transform the illegally organized wildlife "poaching" systems within the Luangwa Valley into something more beneficial for government registered and privileged groups. By the late 1970s, the local practices of lineage husbandry within the central Luangwa Valley were already losing their grips and integrity as necessary resource arrangements. Given the declining and poor health of its chief architect [Kabuswe Mbuluma] and the inflating state of the national political-economy with its effects on rural societies generally, there was a widespread regional demand for reinvestments, if not structural and developmental improvements in the rural natural resource and wildlife commons. As had the colonial government before it, Zambian authorities undoubtedly enlarged and enhanced the openings to these isolated rural "commons" and its wildlife by claiming these natural resources as "state properties" and placing them beyond the "legal" reach, utility, and creativity of those who had earlier husbanded them. Had either government embraced local and regional creative engagements and backed smaller scale initiatives in natural resources, the resulting dynamics between rural people, environmental assets, and the integrity of the land might have turned out quite differently. This local emphasis was not the path of development taken and it might even not been possible to implement then. But the words were in programmatic documents and were used as promotional phrases in the northern hemisphere to garner the political and economic support for its science and tourism.

A precaution is needed. Any governmental intervention, whether abroad or within its own territory, is always uncertain, readily transformed by unforeseen forces and events, whose outcomes, in both short and long term, bear differential costs and benefits for those involved or impinged upon. Even locally crafted regimes have their own dilemmas, pitfalls, and durations as demonstrated by references to adopted means (weapons) and cultural objectives (procurement of protein and protection) in chapters 8 and 9, which follow. Yet crafting limited-resource regimes is undoubtedly a more durable and doable sustainable option than the present quagmire, and at a fraction of the escalating costs in alienation, failure, and poverty. Perhaps beneficiaries would be able to engage in meaningful conversations about ways to structure, enhance or reconfigure a wildlife regime that they understood, and were willing to maintain in culturally imaginative ways.

Notes

1. *Nakudya nga Lesa, nkalamo isha mukule* (recorded 1967).
2. A key dichotomy in Borgmann's philosophy of technology is his use of the contrasting domains of the "thing" and the "devise." According to this theory, technology as a "devise" operates in opposition and negatively to the humane and centering attributes of time-honored "things." See various chapters in Eric Higgs, Andrew Light, and David Strong (2000).
3. The prescriptions recorded for placating other smaller species at Nabwalya varied in behaviors and ingredients, indicating the narrower horizon of lineage affiliation within which most of the meat was distributed.
4. One hunter and his wife consumed zebra meat, yet, as parents, they refused to provide it for their children, whose skins had broken out in rashes from touching its flesh.
5. This porcupine's flesh was *citonga,* a cognate of the Bisa noun for "maize" (*vitonga*). From a conversation recorded in my notes of 9 June 1991.
6. Hippos were scarce around Nabwalya during the 1990s and 2000s because of the widespread drying of lagoons and increased human predation on them during droughts. Hunters exchanged hippo flesh with plateau residents for grains. Local residents referred to a broad, deep bend in the upper Munyamadzi River as "hippo mine," a place from which many hippos were taken during the 1980s. Many of these hippos were killed by outside hunters and ferried to the plateau by carriers. Hippos were also vulnerable to diseases and shrinking habitat.
7. The force behind this prohibition centered on a widespread belief that eland possess a powerful spirit (, see comments by hunters about this species in chapter 5). One hunter elaborated on this spirit's strength as follows: "[It] makes the children and probably the wife die if a hunter is not careful. That's why we must use medicine to make this bad spirit sleep. When the eland is about to die and the hunter is within its sight, it barks as the bad spirit goes to the

hunter or comes in his sleep." I recorded this statement in 1989 after an individual refused to take an eland when it was apparent that he could have done so. Another hunter told me that he killed an eland in the thickets of Kawele in 1986, but he did not approach the carcass to butcher it. When I pressed for details, he elaborated upon his circumstances at that time. His son was sick and his maternal uncle (from whom he would have received the ritual prescription) had just received word that his son in town was fatally ill. Furthermore, his own uncertain standing as a newly returned migrant made him vulnerable, as he felt some within his own lineage might turn against him and inform the game guards or accuse him of some of their troubles ("afflictions"). The day following his kill, other villagers saw the vultures descending on the carcass and butchered it without questioning who "owned it," as they assumed it had died there. The hunter decided not to tell them it was his.

8. Conversation recorded in my Field Notes on 1 April 1989, the day after the event. The hunter noted that when he shot this warthog, the buffalo continued grazing within sight, nor did it move away while he and his friend butchered the warthog. He said that the buffalo even followed them as they returned to their village. Had it been a moonlit night, he claimed that he would have stayed and stalked the buffalo as well. For a chi-square evaluation of one hunter's dreams and his success or failure to secure game the next day, see Marks 1977a: 12.
9. The local name of the banded mongoose is associated with good fortune as in the following adage: *munzulu azulula nzila bwino, apelike izjuko,* which translates as "the banded mongoose makes the path straight and provides fortune." The slender mongoose name of *lukote* is a cognate of the verb *kukota,* meaning "to age quickly." Another embedded meaning is that the banded mongoose is gregarious while the other occurs on its own. Other omens and their associations, along with a table comparing their sightings on hunts and outcomes, are found in Marks 1976: 102–4.
10. Wounding game only to have it escape did not have the same meaning for local hunters as it did for some western "sports hunters" who felt remorse when an animal got away. Unless there were a string of such failures, it was not in a hunter's interest to disclose what had happened during a hunt, except perhaps to his patron in accounting for his expended shot and powder. Local hunters typically explained their failures in terms of spiritual or social retribution. I suspect that a few more buffalo were wounded and escaped in 1988–89 than I was able to record.

11. A stalk usually brings an abrupt halt to a hunter's forward searching movements, followed by an indeterminate time during which the hunter studies site characteristics and the prey's behavior. He may also test wind direction before moving tangentially, directly, or indirectly toward the target. A hunter's stalking movements are calculated and cautious, taking advantage of whatever cover is available. During his searches for game, a hunter remains aware of the direction of the wind and generally freezes upon observing game to assess its awareness of his presence.
12. Casho was an important and active hunter who trained many youngsters during the 1970s. The interpretation here is that his spirit was looking for a descendant to follow in his footsteps.
13. Conversation recorded in my Field Notes for 1989.
14. On 30 May 1989, another hunter spent thirteen hours in the bush, separately and unsuccessfully due to unstable wind, stalking a buffalo group and wildebeest group. He also stalked a group of zebras and a warthog, firing his weapon on each occasion and missing both times. Upon his return to his village, he learned that a cousin had died that day. He attributed this unsuccessful foray to the "spirits" attempting to communicate with him, the substance of which he realized only upon his return to the village.
15. For each of these figures, the unit of measure is a single observation, action, or event. One encounter represents a hunter meeting one group of a particular species, which may contain one or more individuals (a herd for example) during a specified time interval. The unit for stalks and kills or retrieves is also the number (of stalks or kills) for each species.
16. In November 1988, I was invited to attend the *cizungu* (puberty) ceremony for Hapi's youngest daughter. This occasion put considerable stress on his household's finances, especially when it came to feeding all the guests. His wife showed concern about my health and inquired if something was wrong when I did not finish the severely charred meat offered me during a meal. She sent her oldest daughter on bicycle to an adjacent village to purchase meat from a freshly snared buffalo, which she had learned was there. She introduced me the following day to the trapper, who remarked, "So this is the European [*muzungu*] whom you were concerned about feeding properly yesterday!"
17. From my Field Notes of events and interviews recorded during my visit to Nabwalya in 2011. In this interview, I asked this individual if he remembered any conflicts over prior distributions among kin. These paragraphs are his response.

18. I use Janet MacGaffey's definition of "secondary economy" as "economic activities that are unmeasured, unrecorded and, in varying degrees, illegal." As she does, I use "secondary economy" to separate criminal illegal activities (e.g. drug trafficking) from other activities that are legal in themselves (e.g. hunting), "but illegal in that they evade taxes, licenses, or in some other way deprive the state of revenue" (MacGaffey 1991: 12–14).
19. The DWNP issued game licenses and game-management permits on the basis of a quota system skewed in favor of expensive licenses (national and international prices) and against less costly local licenses. The outside demand for buffalo licenses greatly exceeded even the local need. Mpika officials allegedly used this difference between legal supply and demand to require additional favors from purchasers of a buffalo license.
20. In 1988, the Zambian currency (Kwacha) was worth ZK 6 to the US dollar. In subsequent months, it declined rapidly in value, and the following year the Bank of Zambia issued new bank notes to counter the widespread distribution of counterfeit notes in circulation.
21. Bwalya claimed to represent four plateau people as clients and possess their licenses for firearms and supplementary. Although residents doubted Bwaya's possession of these licenses, to my knowledge neither they nor the game guards challenged him. When I asked about their ideas, locals answered in a diffident tone that "one day the animals would get him."
22. The operative rate for a local person to secure the services of a local hunter varied between ZK 100 and ZK 150, depending on the amount of meat retained by the hunter and anticipated difficulties of the expedition. Bwalya contracted to kill these four buffalo for ZK 370.
23. In 1988, after paying licenses, permits, and carrier costs, residents could make a profit of between US $428 and $440 from a hippo carcass (two instances) and US $48 and $123 from a buffalo carcass (five instances). By 1998, license and carrier costs had reduced these financial incentives, yet apparently increased the meat traffic in the shadows (without licenses) especially by the commerical gangs, which operated from villages on the plateau.

Chapter 8

Muzzle-loaders and Snares

Weapons within Their Cultural Contexts

My gun is my future, it produces every time it speaks.[1]
Valley Bisa hunting boast

Recipient societies embellish and invest new objects with different cultural meanings from those held by donor and other groups. In addition, they often incorporate novel items in specific cultural ways to make them meaningful within their lives. In the current age of globalization, rapidly changing technologies, commodity flows, media dialogues, and migrating peoples together with newly imagined lives, individuals and groups alter continuously the face of a similar world by developing the shape and scope of their cultural differences and identities beneath a surface sameness. One grasps the character of these objects through their provincial uses and significance.

Like the bushmeat they produce, muzzle-loading guns rapidly became "focal things" in Valley Bisa users' hands. By adopting and loading them with rich lodes of their own cultural ideas and practices, local hunters domesticated these weapons and made them their own in ways that had little to do with technical mechanics. Upon this device, some men as hunters embedded their cultural worlds of social relations and highlighted their strongholds of masculine privilege, authority, and vulnerability. In doing so, they cloaked their self-serving interests within a different garb, the metonym of lineage welfare. The muzzle-loading gun became an attractive means of investment through which a man established social bonds within his community, offered protection and meat to kin and clients, and mediated between his lineage, their ancestors and the ecosystem. The adoption of this technology overshadowed and temporarily displaced other customary weapons within the local armory.

Later, when the Zambian state targeted the regional bastions of unregistered firearms, valley residents returned to their customary stock of tools, infusing them with innovations and obscure covers to feed and to protect themselves. As a northern interpretation, this return to indigenous technologies, particularly snaring, carries a high price in cruelty to wildlife, yet its deployment has proved difficult for a distant state to control in order to 'conserve wildlife resources' in its own interests. Situated on a tsetse fly landscape which denies the survival of domesticated livestock and surrounded by dense wildlife populations, local residents find state wildlife policies inept and oppressive, given these circumstances and their recent pasts, as well as inimical to their own survival and needs. Officials respond slowly to residents' dangers and complaints and are often corrupt. This chapter describes cultural weapon trappings and local responses to state activities that threatened and changed a once provincial, masculine mainstay of power, privilege, and provision.

Incorporating Muzzle-loading Guns into Bisa Society

The relationships between people concerning property are culturally structured and malleable. In many Central African societies, muzzle-loading guns are not just instruments to kill or to protect property but they have also come to embody a host of indigenous concepts and practices. These reasons were significant in the rapid acceptance of these weapons by Africans prior to colonial rule, as the slave and ivory wars substantially threatened most men and their roles.[2]

This section depicts some of the local cultural embellishments and meanings. First, I establish that these guns belong exclusively within the masculine domain. Women are rarely allowed to touch these weapons, and if they do, the weapon loses its power and productivity. Elder men possess a rich lore of rites and treatments that prescribe causation and perform ameliorative acts when these weapons fail to function. In the past, elders' special knowledge made younger men subordinates, while the latter's productions from the bush trumped those of women from agriculture if only in cultural, but not material, significance. Second, I reveal how muzzle-loading guns played an important role in Bisa iconography of ancestral reverence, call, care, and compassion. Gun 'owners' attributed a hunter's productivity to ancestral oversight and linked this assessment to descendants' behavior. Finally, I weave these cultural threads into the cloth of lineage identity in the social history of a single muzzle-loader. This history intertwines cultural significance with the social calculations and experiences of three generations of users.

A Masculine Domain

Guns are not tools sanctioned for use by Valley Bisa women. In 1989, when I asked a group of men if women ever hunted or used guns, their reactions indicated that my question hardly warranted a reply.[3] During the military sweeps to find and confiscate unregistered weapons during the 1980s, men told bitter tales of wives forced to deliver their guns or face detainment or rape by soldiers. When men mentioned women handling guns, their associations were about ineptness and about spouses' failure to keep household secrets. Numerous prohibitions and taboos, such as abstention of sex prior to hunting forays and the mystical properties of menstrual blood, substantiate the formal polarized gender roles while also sanctioning a prominent place for guns within the overall household economy. Whereas some widows might seek to claim their late brother's gun for the benefit of their children, or to sell in order to get her son released from prison [Ngulube's mother's petition to the chief in chapter 4 for the possession of her late brother's gun to sell] men would interpret such assertions as unwarranted intrusions onto their domain.

FIGURE 8.1. Resident tax collector Lutwika Kunda (sitting), assisted by resident game guard registering the muzzle-loading guns of residents within the Munyamadzi Controlled (hunting) Area, Nabwalya. (Photograph by author in 1966)

With a combination of respect and awe, younger men recall instances of their maternal uncles or fathers using these guns. Such stories are about their introductory incursions into the bush, about learning appropriate gender behavior, and about coping with their initial insights into becoming men. Some adults consciously merge the material and symbolic value of their guns within their social roles. An attentive son of Kameko Chisenga recollects this interlacing of practices with spiritual meanings in these reminiscences about his father.

> My father was a hunter who taught us hunting skills and wanted his sons to learn these customs. He possessed two muzzle-loading guns, and with these he taught us how to trail, stalk, aim, and fire at animals. He also taught us to snare and hunt with dogs. He took me hunting whenever I was free from schoolwork. When we set out hunting, my father led the way. I was behind carrying his second weapon.
>
> My father boasted that his hunting tactics were better than those tutored by others, for his skills were spiritual and acquired from strong ancestors. Sometimes when he was sick, it was because he had not taken the time to go hunting and that his ancestors wanted

him to continue these practices. Then he would take up his gun, go into the bush, and come back with an animal.

When hunting became difficult [when he missed or couldn't get close to animals], he washed his guns with certain roots. He placed these guns on the household rubbish heap [after a ritual wash], then crossed them twice. At other times while going into the bush, my father would lay his muzzle-loader across the footpath and then ask me to cross it with the other gun. Then I walked across the crossed guns followed by my father. He called this activity *kupindwila* [to cause to loosen or open the spirits]. He explained that in the early morning the path was still used by the ancestral spirits from the previous night. He did this practice in the sense of "causing [the spirits] to bring the animals"—by asking the ancestral spirits to bring the animals in our direction. When he did this, we more frequently encountered animals.

Another activity for improving our "chances" was to pass the muzzle-loaders through our thighs and between our legs. When we had walked a long way and not encountered any game, he would pluck the leaves of *kamulebe* [a thorny, fruited shrub] and toss them in the air. Then we gathered only those leaves that fell with their underside facing up. Some of these leaves he placed in his pocket, others we chewed and spat on the muzzle-loader. Smearing the gun was to make it smell like the bush (*kununka mpanga*).[4] These were the main ways in which we got results.

My father used his guns to kill birds [doves, guineafowl] in our field, and to protect our fields from hippos and elephants, as well as our homestead from hyenas. Yet his guns were not just for killing, but also for social occasions. During Christmas and Easter times, he loaded his guns and fired them into the air as a sign of joy. Also at funerals for close relatives, he fired his guns to show respect for the dead. When the coffin was lowered into a grave, he fired in the air as a sign of posting [sending off] the soul of the dead to the other side of the world. My father's guns performed many functions for our lineage.

As the means by which a few men secured wild meat to complement women's agricultural production, muzzle-loaders outmoded spears, bows, and arrows to become emblematic of their mastery within the matrilineage. As a capital investment purchased by migrant laborers, these firearms served as 'placeholders' for those absent from their community. In this capacity, the gun affirmed a presence during a man's physical absence, acknowledging his interest in lineage welfare while retaining his place pending his return. These muzzle-loaders, like the bush meat they produced, were not private, but instead were social possessions valued as lineage property (*mfuti wapacifulo*). Such guns were not for sale at any price.

Firearm Categories, Cultural Creativity, and Rites

Hunters sort their muzzle-loaders into four broad categories on the basis of their length of barrel and size of bore. *Kalobola* features a long gun barrel and a small bore, while *Kabende* has a long barrel and a large bore. *Malike* and *Nsumbwe* are short-barreled weapons, with small and wide bores, respectively. These qualities of length and bore are those which imaginative gun owners or borrowers might use to personalize themselves in raw masculine terms or to exaggerate their superior qualities or accomplishments in challenges with their would be drinking peers during day and night-long beer drinking stupors. Yet, there are many other resourceful ways in which muzzle-loaders feature in the lives and practices of their owners and borrowers. We examine some of these in this section.

Cultural creativity manifests itself in the given names of some guns. One hunter referred to his two muzzle-loaders as *Kete Bashele* (literally 'call those behind)[5] and *Sunga Muchele,* (literally, 'save/conserve your salt'). These names are a subtle (synecdoche) and artistic use of local metaphors and cultural connections. *Kete Bashele,* used for the larger, 'dangerous,' and less frequently encountered mammals, had a larger bore, and inherently instructed its prey to bring their cohorts forward, according to its name. *Sunga Muchele* was used to take the smaller game, and its name, an instruction to have salt handy, anticipated the immediate taking of meat. Names recorded for other Bisa guns include *Cimanya musasa,*[6] *Nakazabwe umwanakazi wacinwa,*[7] and *Kabula.*[8] Another muzzle-loader's name, *Mayo mpapa,* is a shortened reference to the Bisa proverb *mayo mpapa, naine nka kupapa* (literally, 'mother, keep your infant safe [on your back] for s/he will look after you later'). A baby on her mother's back is safe, nurtured, and returns the kindness. A hunter attentive to his weapon heeds his future.

Gun names may also embody stories of its owner's experiences, and even critique state policies and history. Brad Strickland relates how the name of one gun, *Munda waWana* ('field of children') captures ecological history and emphasizes his persistent local identity. As the former owner of this gun, the Kunda grandfather who lived adjacent to South Luangwa National Park, was an important village headman who tirelessly 'poached' and resisted state programs of wildlife conservation. These stories, recalled in just the metonymical mention of the gun's name, relates to this man's pivotal use of his weapon to protect his lineage, its fields and provide meat [the customary ways to sustain village life] for relatives and dependents ["children"] around him, "a community" that he sustained during his lifetime. The telling of this story also criticized the governmental health, education, and agriculture policies that were destroying local autonomy and wellbeing. These themes became implicit in deciphering the story revealed through the mentioning of the gun's name (Strickland 2001).

Some guns have status and possess authority by virtue of their possessor. A Ng'ona lineage member replied in 2006 to a question about the names of his two muzzle-loading guns, one of which he had purchased while a migrant worker, and the other was a recent inheritance.

> My guns have no names because they are for the royal family [*shakwana vyonze shaciaalo*]. Names are not necessary, the name is just "*navyonze*"—everything belongs to us [us as 'one lineage'] just as all the spirits in this land belong to our lineage. We bless people in the land to have successful hunting.

Cultural imaginations also flourish in choosing ingredients for ritual washes to restore the potency of muzzle-loaders. These guns may fail to fire, misfire, or even explode for a number of known, practical reasons. Yet when these problems occur, hunters and others search for answers within the broader contexts and currents of their relationships, seeking answers for the 'why' questions. The presumption is that ancestors monitoring lineage welfare cause these unexpected ruptures to happen. Residents say that petty jealousies and conniving kin can cause monitoring ancestors to withdraw their benign oversight. Ancestral withdrawal and displeasure become visible in hunters' failing to deliver game and even in their misadventures. By interpreting this causal ideology, elders channel the emotions of the young and the inexperienced via these traumatic and transformative events (figure 8.2).

The public acts within villages by which an individual acquired, used, or inherited lineage firearms rarely, if ever, occur now. Such customs included public blessings by elders upon the transference of a lineage gun to a younger man, ceremonies in the bush and in the village celebrating a novice's initial kills, and open displays and acknowledgments of hunting strategies and engagements (Stefaniszyn 1951, 1964; Marks 1976: 86–146). Once a prime topic among masculine peers, one rarely hears hunting or gun songs and stories today. Villagers fear that wildlife agents might use these commentaries to identify suspected 'poachers.' The telling and retelling of stories about wildlife incidents or about the activities of wildlife agents caution everyone to watch what they say publically. A case recorded in September 2001 includes a wife's ineptness with a gun.

> Scouts arrested Kunda Beta at their camp as he returned to his village from employment. Several weeks earlier, his drunk uncle joked with a village scout that he had given his unlicensed muzzle-loader to his nephew Kunda. The wildlife scout played along as the uncle tantalized him, "You game guards, no one can arrest me." When this uncle went to his second wife who lived across the Luangwa River, the scout did not forget what he had heard.
>
> When arrested, Kunda refused to cooperate with the scouts even when told what his uncle had said. When Kunda's mother heard that scouts had detained her son, she removed the gun from its burial in a trench and took it to his father's village for safe-keeping should the scouts decide to search for it. As rumors grew about a search, the gun reappeared unexpectedly at Kunda's house, where his mother instructed the wife to hide it somewhere outside the home. That night, the wife carelessly hid the weapon under a pile of sugarcane. When the scouts appeared with Kunda, they went to the spot where he had buried the gun. As the gun was not there, Kunda asked his mother to give it to him for the scouts had promised to release him once they possessed the gun. While they were talking, a scout discovered the gun in the pile of canes and the ploy turned in their favor. "Friend," the scouts told Kunda, "we were just joking with you before, the real case begins now."
>
> Kunda was handcuffed, tortured again to secure a written statement, and later taken to the magistrate at Mpika for sentencing to hard labor or the payment of ZK 150,000 (about US $32). He served several weeks in prison while his mother visited her relatives to raise money for his fine.[9]

FIGURE 8.2. Kameko Chisenga washes and warms his muzzle-loading gun on the rubbish heap adjacent to his homestead. Note the prescription of herbs soaking in the broken pot (water spilt ground), which he has poured into the gun's barrel to cleanse it. Here he is checking the barrel with the ramrod before placing it atop the pot and herbs to warm in the afternoon sun. This washing occurred soon after the death of his son and his subsequent failure to kill game. (Photograph by author 1993)

Indicators of Unknown "Spirits"

Hunters believe that animal spirits also affect their muzzle-loading guns. Each identified effect has its particular prescriptions and washes; the ingredients vary by lineage and with individuals. A hunter explains the cultural constitution and effects of these spirits.

> There are many ways of washing away bad spirits of animals. We have two types of medicine—*izambo lya nama* [the washing of animals] and *izambo lya vibanda fya nama* [the washing for "bad spirits of animals"]. When you are asleep, the spirits of the animals, which you killed long time ago, will rise up and say, "It's a long time since you began killing us, but these days you will not kill us anymore." If you are not killing animals, you must wash your gun to be able to kill them again. Collect herbs for animals seen by you. Dig the roots of *lungwizi* and take the bark of *lupampa,* place both of these in a pot of water. Early in the morning, take this pot of medicine and place it where you throw away rubbish from your household. Wash your gun with this mixture and pour some down the barrel. This washes away the animal spirits. If these animal spirits wish to trouble you again, *lungwizi* will impale them on its thorns.[10]
>
> If you dream of animals at night, you know that "a bad animal spirit" is making you miss the animals. If you dream of a human being, you know it is from "the bad spirit of a person." To quell these dreams, you dig the roots of *muzolo, citimbe,* and *mwanya.*[11] Put all three in a pot with water and place on the rubbish heap. In the middle of the night, you bring your gun to this site, wash it with the liquid, pour some down the barrel, and take the rest of the liquid and discard it in an anthill.[12]

The most puzzling and intractable afflictions are those believed caused by human agency, a person or an ancestor. Finding abstruse causes usually involves consultations with a patron, dream diviner, or a cult specialist where detailed testimonials to the pervasive powers of observant ancestral spirits emerge in detailed accounts of plausible causes of misfortunes. The following story of problems with a particular muzzle-loading gun illustrates another way in which two hunters sought answers to why these spirits might be affecting one's hunting efforts. This case shows individuals seeking the 'validity' of their assumptions.

> The teller was the younger, more accomplished hunter as well as the narrator. He and his elder Mbokozi, a resident of another village, had an argument about whether the person or the gun, which Mbokozi claimed and was using, was the agent responsible for his [Mbokozi's] lack of success in hunting. On an earlier hunting foray, the two had traded their guns during that hunt and both had made kills. That foray determined that Mbokozi still had power to kill game. On this day, they wanted to see if Mbokozi and the gun he typically used was still "locked up." So on this particular day, Mbokozi assumed the leadership role with his suspect

gun. He attempted five stalks and fired his weapon on a zebra and a buffalo, both of which got away.

My narrator begins his analysis of these two tests with the statement; "The problem with Mbokozi is that he is always unfortunate." The explanation for Mbokozi's extended failure remained centered on his relationship to this particular muzzle-loader that he had "forcibly taken" through intervening in the late owner's expressed wishes. Before his death, Mr. Lilaili, the gun's owner, specifically said that his gun should not pass into Mbokozi's hands and that he intended it for his younger nephew. Yet Mbokosi laid claim to the weapon as a classificatory "brother" who planned to use the weapon until its intended owner reached maturity. Then he would register the gun in the nephew's name. He concludes with: "Today's results show that Mr. Lilali's spirit is still working to tie up the gun [*kukaka obuta*[13]], for Mbokozi has fired so many shots unsuccessfully. *Izambo* [ritual washing of the gun] is not good for human spirits such as this, and Mbokozi must go to the *ng'anga* [diviner] for a resolution. Instead, Mbokozi is hoping to change the blue book [legal registration] and the numbers on the weapons with those of his elder brother. He knows that he has forcibly taken the gun from his young nephew. If the gun remains without killing game while in his possession, then the gun will be kept by someone else until the young nephew matures and can use it."[14]

Hunters may expend considerable time and assets determining the specific nature of their gun's 'affliction.' If associations between weapons and spirits sound bizarre, an interview with an elderly, renowned blacksmith in 2006 substantiates the endurance, prevalence, and nature of conflicts between persons sustained by these lineage weapons.

When there is a lineage gun, all the members expect to benefit whenever it makes a kill. If the hunter is using clever ways of not sharing the meat, the relatives work some magic and these guns break and can become quite dangerous. When that happens, the hunter should sit down with his lineage, for they know already that something is wrong. They may find the reason by appeasing the ancestors through using the gun on a ritual hunt. They will say, if the problem is coming from male relatives, then enable us to kill a male animal, and if from the lineage, send us female animals. The reason becomes visible when the hunter goes out on this commissioned hunt. Until this happens, the affliction can bring death to the hunter, if he fires the gun without this "coming together ritual" attended by all members.

Sometimes a person consults a witch finder to cleanse the gun. Sometimes guns have problems if sold from one lineage to

> another without the relatives coming together and agreeing on its sale. Those who are not happy to have the gun passed to the other lineage will use cursing words that bring misfortune to the gun's user. No peace exists until the whole issue is resolved.[15]

Sons and nephews may bitterly contest the inheritance and disposition of lineage muzzle-loaders, along with other property belonging to their late "fathers." If customary inheritors lose possession of important property, either through disputes or through a court decision, they may resort to sorcery to enhance their claims. The chief and Nabwalya Local Court settled all twenty-five cases of gun inheritance brought to them between 1988 and 1992 'in accordance with customary laws of the area.' Matrilineal succession determined the inheritance of these weapons; all went to brothers or to nephews.[16]

In the early twenty-first century, muzzle-loaders belonging to a lineage had become a financial burden and liability for some keepers. For the gun to remain a legal possession, ownership licenses [called a blue book] needed to be renewed every three years, which was only the beginning of the cash outlays required to legally keep the weapon. To kill animals legally, the gun owner had to purchase additional licenses for specific animals and permits for where they were to be hunted. Each encounter with state officials to obtain these licenses took time, energy, funds, and sometimes presented additional problems. Some residents found these routines tedious and invested their skill in circumventing them instead, if they could. In 2006, a seventy-four-year-old man informed me of his dilemma as the possessor of two muzzle-loading guns.

> I purchased a Kabende type of gun for ZK 15,000 when I worked for safaris. I use it for scaring animals in the fields. I inherited the other from my late young brother. I have not purchased any animal licenses because when we go to Mpika to buy them, we find that all the licenses have finished as other people from outside claim these licenses. Therefore we always renew our gun licenses for nothing because we always think we shall buy animal licenses from the quota. And then we have problems with our blue books. When I last reported to the Mpika police for my gun renewal, I was advised to come back later since the licensing officer had no firearms licenses. [He shows a note dated 21 November 2005 signed by the officer.] This is a case the police have done to me because I have

> paid and they cannot clear me. When I returned home, I couldn't buy any animal licenses [since he lacked a stamped gun ownership certificate]. The officer told me that "we don't have money to buy animals, instead others from the plateau buy animals through residents here."

Whereas lineage guns require the permission of all members to sell, personal guns are not so restricted. Despite the burdens felt by some, there are always other men willing to pay for the prestige that comes with gun ownership. Some employed parvenus seek guns as important symbols of their new status. Local muzzle-loaders become available for sale as their elderly users die, or they are offered during famines when the cash is needed to purchase staples for the lineage. The chief often serves as the broker in these transactions because the weapon is assumed safe from seizure while in his custody and its transfer of ownership becomes sanctioned through his signature. Furthermore district authorities respond to his office. For these services, the new owner pays transaction costs.[17]

Social History of a Muzzle-loading Gun

The social history of this gun, a *Malike* type with a short, small barrel, interweaves its cultural and material strands. Acquired by a young man through a government auction in 1948, this weapon performed for three generations of lineage hunters until its barrel exploded in 1992. Its blue book allowed its owner to purchase a new barrel, and by reusing the trigger and spring mechanism of the exploded weapon, to commission a local gunsmith to reassemble these parts on a new stock. The new gun had morphed from a *Malike* into a *Katobola* type as the possession of the original blue book legally allowed its same registration.

This gun was one of many weapons confiscated by the Northern Rhodesian government in 1947–48 during widespread raids and prosecutions for breaking game, arms licensing, and ivory laws during and after World War II. The region targeted in these investigations was Lundazi District in Eastern Province, across the Luangwa River from the Valley Bisa homeland. As a region bordering Nyasaland, colonial officials considered this district's

residents heavily involved in ivory smuggling and other game violations.[18] Undercover operations resulted in the confiscation of many weapons, which the administration sold later at public auction. At a 1948 auction in Lundazi, Adam Mwaba purchased a small, stocky muzzle-loader for thirty shillings (probably then worth about $5 US).

Adam was then a young man in his early thirties, recently returned to Nabwalya after six years as a migrant laborer in South Africa. He arrived to marry, and elders suggested a cross-cousin living in Chief Chitungulu's area as a suitable wife.[19] While living there in bride service among his in-laws, Adam learned of the auction. He went to Lundazi where he made a deal for a gun with the white auctioneer and his black clerk. He paid with the wages acquired from his labors in South Africa.[20]

Upon completion of his bride service time, Adam returned with his wife and prized gun to his mother's village near the Nabwalya palace. Soon after his return, he became headman, acquiring the name of the village's initial founder. As a youth, Adam had pursued game with bows and arrows, but upon his assumption of the headman role, he had little inclination to hunt.[21] Instead, he told me that the lineage's 'hunting spirit' was passed on to a 'younger brother' (actually a cousin). After several years in bride service, this cousin, Hapi, came to reside in Adam's village with his wife.[22] Hapi spent little time as a migrant laborer—hunting was his passion, and he preferred to spend his time in the bush. As he had no capital to purchase his own gun, he borrowed that of his elder kinsmen. Hapi killed his first buffalo with this gun about 1950, and became its primary user.[23] Hapi also gave the gun its name of *Kabutu*—the term for a premenstrual girl. When I asked him why he had given the gun this name, he replied, "Because young girls know and follow the rules." The weapon remained registered in the headman's name.

In February 1953, Hapi's father and Adam encouraged him to kill a large animal, as all their dependents craved meat while weeding their fields. Hapi's quests after buffalo proved unsuccessful and, after a month of futility, he surmised that someone had 'closed' the weapon. He mentioned this suspicion to Adam, who gave him a prescription to wash the gun. Thus Hapi began a series of ritual washes following the grim circumstances that prompted each of them. Each wash restored the weapon's productivity and affirmed its efficacy.[24]

During the harvest in 1961, Adam Mwaba kept *Kabutu* loaded with a large quantity of gunpowder and two elongated bearings (bullets), expecting to use it against hippos or elephants in his field. He did not fire the weapon immediately and soon forgot about its heavy charge of powder. Later that dry season, he used this gun to kill an impala along the banks of the Munyamadzi River that was distracted by its escape from a pack of wild dogs. As his gun discharged, one of the large bearings lodged lengthwise in the barrel, forcing the chamber to expand and crack near its breech. Hapi dislodged the bearing, hammered the barrel back into shape, and covered and reinforced the breech with a thick strip of buffalo hide.

Subsequently, Hapi consciously kept track of the powder loads in the gun. Still, the gun was often at the blacksmith's shop for mechanical repairs between 1973 and 1977. Hapi remained its main user until squabbles within the lineage caused him to resettle elsewhere with his father. During his long tenure with this weapon, Hapi trained three nephews and four cousins in the intricacies of stalking game, among other bush skills. One of the youths he trained during the 1960s was a nephew who had come with his new wife to reside in Adam Mwaba's village. This nephew, Kandeke, soon became the main hunter for Mwaba and his own father, who lived several villages away.

Pending a visit to relatives in Lusaka in 1983, Adam placed *Kabutu* in the safekeeping of a cross-cousin married to a niece in his village. While Adam was away, military and police officers made a surprise search of households and uncovered the gun. When the cousin could not produce the gun's registration book, the officers seized the weapon, beat him, and held him temporarily in detention. The officers took the gun along with other confiscated weapons for deposit in the police armory at Mpika. Upon his return, Adam went immediately to Mpika police where he presented the gun's blue registration book, told his story about being away, but received no action from the police. After two trips and confrontations in Mpika, Adam returned despondently home, yet he was confident that his gun was legitimately owned and that the police had taken it in error.

Goaded for his failure to retrieve the lineage's gun, Adam finally acquiesced to Kandeke's harangues and gave him the registration book. Kandeke knew a police officer at Mpika and presented him a chicken supporting his request for assistance. Together with the officer, they searched the armory until they found *Kabutu*; the officer replaced its missing hammer and ramrod with those of another confiscated gun. At the rural council offices, Kandeke paid another chicken for a clerk to relicense the firearm. At the cost of two chickens and 'town smarts,' Kandeke had restored the weapon to the lineage. After six months with the blacksmith, *Kabutu* became functional again with a new internal spring and trigger.

Despite Kandeke's intervention, Adam insisted that the weapon remain in his custody. As his hunting reputation grew, Kandeke continued pressing Adam to transfer the registration papers to him. With his own status slipping as age overtook him and with his former villagers joining other headmen, Mwaba finally acquiesced in 1990 to this request. In handing over ownership, he stipulated the gun was lineage property, never for sale.

In 1991, *Kabutu* malfunctioned as it had decades earlier. Kandeke related his misadventures with the gun to Hapi.

> I fired many shots in the past months but killed nothing. I only wounded animals without making a kill. I told my maternal uncle about these failures of the previous months. He laughed and said, "You have neglected *nkombo* [a ritual gourd, named for a guardian spirit]. Find *nkombo* and pray to our ancestors who were great hunters to help you. Why have you forgotten them?"
>
> So I prepared *nkombo* and said these words in front of its mouth: "Please old hunters of my lineage, I want meat to eat. You great hunters of my lineage, be with me. Tomorrow I want to prove that there were great hunters in my lineage. Let me kill an animal. Be in front of wherever I go in the bush." After saying these words, I spat saliva inside of *nkombo* and put it safely hanging in my house under the roof.[25]
>
> Early next morning, I set out for the bush with my friend. Before I left the house, I took *nkombo* and said these words to it, "Be in front of me, great hunters. I want to prove you today."
>
> We saw no animals until we reached Malanda Stream. When I crossed this stream, I saw a male warthog grazing along the bank of Kamusimbite Stream. I stalked it to within forty paces. I attempted to fire at it several times. For at least ten times when I

> cocked the hammer, it fell on the firing nipple with the gun refusing to fire. Then I looked carefully and saw that the gunpowder in the nipple had receded so the spark could not reach the powder to explode.
>
> I took my hunting bag, retrieved some gunpowder from a container, and put some fresh gunpowder on the nipple. Now the gun was ready. The grazing warthog had not gone away and proceeded to come very close. It was eight steps from me now. I fired and the gun sounded. The animal fell down dead without a movement.[26] We cut it up in pieces and went back home. I told my village headman about the animal and about the gun refusing to fire. He said that those were the signs that make you believe in the powers of old great hunters. From that time, I have made a new *nkombo* and I am killing animals without difficulties. SURE OLD HUNTERS' SPIRITS WORK [*sic*].

Early the morning of 9 April 1992, Kandeke was following a herd of buffalo retreating into the hinterland after drinking at the river. The herd headed for cover in a thicket, so Kandeke made a wide circle and placed himself at a streambed downwind where he waited to ambush the animals when they crossed there. As the lead buffalo approached within twenty paces, Kandeke cocked *Kabutu* and pulled the trigger. The gun exploded for the last time.

> I was shocked because I was deaf. When I looked at the gun, the barrel was broke and bent. I called to my nephew accompanying me to check if I was injured. I looked at my arms, there were no injuries. I took off my shirt there was no injury on my back. I sat down, but my head felt very heavy. I instructed my friend to check my head, but he found no injury. I had a small bruise on my eyebrow, but not a very big wound. I then took the gun and examined it. I nearly cried because I loved it very much. The bullet hit a tree and the buffalos went away. Now, I have no gun. This gun belonged to Mwaba for a long time, and Hapi used it. At last, I destroyed it. It has killed a multitude of animals in this valley. This was a good-bye kiss to *Kabutu*. I must now look for another gun.

Initially, Kandeke thought that someone's jealousy had afflicted the gun, causing it to explode. He pushed this possibility in conversations with an aging Adam Mwaba, who asked if the explosion had injured him. As Kandeke was not injured, Adam told him to forget about that explanation. Then he told of its earlier history, including the cracks in the barrel and of its temporary repairs under Hapi. The barrel fissures had ruptured finally from

the heavy charge of gunpowder placed in its chamber intended to kill the buffalo.

Kandeke returned to his field hut with the remains of the barrel and the firing mechanism. He saved the hammer, spring, and trigger to incorporate into the lock, stock, and barrel of another locally crafted muzzle-loader.[27] Kandeke purchased a long barrel from a cousin and took it with the remaining trigger assembly to a blacksmith. The new edition of *Kabutu* looked nothing like its stocky predecessor with its short barrel. The gun now belonged in the *Kalobola* category with a small bore and long barrel. Like the original, the new firearm bore no industrial imprints other than a registration number stamped on its stock by a state agent.

By October 1992, the new weapon was ready for use. With the new gun in hand, Kandeke pushed toward his goal of becoming a local Big Man. When Adam Mwaba died in 1996, Kandeke became headman. The village, which had dwindled to only four homesteads by the time Adam died, increased to thirty-four households by 2004, some of whom were in-laws.

During my visit in 2002, Kandeke informed me how circumstances had changed how he felt about the muzzle-loading gun now registered in his name. It was a lineage heirloom, but he was feeling pinched financially, particularly when he couldn't use it. And no one in the lineage seemed to want it or was able to afford keeping it legally.

> Many of those owning guns have sold them since these weapons have become now a liability. My gun has a fissure on its hammer and I lose when I can't use it to kill animals. I want to give it to a young nephew, but he refuses because the costs of ownership are too high when there is no employment. To renew a gun license costs ZK 15,000 [about $4 US] for a three-year period and a blue book costs an additional ZK 15,000. Yet, I cannot sell this gun because it belongs to the lineage. People will likely return to the older ways, as we have many ways for killing wild animals. Today, younger men seem more interested in fishing.

Kandeke continues to use his muzzle-loader to protect lineage fields, hiring himself out to others who have licenses and area permits, and he seeks local employment when he can. The world that he now struggles in was not of his own making, nor was it one he dreamed or honed his skills for. Instead, he finds that many in the younger generations, loosened of their cultural foundations through formal education, are better competitors in seeking material wealth and employment. They look down upon his bush knowledge, experiences, and past history with contempt and rejection.

This discontinuity with the past was noticeable in the cultural confusion, apathy, and growing poverty I experienced during my inquiries in 2006 and during later visits. Residents still found ways to take wildlife and to supplement their porridge with bush protein, yet very few depended on muzzle-loaders for these tasks. In their struggles with meanings and poverty, many had converted to Pentecostalism, in which they hoped to find answers for their unprecedented questions about identity, survival, and salvation. Among these was Kandeke.

Return to Earlier, More Silent Technologies

Given the current liabilities in possessing firearms, residents have resurrected earlier technologies to put meat into their relish dishes. Game pits reappeared around gardens in unsuspected places while wire snares were deployed covertly throughout the GMA. These older methods were more than just another means to procure meat; they were subversive subsistence acts reasserting local customary entitlements to wildlife. Through these methods, some residents affirmed their identities of place, which in turn energized their capacities under adverse circumstances. Scouts and state employees also utilized these practices when the government fell behind in delivering material support and salaries, as well as to enhance their local coffers.

The Callousness of Conventional, Centralized Wildlife Management

The meager incentives under the 'community-based' wildlife program have largely failed to culturally engage most GMA villagers into conserving wildlife. With the exception of the comparatively few on official payrolls, whose loyalties might be suspected, most residents silently have opted out of the state program and depend instead on their own ingenuity and shrewdness to get by (Hirschman 1970). During 2006, residents discussed snaring within the context of two issues: procuring food and protecting fields from wildlife. Both subjects have local and official versions; the former account concerns mostly small game and protection, the officials focus on the larger mammals, particularly elephant, buffalo, and hippo. Residents worry about their livelihoods—about access rights, welfare, and protection, while outsider interests converge on 'poaching,' 'problem animals,' and revenues. The government had previously assumed some responsibility for 'crop protection' as it employed elephant control guards to kill persistent marauders in the fields and even hunted down those animals that had killed people.[28] Under ZAWA, wildlife scouts seldom killed animals allegedly raiding fields. If scouts respond at all, they only temporarily scare the marauders away.

The sector-in-charge of the Munyamadzi GMA Wildlife Unit repeated these ideas in his summary of their efforts for the warden in 2002.

> During January to July, the sector was receiving disturbing reports over buffalos, hippos, bushpigs, monkeys, and elephants raiding people's fields. Effects (sic)were being made to control the problem animals however it was not all the people [who] appreciated our work. Most people once they report, they expect any animal, which will enter their field should be killed, which is not [our] case. The office of Sector-in-Charge worked according to what is expected to be done, not to work to please the residents (sic). Some people had their fields in grave yards and expected officers to be there guarding their fields at night whereby there were no houses. . . . Some reports were being followed and some were not due to some management problems [lack of fuel for vehicle, manpower on anti-poaching operations]. Most of the people attacked were themselves to blame. Most of the attacked by crocodiles were fishermen and some were doing careless bathing in the Munyamadzi River. The Sector-in-Charge was very careful handling such cases. Very few animals were killed on control.[29]

The report mentions that wild animals attacked fourteen people during 2002 and that six had died. The official writes that the unit killed eight animals, which included one elephant ("damaging people's houses, raiding fields"), two hippos ("troubling people when crossing rivers"), two buffalo ("previously wounded"), and three monkeys ("troubling people on new shooting plants"). Behind the official proactive role to 'protect' wildlife lies an institutionalized and insidious market approach for conservation based on generating revenue ('paying its own way') oriented toward foreign wildlife values and privileging tourist interests. These are not local values nor directed toward enhancing local involvements in conservation.

"Snares Don't Make Noise Like Guns Do"

Snaring has always been part of the local arsenal, but it became prominent again during the 1990s under the long prosecutorial shadows of ADMADE. The deployment of snares was a rational choice by those who recognized the liabilities of continuing to use muzzle-loading guns. Snares have several advantages over firearms. They are quiet whereas guns are loud and draw attention. When a man checks a line of snares, his intended activities are inconspicuous and camouflaged, fitting into his bush activities more so than if he were hunting with a visible gun. When men wander through animal habitats to cut poles, collect various forest materials, or fish, they typically arm themselves with only an axe or spear. These utilitarian tools are unlikely to arouse scouts' or others' suspicion. Moreover, snares are not large capital or emotional investments and do not require a protracted apprenticeship to deploy. Trappers rarely suffer debilitating injuries. The materials are readily available. Snare sets require local knowledge and perceptive observations, but there is little subordinative training involved and the practice is difficult for elders and even authorities to control. If wildlife scouts find a snared animal, ownership is difficult to prove—nobody willingly claims a snare or a catch that is in scout hands.

Yet snaring carries some of the same cultural baggage as other forms of taking wildlife. Users apply prescriptions (magic) to entice animals to enter the noose, to protect snared prey from theft and from predators, and to hide snares from detection by others, particularly wildlife scouts. Trappers expect certain reciprocities, mainly silence and appropriate concealment, when they share their bushmeat. Like a hunter, the trapper develops his strategy over the course of his agenda as he traverses the bush terrain. While moving about and scanning for bush products, a trapper searches the ground for recent animal signs and movement patterns for likely sites for his snares. Once he sets them, the trapper remains sensitive to dreams, the dispositions of neighbors, and indicators of success or danger. Trappers, like hunters, constantly monitor scouts' movements.

In 1989, Chunga Malembo, a near neighbor, a successful trapper, and an acquaintance of a mutual friend, was eighteen years old. His father's brother taught him when he was nine years old about the intricacies of snaring. Since then, he claimed to have snared a host of animals from a hippopotamus to a grysbok, including a lion and a hyena as well as a monitor lizard. He locally raised chickens and exchanged them for his snares during visits to a village on the plateau. The following entries suggest his tactics during a three-week period. At the time he possessed thirteen snares.

> *16 April 1989.* I set two additional snares, a total of four for buffalo. I have gone to check these snares every day for the past week. Buffalo in this area seem to sense the presence of my snares, for they continue to pass nearby. I dreamed earlier this week that a large buffalo herd was located at Maida, but this dream did not materialize in my taking a buffalo. I will continue checking for two weeks before considering another site. Many impala are in this area as well. I am thinking about placing more snares at Maida. I have not yet changed the location of the earlier sets.
>
> *26 April 1989.* I now have nine snares in the bush. All are set for buffalo and zebra. Today, I snared a buffalo, but somehow it uprooted the tree to which the snare was anchored. I suspect this buffalo is still carrying my snare as I did not find it discarded. Last night, I dreamed that a person was dying but he came back to life and afterward went swimming with me. At the river, this man drowned and I survived. When I awoke this morning, I told my friends that I was sure that I had caught something in the bush. When I went into the bush to check, I found that the buffalo had escaped. I also trapped a warthog, which also escaped, on the other side of Bouvwe plain.

[Responding to my question if he had consulted anyone about his dreams, he said,] No, I never consult about my dreams. This dream informed me that game was snared, but not secured. So far, I have not changed the location for any of my snares. Buffalo and other species are still passing nearby.

3 May 1989. This week, I set one more wire, which now makes ten. On 29 April, three of my snares were found and taken by the game guards [wildlife scouts]. These guards sighted my snares while using the road near Maida. Earlier, I had trapped a zebra there, but the wire snapped as it struggled and got loose. This occurred on 1 May. The previous night I dreamed I was pulled from a tree I had climbed. No additional animal was snared on that day. I changed the position of three wires and set them near Mutantangwe stream. I changed their location because the site where the zebra struggled filled the surrounding area with telltale signs that were sensed by other animals passing near there. Every day, I find signs of zebras passing and grazing near my snares. Now I have set one wire for smaller animals like impala and warthog.

Trappers can be as dependable at supplying bushmeat as hunters who used muzzle-loaders. Two other records of their yields—one occurring for a few months in 1997 and the other during my residency in 1988–89—indicate trappers' reliability. During September and October 1997, four men took ten mammals by deploying fourteen snares: four buffalo, three impala, and one each of warthog, puku, and kudu. They procured this meat even after allegedly losing some of their snares through theft or removal by wildlife scouts. In 1988–89, I asked eight individuals between the ages of nineteen and seventy-five to keep weekly records of the wildlife caught in their snares over an eighteen-month period. They snared fifty impala, seventeen warthogs, eleven buffalo, six wildebeest, five bushbuck, five puku and two inedible species (lion and hyena) which together totaled 4,290 kilograms of edible meat (carcass yield). This harvest was comparable to the meat of 21 buffalo or 237 impala carcasses. These trappers shared their smaller takes with close kin and sold portions of the larger carcasses to employed residents or to those on the plateau (Marks 2001).

Snaring a buffalo presents a greater risk than taking smaller mammals. Buffalo are dangerous when caught and rarely expire quickly when snared. Killing a snared buffalo takes effort and a plan, for the process may extend over several days if the animal is not dispatched quickly with other means. The large size of a buffalo's carcass adds another level of liability. To dispose of the evidence requires its trapper to distribute the meat quickly over many households, some of whom may not be close kin, appropriately discreet, maybe even passersby. If the carcass is not flayed swiftly, the site might draw vultures, whose descending flights alert scouts.

Although snaring may be as productive in some ways as hunting with guns, its impacts on wildlife populations are more devastating and visible within the short term. Snares are non-selective, taking males, females, and young indiscriminatingly, and are thereby capable of causing rapid decreases in wildlife numbers (Freehling and Marks 1998). Conservationists describe snaring as cruel, illegal, and subversive to their cause and remain dead set against snares. The tragedy is that their predetermined evaluations become barriers to inquiries that might disclose the cultural and political issues behind the (re)appearance of this technology.[30] Such studies would reveal residents' attempts to maintain their identities within a capricious environment and to use their time-tested expertise and means to tap into an abundant protein source to survive. Moreover, wildlife managers have missed the obvious message that many GMA residents have chosen to forego the meager and unreliable options allocated under the state wildlife agency's program. The official strategy of upping the ante through increased brutal enforcement just hasn't worked here. Residents are well aware of the program's closing noose on their welfare and have not been given a meaningful role in the program's processes.

Both social purposes and cultural values shape recipient societies' uses and modifications of new technologies and concepts. Valley Bisa men adopted muzzle-loading guns, both physically and symbolically, to creatively enhance their new opportunities under a somewhat distant colonialism. Once incorporated, muzzle-loaders became depositories of dominant masculine aspirations and, in the hands of a few patriarchal elders, were used to enhance their authority over women and younger men. When this local management system became criminalized by the state, other villagers redeployed their creative imaginations and recreated former and indigenous technologies to cope under the new challenging circumstances. As a 'democratic' technology, readily modified to fit local routines, skills, and subversive calculations, snaring has cultural energy and ingenuity behind it. Unlike the largesse of past patrons, who with their guns supplied and supported clients in state-mandated large villages, trappers seem to fill an appropriate strategy in the shadows of a scaled-down political economy of fractured settlements and individual struggles.

Notes

1. *Mfuti yanji nimawezela, Yenda ne kawele* (recorded 1973).
2. I base my assertion on men's vulnerability in a rereading of Martin Chanock (1998) and in a rethinking of my own experiences. Men probably suffered higher casualties under the warlords as women were often the main targets of capture; many younger men were captured for export. Colonial taxation forced most men into migrant labor for long periods.
3. Their answers included that God did not create women to hunt but instead dictated their household chores, that women have no rights (*inzambu*), or that the spirits would not allow women to survive alone in the bush. Women (sisters) might prompt their men to secure meat and to trap, but, in my experience, they never actively engaged in hunting with guns or snares. But women did occasionally scavenge from closeby predator kills and brought back from the bush, young animals killed by the dogs accompanying them, when they went to collect firewood.
4. The falling of leaves symbolically connects to the expectation of encountering game in this way. If most leaves fall with their undersides to the ground and most mammals flee from hunters, then gathering those leaves falling upside down (unexpected, unnatural) brings prey running in the hunter's direction.
5. As a reader of this chapter, Mwape Sichilongo noted that the name given this gun might have an additional meaning beyond that mentioned in the text. Since it was loaded for larger game, its name might be a summons for relatives to remain on notice for service as butchers and carriers should such a large mammal be found and killed.

This name may have had such an ambiguous reference, but this latter interpretation was not that given by the owner of the weapon.

6. *Cimanya muzaza*—literally [everyone]' *construct a butcher table,'* perhaps as a lineage consensus so the ancestors will deliver the game [my interpretation].
7. *Nakazabwe umwanakazi wacinwa*—literally "woman with a big mouth." The explication included the following attributes: "when a wife explodes, you know where she is and where there is [no] relish. A woman is audible at all times and means trouble; a gun speaks rarely, yet produces much."
8. *Kabula*—literally "a person without lineal sisters." The explanation was that "a sister of a different 'womb' is never satisfied with what her 'brother' provides her."
9. About the same time, another arrest for an unlicensed gun occurred in an adjacent village. A fight broke out between two young men at a beer party. In front of an audience that included wildlife scouts, the loser shouted about a gun in the possession of the other. They were both arrested, so disturbing village life that many people slept in the bush for nights fearing impending raids. People ran away whenever the wildlife vehicle approached as they might be asked to give evidence or be arrested for illegal and suspected activities. Almost any man could be framed, as those who didn't hunt with guns probably took game with snares. The chief ordered the ADMADE unit leader to visit the villages to reassure residents that they intended no searches and that he considered this case closed. This incident was recorded in my field notes [FN 8 August 2001].
10. *Lungwizi* (*Mimosa pigra*) is a sensitive bramble that grows along the edges of lagoons and drying pools. When touched its leaves fold on its acacia-like thorns. *Kupampa* is a transitive verb which means "to pile or stack in layers" (such as concentrating problems in one place where they can be resolved).
11. *Muzolo* is *Pseudolachnostylis maprounefolia, citimbe* is *Piliostigma thonningii,* while *mwanya* remains unidentified.
12. The prescription holder did not elaborate on these items. *Muzolo* has white bark (symbol of good fortune), and some wildlife relish its fruit. The tree provides a curative function in many types of prescriptions. *Mwanya* is a white, slippery root and is used to placate the pulp cavity of elephant tusks (cf. Marks 1976 [2005]: 132–43; also my Field Notes for 21 June 1989).
13. *Kukaka obuta*—literally 'to tie up the bow.' The retention of these terms suggests that gun hunters reinterpreted and used earlier ritual elements in solving their current dilemmas.

14. These attempts of these two hunters to resolve their dilemma over the dilemma that Mbokozi are recorded in my Field Notes for 27 March 1989.
15. Interview 8 October 2006, Kalimba. This elderly blacksmith died in 2007.
16. From my review of all court cases heard at Nabwalya, 1988–92. See also Chanock (1998: 65–66) and references in his footnotes for Northern Rhodesian (Zambian) cases.
17. The sale of legal muzzle-loaders occurs infrequently. In 2006, two muzzle-loaders in the chief's custody were pending sales to safari workers for ZK 320,000 ($92 US). This cost did not include the commission.
18. Colonial officials confined these investigations and raids mainly to the Senga areas, where they managed to convict, imprison, or depose six of the seven resident chiefs. In addition to these convictions, the government seized a large number of rifles and muzzle-loaders. For a summary of this case, see Lundazi District Notebook, volume 2, in the Zambia National Archives, KST 3/1.
19. Chief Chitungulu belonged to a different ethnic cluster (Senga) east of the Luangwa River in Lundazi District, Eastern Province. These formal administrative boundaries meant little to central valley residents who relied on their broad-based kinships to cope in times of hardship. Adam's marriage to a cross cousin in this chief's territory indicates the breadth of his lineage's range, as his father had come from Chitungulu's village to marry a wife in Nabwalya decades prior.
20. During a conversation in 1988, Adam told me that the thirty shillings was an inflated price. At auction, he claimed that a European official sold him the gun for one pound (twenty shillings). When he appeared the next day to claim the weapon, the African clerk charged him an additional ten shillings and 'pocketed this extra charge!'
21. Adam was born around 1914, soon after his father left his pregnant wife to carry war loads for the British fighting the Germans during World War I. In the 1930s, he used bows and poisoned arrows to kill many smaller bucks as well as a wildebeest and an eland. These kills followed the normative pattern of submission to elders for ritual procedures and prescriptions. Adam used these same rituals and prescriptions to placate the few kills he made with his new muzzle-loader. Interview with author, 21 August 1973.
22. Hapi's father was married to a 'younger sister' of the headman and spent some of time in Mwaba's village. As a distinguished descendant in a long line of hunters, Hapi's father trained many younger men in the 1950s and 1960s. A cousin gave his father his nickname, *Mutelele,* after a wild, edible weed that grows prolifically in culti-

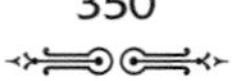

vated fields. Like many hunters, the father despised agriculture and complained constantly of backaches from having to cultivate. Hapi is the same individual who has appeared elsewhere in this book. Interview with author, 7 August 1973.

23. As lineage property, this gun remained in the keeping of its headman. Interview with author, 7 August 1973.
24. The following account by Hapi Luben describes a ritual wash together with the reasons for its use and the frequency it was performed between 1953 and 1970. I recorded this sequence in response to my inquiry with him about the years and the circumstances under which he had washed "Kabutu."

> **Prescription and its use:** Cassava leaves [a domesticated plant from the village] gives power [*vizhimba*] to a potion mixed with the ground roots of *lungwezi* [*Mimosa pigra*]. Every wild animal either treads upon or eats this shrub, which recoils when touched. This mixture has two purposes: to cause the animals to approach or come nearer (*'ifyo ibomba kupalamike inama ne kufunya cibanda pamubili'*) and removes a bad omen or "bad spirit" from the hunter's body. The hunter places this mixture in water and washes the gun at the village rubbish heap in the late afternoon. After washing, he leaves the gun "to warm" in the rays of a setting sun. When dry, the hunter retrieves the gun for use early next morning.
>
> **Frequency (Month) and Circumstances of Use:**
>
> **1953 (March)** I asked my maternal uncle (*bayama*) for a prescription to wash the gun. He gave me a prescription [above] as I had fired this gun five times previously at game without success. People in both Mwila's and in my father's village were expecting meat and a big carcass from me. Three days after washing, I killed a male buffalo at Kapili Ndozi. Subsequently, I was successful for the rest of this year and the next and killed at least an animal every month.
>
> **1955 (August)** I washed the gun after experiencing unspecified failures. After four days, I killed a male warthog at Ngala lagoon, close to the village.
>
> **1958 (March)** My relatives wished meat and I went a full month without success, so I washed the gun again. The morning after this washing, I killed a cow buffalo at Nyanvi stream (eight kilometers from the village). The explosion that sent the bullet to kill the buffalo also burst the gun's stock. Consequently, we sent the gun to a blacksmith to replace the stock. Mwaba paid thirty shillings for refurbishing the weapon. In the interim, I used my father's extra gun, but I never washed his weapons.
>
> **1959 (February)** My father's gun exploded while I was using it, and the explosion almost killed me as the housing chamber and nipple blew past my head. *Kabutu* returned the following month and I used it for the rest of that and the following year. Both years were very good for bringing down animals.

1961 (January) I washed the gun after two weeks without killing anything. In the week after this wash, I killed two buffalos, a bull on Kapili Ndozi and a cow in Kawele thicket. Both were killed on the same day and from the same herd! I killed the bull instantly, then reloaded the gun on the run while following the herd. When the herd gained a thicket and cover, I shot the cow.

1965 (January) I washed the gun after three days without making a kill. On the fourth day after this washing, I killed a bull eland at Mwana waMbouve plain. My father helped me ritually treat this "dangerous animal."

1970 (April) I washed the gun and the following day killed a sub-adult cow elephant (possessing one and a half feet of ivory) at Nyanvi Stream. I placed the tusks in a tree in the bush. This was the last time I washed my uncle's gun.

25. Although not mentioned in this account, supplications of beer, sorghum, maize flour, or even blood are often placed in the gourd's opening. Spitting personalizes the supplications.
26. That this warthog did not flee but continued to come to the hunter had additional significance, because the behavior was neither normal nor expected. Kandeke saw it as a sign that his guardian spirit was answering his plea. His ancestors provided this animal to feed the lineage and to reaffirm their faith in Kandeke as a provider. *Inama zya mipazhi* is an animal sent by the spirits.
27. In 1993 ,Kandeke presented me with a section of *Kabutu*'s exploded barrel.
28. These elephant operations had their abusers as well as beneficiaries. Colonial game officers and some administrators wrote that Africans complained about garden raiders when they could find no signs of recent damage. Africans only wanted meat and always tried to sucker itinerant officials into providing it for them. Game Ranger Poles's reports noted that 'his' elephant guards often shot the largest, big-tusked elephants and at long distances from any fields.
29. Senior Wildlife Police Officer, Sector-in-Charge, Kanele Sector, Munyamadzi GMA 2002, Annual Report 31 December 2002.
30. A paper by Lewis and Phiri (1998) focused their criticism upon a competitive community-based wildlife management regime LIRDP (Luangwa Integrated Resoure Programme) rather than analyzing snaring responses generally as a reaction to the national community-based program ADMADE (Administrative Management Design), which they both helped to manage and implement. From its beginning the Munyamadzi GMA was administered under the national ADMADE program.

Chapter 9

Buffalo Mystique

Protein, Privilege, Power, and Politics

In killing a buffalo, the game guard likens it to his mother.[1]
Valley Bisa proverb

In their insightful essay "Goodly Beasts, Beastly Goods," John and Jean Comaroff (1992: 127–154) argue that among the Tshidi Barolong, a pastoralist group in Botswana, men spent their time thinking, strategizing about, and if they could, investing in livestock. Under colonialism and the political economy of southern Africa, most Tshidi Barolong men were forced into migrant labor markets, becoming ensnared in an arduous cycle of poverty and dependency, experiencing powerlessness in a world dominated by cash and low wages. For these men and women entrapped respectively in vicious cycles of outside labor or homeland farming on degrading soils, neither prospect provided sufficient subsistence or welfare. Consequently, investments and care of livestock turned into dreams of self-determination, an illusive hope for opting out of iterative servitude and wage labor through speculative purchases of cattle deposited within a degrading and chancy homeland. Livestock ownership was a social strategy, a fantasy from a past that few men could ever realize. Stocked and structurally enclosed beyond the practical boundaries and influences of women and detractors, cattle became the proverbial standing for the political competitions among elite and residential men. Through their domestic chores, and cultivation and gathering tasks, related women enabled these privileged livestock owners' contentions.

For the Tshidi, cattle were the dominant form of material and symbolic possession that could increase in value given good fortune in space through time, could hold its value against other inflating commodities as well as represent an owner's identity individually through ties of alliance and patronage. These domesticated mammals became the media through which few men crafted their cultural biographies, reflecting a patrilineal individual's ability to organize others, to extend and sustain support, as well as radiate a personal presence, even during their absence. "As a focus of everyday activity, cattle were the epitome of social and symbolic capital: the capital . . . that linked a material economy of things to a moral economy of persons, and so constructed a total economy of signs and practices" (Comaroff and Comaroff 1992: 144). To paraphrase further the thesis of these authors, the "beast" became a mythical "symbol of economic and cultural self-sufficiency" embodying a haunting vision of turning cash earnings into a productive herd, thereby becoming an influential actor within a promised land by successfully rupturing the seemingly inevitable "cycle of migration and want" (p. 148). Once money was invested in cattle as a protected resource ("goodly beasts") as a wager and legacy, it could be reconverted to transient cash ("beastly goods") only under dire circumstances. Although rarely realized, this dream was a powerful representation of a former world, "a mythic society in which men were men, in which women did not struggle alone in the rural wastelands, in which the control of social vitality was ultimately ensured by the goodly beast" (p. 149).

In his critique of a World Bank report on the Thaba-Tseka project of a pastoralist group in Lesotho, Ferguson (1994) explores the provincial barriers to this donor's livestock/range management goals. Ferguson's analysis indicates that development discussions and projects often generate "less-developed fantasies" of their own of unfamiliar places, where their goals for stated "improvements" often clash with local cultural practices and eventually fail. He suggests that the persistent effects of unsophisticated inputs of external funds, expertise and materials are not the achievement of program goals on the ground, but the expansion and reinforcement of bureaucratic management into new domains. To explain local cultural resistance to this Lesotho livestock project, Ferguson elaborates upon a "bovine mystique," a similar tragic vision to that described by the Comaroffs above. For the Basotho, cattle were exclusively a masculine possession, protected as a separate economy by cultural precepts, and deployed as individual investments in this domain of understanding and meanings. Despite the pretensions of the Lesotho project to improve efficiency in livestock management and to bring the "beasts" to market, cattle remained the local currency for storing wealth and welfare, which for their owners had proven more resilient for sustaining their identities.

I employ these scholars' insights while transferring their discernments from patrilineal societies to a matrilineal one, from southern African pastoral societies to a marginal subsistence agricultural society that earlier incorporated hunting of large mammals within Zambia's central Luangwa Valley. I also shift from domestic (bovines) to wild (mainly buffalo) animals as the "primal beast" for matrilineal investments in social structure (lineages) and symbols (feelings) and from herders to hunters, who are deprived of livestock through the fly on their premises. I use the trope "buffalo mystique" to designate this Valley Bisa cultural conundrum, at least for conservationists and planners, who constructed the Zambian "community-based" wildlife management program thinking that local rural hunters, with their practices grounded in real lives and places, would readily convert to their logic, plans and models. Planners' efforts may have changed that world, but not in the context or with the content originally intended. Promoting "wilderness" built upon a "safari fantasy" was not just a lost opportunity for building allies and learning about reconciling cultural differences; moreover, "community-based" has become a contradictory and expensive illusion, which has yet to achieve its declared goals of improving human welfare and sustainable conservation.

In the previous chapters, I argue for the centrality of bushmeat (mainly buffalo) as a prevalent wild product and conceptual force on a landscape where the presence of tsetse fly precludes the keeping of livestock. I contend that the cultural production of wild animals in Valley Bisa society, its instructive idioms, venerated verses, the need for protection of lives in addition to properties against predatory and depredatory "beasts," the uses of bushmeat as food in forging and maintaining social ties as well as its cultural currency as symbols of health and wealth (*"goote milile"*) demarcated a revered and provincial place in the minds and activities of valley residents. I claim that these cultural factors, including lineage elders' allocation for most of its men as migrant laborers to meet their tax as well as cash requirements were the significant elements, rather than geographical distance or marshaling the essential foot-power necessary to ferry a perishable product to market on the plateau, which retained the production and consumption of buffalo and other wildlife within the lineage arena.

Within the imperial envelope of colonial rule and under the hegemony of a long-lived chief (Kabuswe Mbuluma), a few residential young men, encouraged by their descent groups created parochial identities as lineage hunters, submerged their ancestral "calling" within the ideals of lineage protection and maintenance, embedded their practices within traditional idioms, founded their productive arrangements based on gender and generational alignments while remaining within their homelands as their matrilineal "brothers" became entrapped in the cycles of migrant labor and its paltry pay-offs. The remaining few men as local hunters and elders employed outdated muzzle-loading guns of their own, or inherited from their late, maternal "brothers" or "uncles," who had purchased them with their meagre migrant wages, in pursuit of their provincial "games" (pun intended). The uses of these muzzle-loaders, even as placeholders for absent men, provided an explosive claim extending the ascendancy of men's work and influences in space and through time as hunters covered the weapon's and their own vulnerabilities in abstractions and rituals. The cult of the younger lineage hunter, epitomized as "beer drinking and hunting," became a youthful dream of manliness and local autonomy compared with its alternative of becoming an itinerant laborer, whose experiences were shot full of confusions, humiliations and tragedies while subordinated to foreigners, whose rewards in cash, materials and skills were quickly dissipated upon returning to their homeland. Whereas hunting was attractive, comparatively few youths were enabled to secure essential social and material supports, survive its surprises, and achieve its celebrated standing throughout their lifetimes. From my reiterative accounts and observations spread over five decades, what remains to show is how, for three generations of men, this hunting and their wildlife products were significant as a "total social phenomenon" in the making of local personalities and materials as well as in their relations and distinctions. These men spent much of their lives within the shadows and political intrigue of the Nabwalya chieftaincy. My account seeks to situate and connect their stories chronologically and strategically, thereby revealing their changing personal relationships as well as the strategic roles for bushmeat within local and national frames.

A species long considered as a proverbial disease threat to colonial livestock enterprise, cape buffalo became the main targets for local hunters and the preferred staple in their provincial subsistence economy. Buffalo, with its large size and accessibility, offered a dependable and abundant source for feeding the stomachs and the imaginations of residents and hunters alike. For local hunters, buffalo pursuits became a cultural threshold to "a certain type of local man," one who epitomized the virtuous masculine thickets of potency, patronage, protection, and political savvy. Buffalo meat, secured mainly by younger men, was the valued cultural currency upon which a few astute Big Men, who controlled the guns, built their reputations as patrons (see chapter 6). For a few local men, the hunting of buffalo grew into a meaningful quest for sustenance and structure as young and old, men and women wrestled to obtain the "good life" (figure 9.1) within an otherwise parsimonious and insulated landscape.

While he lived (1900–1984), Chief Nabwalya [Kabuswe Mbuluma] and his counselors provided the local authoritative and policy structures to which most Valley Bisa adults responded. Zambian independence in 1964 and the chief's declining health and death in 1984 gradually eroded his authority. The ensuing contentious interregnum, during which a new chief [Blackson Somo] was confirmed in 1990 and a new Zambian president elected in 1991, saw the state begin to implement its new "community-based" wildlife program and structures within the Munyamadzi GMA.

FIGURE 9.1. Buffalo bull killed and butchered adjacent to the cultivated fields of Luba Village. This kill was assigned to a local hunter by Chief Nabwalya, who is standing out of view with his retainer (right), directing the butchery as the assembled youthful carriers compile the chunks of meat to take it back to village households. (Photograph by author in September 1973)

This new program initially depended on donors and reflected urban interests and revenue generation rather than promoting rural livelihoods and addressing local contingencies. Chief Blackson Somo continues to struggle with these contradictory weights on his shoulders.

The Bull's Eye: Efflorescence of Clientage Based on Bushmeat

1950s: The Flowering of a Cultural Tradition

[In 1953, the British protectorates of Northern Rhodesia and Nyasaland were joined with the former "self-governing" European colony of Southern Rhodesia into the Federation of Rhodesia and Nyasaland. It was created as an alternative to the newly independent African states to the north and the white-dominated governments to its east, west, and south.

In the late 1950s, African majorities increasingly protested against white minority rule within the federation, leading to the banning of African political parties and eventually to a declaration of an emergency, during which Kenneth Kaunda, a nationalist, was jailed before he eventually became Zambia's first prime minister in 1964.]

After I got to know them, I discovered that the five headmen of the villages around Nabwalya's palace on the Munyamadzi River had spent at least some of their earlier years as migrant workers throughout southern Africa.[2] I place these elders and their affiliates on the 1950s landscape as background to my studies, which began in mid-1966 and have continued intermittently during each subsequent decade. These headmen, who had consolidated their reputations,[3] possessed one or more muzzle-loading guns loaned mostly to younger men for hunting[4] (Marks 1979a: table 1, p. 60). As keepers of the guns of their respective lineages, these elders hunted to assert their potency as protectors or to seek animal protein when their subordinate hunters were elsewhere. The elders' main responsibilities were to assure harmony in village life, to attend to dependents' concerns, to implement the chief's dictates and inform him of events, to protect lineage properties, and to oversee the judicious distribution of lineage prosperity. Younger men within the same or nearby lineages (in-marrying) mostly performed chores as assigned; a few of them joined older men as assistants and carriers on hunts.

Patrons supported younger men by supplying them with weapons and materials, facilitating their access to women, and legitimizing their progression of demonstrated hunting success through interpreting the esotericisms of lineage rituals. Elders also paid a subordinate's taxes, protected him against outsider (often chief's) demands, and assisted in meeting monetary and household needs. While attending local school or before embarking on their first journeys into town, some boys assisted in bush forays. For them, these bush trips were their first and most compelling engagements with the adult masculine world, which some sought to emulate later.

European staff of the colonial Game and Tsetse Control Department suspected Chief Nabwalya and those around him of supporting "poaching" and thereby significantly depressing "the carrying capacity" of the valley's wildlife. Because some prominent administrators did little to support the Game Department's rural programs, the game ranger at Mpika surmised that the chief was in league with these other district and provincial personnel. In their official memos, provincial officers showed little concern for buffalo other than describing it as a "pest" and as a danger to local people. If administrative reports applauded the pluck of African hunters, Game Ranger Eustace Poles's reports from Mpika graphically described local hunter deaths by buffalo as apt lessons of degenerate residents and for others not taking his admonitions about presumed poaching seriously (Poles Mpika Tour Reports 1947, 1949, 1951a, 1951b, 1953[5]; Boyd 1992).

At Nabwalya during the 1950s, each reigning headmen resided with his followers within a village, observable as a discrete cluster of households to facilitate the keeping of official records. The biggest patron was the chief whose powers and resources affected all villages and settlements within his territory. He was married to multiple wives strategically placed throughout his chiefdom, received tribute (*mitulo*) in wild and domestic products from residents as well as a government salary, operated a store, owned a Land Rover, and as the "owner of the land," held political and spiritual hegemony. Since the wealth and health of a reigning chief symbolically reflected the welfare of the land and its inhabitants, he officiated at many provincial rituals, held court as the local arbitrator in lineage disputes, and possessed a pharmacopoeia to restore health and to achieve success in sundry endeavors. If a patron killed a state "protected" mammal (elephant, eland, rhino)[6] or needed some esoteric hunting ritual or resolution, he likely obtained these means from the chief. The chief sanctioned the killing of these ritually marked species and, in any case, expected tribute in meat to learn of their demise on his terrain.[7] He could also act as a protector in foiling investigations of an illegal kill by game guards and was acknowledged and suspected of doing so.

This chief never lacked meat or beer as he depended upon his many wives for the latter and secured the former from various subordinates. Goliat, one of these subordinates and the son of the chief's sister from the same "womb," was described to me as driven by a strong "spirit," and he allegedly possessed powers for venery (both in the hunting and womanizing senses). Although he was an expected contender for the throne, the tasks of farming and other responsibilities ("keeping the land prosperous") were never among his ascribed characteristics. He spent days in the bush without advising others of his movements and would reappear again just as suddenly. The animals he killed entered the households of his two wives and the palace. The chief appointed peers to monitor Goliat's whereabouts and gave his abnormal habits as the reason.[8]

Beyond his relatives and subjects, the chief's constable (*kapaso*) added to the palace larder. This was the role that Jackson Katongola so adequately filled this decade earning the nickname *wabukazha* ("meat lover") before he advanced in the district and provincial hierarchies of constables. Other aspiring younger men also eagerly hunted for this chief. Among them was Hapi Luben (remember him?), who reportedly killed many buffalo with the chief's government-issued Martini-Henri rifle. In 1959, he killed three buffalo on the same day, almost at the same time, and in the same place. The buffalo herd was apparently alerted and facing him when he killed the first one. When those remaining became aggressive and charged, he shot two more, forcing the herd to disperse and saving himself from being trampled. Such bravery and sequential killing increased Hapi's local reputation. That reputation worried his father Luben, who at the time was struggling to keep together an array of kin that he had assembled a few years earlier in establishing a new village.

Hapi was neither a member of Luben's lineage nor a resident in this village at that time. Another young man Lowlenti (remember him from the bush encampment in chapter 4) belonged to Luben's lineage and lived at times in his maternal uncle's (Luben's) village. Lowlenti, like Hapi, never "owned" a gun. Borrowing a gun was no problem for Lowlenti despite the fact that four muzzle-loading guns, two of which belonged to Luben, had exploded in his hands during this decade. As a result, Lowlenti was missing a thumb and index finger from his right hand, which merited his nickname *minwe* ("fingers"). Another moniker for Lowlenti was a synonym for zebra, which referred to his stubbornness and idiosyncratic behaviors. Yet throughout the decade, Lowenti reputedly killed "many" buffalo for his patron Luben and for in-laws in an adjacent village.

One day while accompanying Luben after buffalo, Hapi was not given an opportunity to stalk, even as his father's stalks had repeatedly failed, until later in the hunt. Their hunt was troubling, as the villagers were famished since their fields had flooded. Eventually Luben handed his gun over to his son hoping that he would make "an impossible" shot at a single impala. Hapi killed the animal as his fortuitous shot saved an otherwise "meatless" return to the village. Luben had expected Hapi's shot to miss and commented afterward that the kill was too small to distribute to the whole village.[9]

When I met him for the first time, Kalaile Muchilibala, the local court messenger and the patron for a cluster of settlements north of the Munyamadzi River, expressed his need for recruiting younger men to marry within his village. He was a migrant laborer before becoming the local court messenger. In that capacity, he purchased a muzzle-loader in 1957 and, before his retirement in 1960, acquired a shotgun on a letter of good conduct from the chief. In his words, these guns were his "pension plan," as their primary function was to lend to younger men. Kalaile was rather elderly by the time he killed his first mammal, a male impala, with his muzzle-loader. A younger nephew helped him treat the carcass and his scattered settlements performed the traditional *vizimba* ceremony combining beer drinking with feasting. He intended through this performance, an earlier tradition, to demonstrate his power as a Big Man to attract younger men as marriage clients.[10] He spoke then of his need "to retain his village" in metaphorical terms of mending an old and frayed cloth (*wazapuka*) by attracting new threads (in-marrying younger men).

1960s: A Decade of Individual Accomplishments and Contributions

[Northern Rhodesia, a prime area for demonstrations and political turmoil particularly in 1960–61, led to the dissolution of the Federation of Rhodesia and Nyasaland at the end of 1963. In January 1964, the newly elected prime minister of Northern Rhodesia, Kenneth Kaunda, crushed the dissident Lumpa, a religious cult under Alice Linshina in the Chinsali District (northern Luangwa Valley). He then led Zambia to independence on 24 October 1964. He served as Zambia's first president

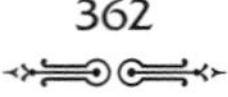

until the 1991 elections. Southern Rhodesia remained ruled by a white minority and faced a long and costly nationalist revolt following its unilateral declaration of independence from Britain in 1965. Southern Rhodesia's relations with Zambia were hostile in heated rhetoric and in cross-border incursions by their military, until Zimbabwe obtained its independence with majority African rule in 1980.]

Some of Kalaile's concerns about prospective recruits eased when Kameko Chisenga, of the chief's lineage, married a young woman in his village. Returning from work in Southern Rhodesia, Kameko struggled to find a place for himself on the local scene. Beyond "hawking" goods and serving as the local Catholic catechist, he practiced his hunting skills. Initially he used his father-in-law's gun, and a year later he acquired a barrel from the chief (his uncle).

Late one evening during 1962, Goliat took his own life with his muzzle-loading gun. According to a nephew, Goliat had forced himself on a woman while her husband was away. This woman reputedly placed a "long-acting medicine" on his body that caused Goliat's eventual demise. With this death, his strong, haunting spirit searched for another relative to possess; whom it captured depended on who was asked.[11]

In the early 1960s, Jackson Katongola departed for a district posting, depriving the chief's palace of its master hunter and a dependable subordinate. Yet Jackson returned frequently to the palace while on tour with government officials. During these visits and in his capacity as official escort, Jackson killed buffalo and other game to fill the chief's, his relatives', and his employee's larders. Morgan Kabongo, another accomplished hunter and *kapaso* from another chieftaincy, replaced Jackson as the chief's constable.

Later in the decade and perhaps more significantly, safari operators within GMAs were required to distribute the meat from their kills to residents. When safari clients killed an elephant, rhino, or buffalo, these operators loaded their vehicles with fresh meat and dumped flayed strips or carcass portions at the palace for the chief to redistribute.[12]

Hapi Luben expanded his own social horizons as a seasonally employed tracker with Luangwa Safaris. In the offseason during the rains, Hapi supported three wives and their children through his exploits in the bush. During these months, he mainly hunted and attended beer parties, for, with cash and favors, he hired others to work in his wives' fields. Borrowing others' guns, Hapi dispersed his hunting prowess among his various patrons including the chief, his maternal uncle, his father, and the headmen of the villages where he was married. His success and reputation attracted youngsters who joined and later tried to emulate him. During two decades, he trained six relatives and an in-law.

In 1964, Hapi upped his record take to four buffalo in a single day. He had been given a rifle and sent along with the chief's son, Poulande, to secure a buffalo for the visiting district secretary (DS). Hapi and Poulande came upon a spooked herd in an open glade. As the buffalo stampeded toward him, Hapi shot the first at point-blank range; then, from atop this carcass, he killed three others in rapid succession as the animals continued to rush forward. By local accounts, this feat was laudable, impressing both the chief and the DS.[13] Such exceptional triumphs paradoxically contributed to Hapi's relational problems, as later this decade his kin were to accuse him of witchcraft (chapter 5).

Lowlenti's fortune waxed and waned as well. He spent much of the decade drinking and pursuing buffalo and small game. Once Lowlenti tried to short-change his uncle, Luben, when he borrowed a gun with intent to kill a buffalo but instead sent him only the hindquarter of an impala. This token amounted to an insult, for not only had Lowlenti killed both a buffalo and an impala, but sent the owner of the weapon tribute from the "lesser animal." Those borrowing guns were not sanctioned as the sole distributors of the meat produced.

This additional confrontation with Lowlenti added to Luben's growing problems.[14] Luben's second wife died unexpectedly as had his first, a lineage "sister," several years earlier. This coincidence caused his in-laws to charge that, as a wizard, he was employing the labor of their dead sisters and their children in his searches for wealth. This accusation was a serious affront on top of his other problems. This charge left Luben with few options but to leave. In 1970, Luben petitioned Chief Mwanya for a plot of land east of the Luangwa River and went there to resettled there with his wife, a nephew and his wife and children.[15]

According to my records of 1966–67 (table 9.1), Hapi was the most prolific local hunter, killing seven buffalo, an eland, an elephant, and smaller bucks totaling more than 5,100 kilograms (butchered weight). Most of this meat went to the two villages where he had wives and to the palace. He also dispensed meat and other unspecified favors due to his position with the safari firm. Yet, Hapi's take was only half that of the resident game guards, who in 1966–67 killed six elephants in addition to their monthly rations of other game. Some of these elephants and other kills were at the chief's discretion and insistence.

As Hapi's peer, Lowlenti killed six buffalo and four smaller species in addition to salvaging three buffalo from lions. His take totaled some 3,100 kilograms, which was mostly claimed by his patrons. Both he and Hapi accounted for 38 percent of all hunting time recorded during that year (610 hours). Luben killed one buffalo, retrieved another carcass from lions, and wounded three more. In addition, he butchered a waterbuck. Residents in Chibale's village claimed twelve animals including five buffalo. Although Chibale (chapters 4 and 5) did not kill these animals, he was the main distributer of this bushmeat.

Throughout the decade, the chief actively killed elephants, mainly at night in his wives' fields throughout his chieftaincy. Provided the game guards were notified of a kill within forty-eight hours, the wildlife code sanctioned this activity as protecting personal property. A tax collector attached to the palace killed some eighteen smaller animals (about 600 kilograms) with his

TABLE 9.1. Wild meat as main constituent of daily relish dishes recorded as quarterly percentages in three villages, Nabwalya Study Area, 1966–67. (Source adopted from Marks 1976: table 7, p. 48)

	Aug–Oct 1966	Nov 1966–Jan 1967	February–April 1967	May–July 1967
Number of Relish Dishes sampled	761	1726	1822	2387
Dishes containing Wild Meats	351	477	549	1022
% containing Wild Meats	46.1	27.6	30.1	42.8

(Source: information from Marks 1976: 48)

shotgun. Most of these proceeds went to his local wife, her in-laws, with some portions sold during his trips to Mpika.

1970s: The Impacts of a Declining National Economy

[The decade began with a revision in Zambian wildlife policies supported by the United Nations Development Program, with the incorporation of additional Bisa land (Chifungwe Plain) into the South Luangwa National Park. It ended with a worsening national economy, with a demoralized and weakened wildlife department (now the Department of National Parks and Wildlife Services [NPWS] under a Zambian director), and with widespread assaults on the national wildlife estate. The conflict with Southern Rhodesia resulted in the intermittent closure of Zambia's southern border in 1973, posing severe problems for the international import and export of goods. The energy crisis dramatized the national need for additional regional connections, including the building of an oil pipeline and railroad through to Tanzania and the African East Coast. These additions to the national debt, international loans, and the rising costs of materials and transport contributed to the general economic malaise. At the regional level, increasing costs for basic commodities together with ineffective enforcement of the new wildlife codes opened the valley to foreign trafficking in elephant, rhino, and buffalo for ivory, horn and meat.]

With the exception of Luben and later Hapi, the same patrons and cadre of hunters remained from the 1960s. When Luben settled across the Luangwa River in 1971, most of his former clients settled in villages near the palace. An elephant reportedly killed Kapampa, an elderly headman of a settlement north of the Munyamadzi River while he was collecting reeds in August 1973. Years later, his nephew told me that Kapampa's relatives suspected him of being a witch, had probably killed him and left his body in the bush (Marks 1979a).[16]

With the suicide of his nephew Goliat, the chief became the sole survivor of his lineage ("womb") and increasingly distant from most other Ng'ona members around the palace. Consequently, he depended primarily on the support of his sons.[17] Morgan Kabongo, as the chief's constable, together with the chief's sons became the main gatekeepers around the palace. They did the chief's bidding as well as his hunting.

Born in 1937 to a wife of the chief, Poulande killed his first buffalo in 1962 as Hapi's understudy. His uncle, Goliat, provided the prescription to treat this carcass and assisted with the *vizimba* celebration of dancing and feasting. For most of the 1960s, Poulande drove the chief's Land Rover and was often away on commercial missions to the Copperbelt. As a driver, Poulande was responsible for stocking the chief's local store and for loading the vehicle with local bush products including bushmeat, chickens, smoked fish, and tobacco for sales in town. In 1971, the chief insisted that Poulande remain in the village to look after the latter's sick mother. Although Poulande had married three women, he now spent most of his time with his first wife and his mother near the chief's palace. Another of the chief's sons became the subsequent driver of the vehicle.

Poulande resumed buffalo hunting in December 1972 with the chief's rifle, and by September 1973, he had killed three bulls and two cows. Hapi also killed a buffalo with the chief's rifle in 1972. In the process of carrying this meat, the game guards caught Poulande and took him to the magistrate's court in Mpika where they charged him with illegal possession of game meat.[18] The chief temporarily paid the fine, but at the time he could solicit only a small pittance from Hapi, who had lost his safari job. When I briefly employed Hapi in 1973, the chief summoned Hapi and reminded him to pay this debt.

Through September 1973, hunters around the chief's palace, including Poulande, Hapi, and Morgan, accounted for eighteen buffalo (some 5,910 kilograms) in addition to numerous smaller bucks and warthogs. During this period, the chief killed at least one elephant (a cow shot in a wife's field), received several truckloads of meat from safari operators, and played host to several government officials and chiefs, who often came to hunt or collect bushmeat. Beyond that taken by officials and chiefs, this meat (estimated over ten thousand kilograms) served as part of the chief's customary largesse and hospitality. During that dry season, the palace transported some of this meat and sold it on the plateau and beyond (Marks 1977b: 259).[19]

The chief's presence was not necessary for the palace to end up with the bulk of meat from a nearby kill. In the early morning of 10 August 1973, Hapi killed a subadult bull buffalo near Kapili Ndozi.

That afternoon, seven of Hapi's relatives, in addition to six from the palace, assembled at the kill site. Poulande and Morgan flayed the carcass and took much of the meat to the palace. Hapi insisted that none leave the kill site bearing a grudge from the amount of meat they received. On 17 August in the presence of both Morgan and Poulande, Hapi killed a female buffalo with the chief's rifle in nearby Bemba Stream after Poulande, with the same weapon, had missed a buffalo in the same herd at close range. Assembled at the butchery, eight men, three boys, and three women hauled this meat to the palace.[20]

The largesse of headmen and their hunters was small in comparison with that of the chief. By September 1973, residents in Chibale Village accounted for four small mammals (335 kilograms) in addition to the three buffalo killed by Lowlenti (1,098 kilograms). Paison killed two buffalo and a warthog (785 kilograms) while Kameko, still residing north of the Munyamadzi River, claimed one buffalo and many impala (1,225 kilograms). Hapi himself killed four buffalo and smaller game (1,605 kilograms), some of which went to the palace.

When I returned in August 1978, Hapi had become his father's client and resided east of the Luangwa River. Poulande replaced him as the palace's main hunter. Earlier that year, Poulande began to kill elephants, a species still on license in Zambia and increasingly slaughtered illegally as well as on license (Marks 1984: table 4.5, p. 101). By August, Poulande had killed at least four elephants (seven thousand kilograms) while acting as a hired guide. In addition, he claimed six buffalo (two for a visiting Bisa chief) and several warthog and impala. This take alone amounted to some 9,500 kilograms of bushmeat. The chief killed an elephant and a buffalo and received truckloads of meat from the safari operators. Three other local hunters killed a buffalo each.

In 1978, Nabwalya was awash in development activities. A few years earlier, the government had shifted from resettling the Valley Bisa to their development and to accommodating their presence between the two major Luangwa Valley national parks. For most of 1977, the district supplied a large labor force at Nabwalya to build a clinic and court, to refurbish the school, and to dig wells. Although not without precedent, this contingent of outsiders and their movements gave some residents incentive to participate at the source in supplying bushmeat to those on the plateau and elsewhere. The impacts of these activities upon wildlife were apparent in the frequency with which I observed some species during a month's visit. The numbers of elephants, buffalo, and zebra were noticeably down from counts in earlier years. I saw few buffalo, and local contacts told me that only two small herds remained within hunters' normal walking range (Marks 1984).[21]

Given the deteriorating national economy and the developing ivory, rhino horn, and bushmeat "businesses," some local hunters found new opportunities to "cash in" with some of their game pursuits. *Sokola* ("tooth extractors") referred to those hunters who removed only the ivory from elephants or horns from rhinos while leaving the carcasses to rot in the bush. Attributed mainly to the foot gangs of hunters and their many carriers operating from villages on the plateau, this routine was a response to the high prices that ivory and horn fetched and also a strategy to evade the risk of capture by armed wildlife patrols (Milner-Gulland and Leader-Williams 1992; Leader-Williams and Milner-Gulland 1993). Some residents expressed this behavior as "cash and carry," an idiom from town originating in the supermarket, but in the face of unemployment and long-distance travel to urban centers, they switched course and participated in these local activities.

Meat became detached from its earlier social and symbolic moorings. Cheap, unregistered muzzle-loaders flooded into rural villages to facilitate the regional bush enterprises, which undermined the control that elders had over these weapons earlier. For younger men and others subordinated under the earlier elder-dominated cultural order, the pursuit and selling of wildlife products became their means of acquiring "respect" and material things they lacked. Increasingly, meat was dried, smoked, trekked to the plateau, and sold. The paved roads and railroad provided swift transport links to markets within Zambian cities. Coupled with this burgeoning informal marketplace, hard times eroded the efficacy of the earlier hunting rituals, dissipated the normative attention formerly paid to lineage welfare, and deepened the trenches of behavior separating young from old, men from women. Yet beliefs in hunting prescriptions remained persuasive, as did those in witchcraft and the abilities of a few to affect the welfare of others. With their seniority and much of their cultural worlds turned upside down by "undisciplined" youth, many elders were besieged and were accused of being "witches" and withholding their fortunes. The decade ended in transition—a harbinger of insecurities brought to the fore by insidious human and environmental changes.

On Target: Cultural Turmoil in the Wake of the New Economic Order

1980s: A Decade of Uncertainty and Tension

[The 1980s included persistent crises and transitions within the national economy and the wildlife sector, which were reflected in Valley Bisa cultural and social life. Employment remained difficult for residents to find, inflation was rampant, and necessities expensive, if indeed available. Tenacious droughts put valley livelihoods at risk, prompting frequent famine assistance from government and others; most intended relief was late in arriving within valley villages. Local residents turned increasingly to their surrounding resources, particularly wildlife and other bush products, to meet their needs and to barter for other goods. The NPWS was in shambles and in transition for much of this decade. At the beginning of the 1980s, NPWS was financially strained with many of its field staff unsupervised, unpaid for months, and demoralized. Some staff engaged in the illicit wildlife trades of ivory and meat. Yet by the end of the decade, international donors and conservation organizations provided new resources, developed a "community-based wildlife management" program, which used revenues from safari hunting to "beef up" faunal protection (predominantly through anti-poaching) and support for infrastructural improvements. The new wildlife program added to the cultural changes and traumas already on the ground. Whereas many local patrons and their clients had benefited temporarily from informal exchanges with outsiders in the past, the new wildlife structure had a different scaffolding. Under the new regime, government forcefully intervened as the new national economic patron, while global funders, safari hunters and national participants became the transient beneficiaries of the declining wealth and health within the GMA.]

The fifty-one-year reign of Chief Nabwalya (Kabuswe Mbuluma) ended with his death on 28 April 1984. Unfortunately, the chief's poor health and lineage squabbles left Valley Bisa cultural life in limbo for much of this decade. Many residents remember Kabuswe Mbuluma for the many occasions on which he defended local ways against outsiders, particularly those in the game department and more recently the NPWS. His death began the lengthy and litigious processes of choosing a successor, bringing an abrupt end to the former social ties that had held the chieftaincy together. The appearance of contenders and claimants to his properties and wives added a dismal dimension to a long drought, which had already affected the lives of many inhabitants. Residents spoke about those closest to the dead chief in demeaning terms and implicated them in his death. Belonging to different lineages, the chief's wives and children retained little or nothing from his accumulated wealth, as they were preempted by those in his lineage. Their status changed from royalty to commoner overnight. In many ways they were hit the hardest.

The death of the chief displaced some hunters associated with the palace, most noticeably the chief's sons who lost their privileged standing. Poulande had anticipated this scenario earlier and had taken up residence with a wife some distance from the palace. Death threats, allegedly by some Ng'onas, and encouragement by district officials distanced the late chief's younger sons, who had hoped that the state would honor their father's wish that they inherit his belongings.

Another son, Edson, who served as the ward counselor on the Mpika District Council, reportedly informed an official in Mpika that game guards harbored elephant poachers in their camps. These guards heard about the report and, late in 1984, they intercepted Edson bringing a warthog carcass in from the bush. Possessing no license and carrying an unlicensed gun, Edson was convicted by the Mpika court and spent a year in prison. More importantly, his conviction opened his district position for occupancy by Ng'ona clansmen.[22]

Retired elders and unemployed younger men drifted back to this valley chiefdom as the national economy declined and the currency inflated. Some became active in local affairs, others retained and deployed their possessions and skills as they could. Among the elders were Philip Njovu and Mathew Sokwe, both retired miners and important members of an organized Valley

Bisa welfare group on the Copperbelt that helped organize urban resistance to Valley Bisa resettlement in the 1960s. Njovu was a close associate of Blackson Somo with whom he worked to press genealogical and legal claims so the latter could become the next chief. Mathew Sokwe brought a rifle and added to the welfare of his relatives. Both Sokwe and Njovu brought personal grudges that, as the new chief Somo became officially appointed, separated them in ways reminiscent of the earlier antagonisms between the late Kabuswe and Luben.[23] Among the younger men, Mulenga Morgan returned to the valley to marry and become a local hunter in the image of his grandfather (Luben) and uncle (Hapi). Kunda Jacobi remained mostly in the valley, married, and later trafficked in trophies and bushmeat.

Besides Kabuswe Mbuluma several prominent patrons passed away during the 1980s, including an enterprising storeowner, Frank Mbewe, who was killed in 1981 by an enraged elephant as he was bicycling on a path east of the Luangwa River. He had gone there to obtain items for restocking his store. His matrilineal kin inherited his rickety shotgun and a rifle purchased earlier from his father.[24] Headman Paison died in 1984 after several years of failing health from tuberculosis. Jackson Katongola retired from work in Kasama and briefly became headman of Paison Village. He dealt briefly there with Lowlenti's contentions and idiosyncratic ways. Fed up with the animosity he engendered among kin along the Munyamadzi River, Lowlenti moved to the Mupamadzi River where he married a second wife. In this new location, he continued to hunt with borrowed guns; unfortunately, the game guards caught him with a cousin's unregistered gun and fined him. Three additional muzzle-loading guns exploded in his hands. In 1988, the venerable headman Chibale passed away after clearing himself of a cousin's charge of witchcraft.

As the times changed in favor of younger entrepreneurs, elders found themselves increasingly marginalized and sometimes victimized by youthful ambitions. They were humiliated at beer parties, beat, and accused of causing harm to others through witchcraft. Younger men, impatient to acquire the good things in life, repudiated their former subordinate roles. They possessed the energy, time, numbers, and conviction to thwart the lineage system with a more individualized youthful ethos that included visions of a different and more affluent future. During the 1990s, much of this youthful energy became circumscribed, dissipated, or was driven elsewhere as the new chief Blackson Somo began mending frayed alliances, demanding loyalty. The unconventional and reckless ways of youth also had a downside as many became victims of HIV/AIDS.

In retrospect, my return for a year of studies at Nabwalya in October 1988 occurred during the lull before the coronation of the new chief and the inception of the "community-based" wildlife program. I sensed four cultural currents immediately. First, residents felt repressed and abused from military and police sweeps through their villages intended to uncover unlicensed trophies and firearms. Second, the widespread uses of these locally manufactured firearms and the international consternation over the slaughter of elephants were the "official" pretext for these military sweeps. Third, the availability of muzzle-loading guns in youthful hands indicated the erosion of elderly authority and suggested serious fractures within community life. Finally, I detected a significant lack of cultural vibrancy, the result of protracted famines, the limbo of a prolonged interregnum, and the lack of leadership.

No longer legal game in Zambia, elephants were rare and protected when convenient. An elephant killed by a local hunter in defense of his fields in 1988 led to intense questioning by the game guards. Local hunters in their prime continued to kill buffalo, yet they began to shift toward smaller prey that were more readily hidden. My records for 1988 show that Mulenga Morgan killed five buffalo and twenty smaller species (2,719 kilograms) while Kunda Jacobi accounted for seven buffalo and fifteen smaller animals (3,801 kilograms). A recently retired individual in Chibale Village killed two buffalo and four warthogs (912 kilograms). All hunters in nearby Paison took five buffalo and ten smaller animals (1,993 kilograms), with most of this buffalo meat smoked and sold on the plateau. Kameko and his sons accounted for 1,586 kilograms of bushmeat mostly from warthog and impala, two zebras, and an immature bushbuck caught in his field. Appointed as interim chief, Blackson Somo inherited the former chief's (Kabuswe Mbuluma's) rifle and other guns and occasionally took game himself. The one buffalo and two smaller animals recorded for him (487 kilograms) were tokens of the reservoir of bounties he tapped from others. Meat appeared at his doorstep from safari operators, game scouts, and residents anticipating reciprocity in other forms.

The total meat produced by all hunters and trappers in 1989 was similar to that recorded the previous year. Again the animals attributed to the chief (two large and three small mammals totaled 948 kilograms) were minimal figures of his real bounty. Kameko and his sons took thirty-five smaller species, mostly impala and warthogs (1,733 kilograms). The head teacher claimed licenses for one hippo and four buffalo, had a hunter kill three and a substitute wildebeest (1,967 kilograms), and sold all the meat on the plateau. Still courting his chances of becoming a headman, Mulenga Morgan killed four buffalo and twenty-four smaller mammals (2,423 kilograms) and deftly distributed the meat to repay for labor to extend his fields. Some distance away, the more commercially oriented Kunda Jacobi killed eight buffalo and nine smaller species (2,959 kilograms). He sold most of this meat locally to government workers. Two other hunters in Chibale Village killed another two buffalo and one each of bushpig, impala, and grysbok (683 kilograms). Four hunters in adjacent Paison Village killed six buffalo and eight smaller species (2,402 kilograms).

Trappers using snares increasingly contributed to the meat in households. Their takes are difficult to measure, yet their production was comparable to that of some gun hunters. My records for ten trappers showed that they took four buffalo and nine smaller species in 1989 (1,784 kilograms). The Nabwalya wildlife counts began to progressively show these erosive effects of non-selective snaring, particularly for smaller wildlife species previously found in the environs of villages.

In 1966–67, my general surveys of what residents in three villages ate assumed that available bushmeat was evenly distributed among village adults and settlement households. Calculations (table 9.1) showed bushmeat averaged 43 percent as a major constituent in all relish dishes for that year, with higher percentages during the late dry and wet seasons. In 1988–89, an intermittent survey of household relish dishes found that bushmeat was more unevenly consumed (table 9.2). During this year, wild meat consumption was highest in households in which a relative hunted or snared (KT, JF, MM, CH, DK, RC) than in those where they did not (LK) or in a female-headed household (KF).[25] The "community-based" wildlife program ADMADE began this same year, and would accentuate this differential in access to bush products and further widen its chasms in household welfare between those with and without employment.[26]

TABLE 9.2. Main composition of relish dishes for households during days individuals were surveyed. The relish dish is classified according to its main source, Nabwalya Study Area, 1988–89. (Adopted from table published in Marks 2014: 89.)

House-	Number			Meat	Vegetable		
hold	**of Dishes**	**Wild**	Fish	Domesticate	Wild	Cultigen	**?/Beer**
KT	38	20	1	5	–	5	4/3
JF	39	8	9	1	3	12	4/2
MM	28	13	4	1	3	2	3/2
CH	30	10	6	–	5	9	–
DK	43	9	4	4	8	14	3/1
LK	43	3	3	1	12	22	2/–
KF	40	7	1	1	13	18	–
RC	40	16	4	1	9	10	–

* Relish Dish is Classified According to its Main Source
Source: adopted from table in Marks 2014: 89.

1990s: The Transition to the New External Order

[With worldwide prices for copper, its major export, depressed, Zambia by the 1990s was ranked among the world's highest per capita foreign debtors and found itself increasingly dependent upon foreign aid for much of its recurrent administrative costs. In the 1991 national election, Dr. Kenneth Kaunda and the UNIP party lost to the Movement for Multiparty Democracy (MMD) led by Frederick Chiluba. President Chiluba was to govern for the rest of the decade and bring about fundamental changes in the political economy of Zambia, including structural changes in NPWS to become a semi-autonomous ("apolitical") agency, the Zambia Wildlife Authority (ZAWA). Donor funding, materials, and expertise were to affect the tempo of changes on the ground that effected both human relations and wildlife populations within the Munyamadzi GMA. ADMADE, the new state-sponsored wildlife program, began recruiting local wildlife scouts and training them for anti-poaching operations in the Munyamadzi GMA in late 1989. Officials assumed that rural residents under this program would forego their customary

dependence on wildlife, thereby allowing wild populations to increase so that the "community" would benefit from the higher revenues generated by the tourists, who paid to hunt them as trophies or to observe them during vacations. As this program foreclosed on customary wildlife entitlements and endowments, it disenfranchised most local hunters, as younger men, with some education and connections, were employed to enforce the new national regulations. National planners placed their bets in cash by allocating a slightly larger stream of funds (40 percent of fees assessed on safari hunters seeking GMA wildlife) to enforce the new rules (mainly to supply, equip, and pay salaries for "anti-poaching" scouts) than what they paid to improve the welfare of all the thousands of GMA human residents, who received 35 percent of these funds. Yet the accounting was never transparent or certain. Planners assumed that the high initial enforcement costs would decrease as villagers realized the "benefits" from their program and abandoned their earlier practices. Yet these enforcement and human costs increased steadily with time.]

Given the change in locus and the intensity of wildlife enforcement during the 1990s, every resident man became a suspect "poacher," forcing the possession of bushmeat undercover and into the shadows. Hunters devised new tactics to cope with scouts and snooping neighbors. Residents used guns only under favorable circumstances. Many households deployed snare technologies underscoring the importance of wild animal protein for their survival. The use of snares was a way of protesting the new wildlife regime, which was undermining local welfare, as it represented an affirmation of local identities based in previous presumed rights of access.

The 1990 records show that local hunters, local officials and wildlife scouts around Nabwalya took some 20,132 kilograms of bushmeat. That year, wildlife scouts made most of the larger kills (buffalo, hippo) and accounted for 55 percent of this biomass; local hunters and two local officials "harvested" 9,026 kilograms, about 45 percent of the total game meat recorded. Local hunters took most of their animals (75 percent in biomass terms) during the first six months when the lush vegetation covered their excursions. They used their contacts and time to decipher the orientation of the new regime.[27] That dry season, the new subauthority employed most adults (mainly men) as carriers to move famine relief from the plateau to the valley.

In 1991, NPWS authorized and the Munyamadzi Wildlife Authority in Mpika recorded that "as a measure of strengthening revenue generation base for Munyamadzi and as a measure of good wildlife management by reducing excess wildlife species to bring them in line with habitat potential, a culling programme (sic) will take place this year on a date to be announced." The authority then listed 99 wild animals for "cropping" that year including eighteen hippos, nine buffalo, forty-eight impala, ten wildebeests, eight warthogs and two each of waterbuck, bushbuck and puku, for a total package of species estimated at butchered weight of 24,340 kilgrams (Minutes of the Munyamadzi Management Authority, 17 May 1991). The list appeared a-made-to-order for an export market as the authority could expect reduced prices on bushmeat while on an official visit to the valley or have the meat delivered to them the plateau at subsidized rates. Yet this demand for bushmeat greatly exceeded local capacity in transport and supply. A month earlier, the Munyamadzi Subcommittee at Nabwalya noted that the local unit "hasn't adequate funds" for the culling project and had requested funds be diverted from the 40 percent management funds "to provide protein to the local people and diversify income for the unit" (Minutes of Munyamadzi Wildlife Management Sub-Authority Committee, 6 April 1991). The subauthority spent money and time to build two sheds accommodating this initiative, yet the quantities of wildlife culled this year remains largely unknown as the records in my possession indicate that the figures were juggled.

As 1991 was a national election that brought about an abrupt change in administration, the culling program was hastily implemented and open to outside manipulation and shady deals by many occupying their new roles. Even newly appointed chief Blackson Somo had reservations about the capacities of his culling staff to deliver, for, in his report at an ADMADE Chief's workshop (1992), he appealed for "experts to run the [culling] project" as he had noticed that with wildlife scouts and officers selling the meat there was "a big loss after comparing cash and the animals killed." He and the subauthority suggested that the numbers of outside hunters and buffalo allowed on permit in the Munyamadzi GMA be reduced by half so more could be taken by local residents. In addition, the chief himself proposed that the "community share" of safari hunting revenues be increased to 40 percent, the same as that given wildlife management.

In 1992, the culling station at Nabwalya was allocated a license of one hundred and six wild mammals, yet only accounted for thirty-seven killed. The unit killed four less wildebeests in exchange for four buffalo and also received nineteen other species (one hippo, six buffalo, four wildebeests, five puku and three bushbucks). A Zimbabwean professional hunter, who leased the southern GMA hunting concession, offered to pay a bonus price to the SubAuthority Committee for each of the species of district game licenses he could obtain from the committee's quota. He also offered to deliver each carcass to the culling station where scouts would butcher them into smaller packets for local sale. Promoted as a "win-win" bargain, the local subauthority accepted his offer. Thus began a series of informal, sometimes unscrupulous deals between a few privileged insiders and outsiders that, while mutually profitable for both parties, remained off record. The big winner in this deal was the safari outfitter, who was able to extend his hunting clients daily charges and fees as they "cropped" additional animals at bargain prices. The Nabwalya culling station supervisor documented that this professional hunter delivered small portions of carcasses, some other outsider demands, privileges and conflicts as well as theft, donations, and the unscrupulous behavior of the Unit Leader, who spent his time in Mpika rather than in the valley. During 1992, the supervisor recorded that the subauthority received ZK 263,740 (US$ 1,691) of which ZK 80,070 (US$ 513) came from the carcasses delivered by the professional hunter. He showed a profit of US $978 for the year after subtracting the wages for nine men for four months without mentioning vehicle and delivery costs.

Besides salaried government employees and plateau residents, few Nabwalya residents purchased meat offered at "subsidized prices" supposedly processed through the local culling station. "No local demand in the valley but very high demand on the plateau" is how the ADMADE unit leader described this situation for me. "His significant development initiative" allowed him to use the community vehicle to truck tons of undisclosed bushmeat to Mpika for higher prices. The subauthority absorbed the opportunity costs for repairs to its vehicle and suffered unspeci-

TABLE 9.3. Types of wildlife taken by residents and non-residents and its estimated dispositions (kg), Nabwalya Study Area, various years 1966–93.

Year (months)	People*	Large Mammals Killed (Elephants)†	Small Mammals Killed‡	Total Known Kills	Biomass (kg)**	Biomass remaining on site/ Estimated exported	Sources of Information
1966 (8)–	LH=12	39 (1)	51	91	16,535	27,416	Marks (1976);
1967 (7)	O=3	2 (6)		8	10,883		residence 12 mo.
1972 (7–12)	LH=4	2	6	8	883	883	Hunter recall; Interviews 1973
1973 (1–8)	LH=12	34 (2)	44	80	15,460	15,460	Hunter recall;
	O=5	4 (7)	1	12	15,413	15,413/?	residence 4 mo.;
	S	67 (18)	169	254	82,199	82,199/?	NPWS data
1977 (12)	LH=1	9	1	10	2,737	2,737	Hunter recall; Interviews, 1978
1978 (1–8)	LH=5	11 (5)	13	29	13,298	13,298	Hunter recall;
	S	48 (47)	195	290	140,146	140,146/?	residence 1 mo.; NPWS data
1988 (1–12)	LH=15	22 (1)	93	116	14,173	14,173/?	Hunter recall;
	T=10	2	6	8	1,409	1,409/?	Residence 3 mo.
	O=3	5	2	7	2,967	2,967/2,967	
1989 (1–12)	LH=10	17	54	71	6,283	4,683/1,600	Hunter recall;
	T=10	4	9	13	1,784	1,584/200	weekly interviews;
	LO=7	7	40	47	4,221	2,441/1,780	residence 9 mo.
	WS	2?	3?	5+	676+?	676+?	
	O=6	5	2	7	2081	– /2081	
1990 (1–12)	LH=5	7	17	24	6,846	6,846/some	Hunter records;
	T=3	1	6	7	821	821/some	diaries;
	LO=2	2	25	27	1,359	903/456	WS records
	WS	20	12	32	11,106	6,026/5,080	
	O	3	10	13	1,500	– /1500	
1992 (1–12)	LH=3	7 (1)	101	109	10,699	5,574/5,125	Hunter records;
	LO=1	0	28	28	1,080	1,080/most	diaries;
	WS	16	21	37	9,509	1,372/8,137	WS records
1993 (1–12)	LH=4	9	48	57	6,356	6,356/some	Hunter records;
	LO=1	1	33	12	1,472	1,472/ –	dairies;
	WS	?	?				WS records

* Local Hunters (LH), Local Officials (LO), Others (O), Safari (S) Wildlife Scouts (WS), Trappers (T). Others include wildlife personnel, government officials, and visitors; WS includes wildlife scouts; Safari include all kills made within the Munyamadzi GMA, some of which was distributed through the chief's palace at Nabwalya and elsewhere. Included in safari totals for 1973 and 1978 are all edible mammals killed and supposedly distributed throughout the GMA (author's correspondence with NPWS and NPWS records).

fied losses for the unit leader's undisclosed revenues from his personal outside meat sales. Sanctioned by NPWS headquarters, this scheme became a cover under which scouts and other prominent figures took scores of buffalo and lesser bucks, yet the records never revealed the full amount of revenue received nor pretended to show how the proceeds enhanced prospects for conservation, development, or local welfare. Common knowledge indicated that the unit and local leader siphoned off these proceeds into their own pockets while the community paid for fuel and vehicle maintenance.[28] Later during a grim famine when the NPWS authorized the subauthority to kill one hundred hippos to sell or exchange for grain on the plateau, the project fell far short of expectations because of costs, lack of transport, and theft (Marks 2003).

In 1992, my notes show that local hunters killed eight large and 137 small mammals (11,779 kilograms). Kunda Jacobi took all of the large mammals and most of the smaller ones (8,821 kilograms); three other hunters took forty-seven mammals, all small. In 1993, Kunda reported a kill of ten small mammals during the first seven months before disappearing from local view. During that year, Mulenga killed or scavenged eight buffalo, most of which went to feed those who cleared and weeded his fields, in addition to thirty-five smaller species (4,088 kilograms). Kameko scavenged the carcass of a lion-killed buffalo (estimated two hundred kilograms) and thirty-three smaller species (table 9.3).

During a return visit in 1993, I realized that surveillance by village scouts together with villagers' experiences with ADMADE had significantly changed local values and their rapport with outsiders. Residents were suspicious of all inquiries, particularly about wildlife—even the scouts became coy in wildlife matters. In many ways, trust had eroded for nearly everyone, necessitating a shift in my objectives and different tactics for my studies.

† Large mammals include buffalo (mainly) but also hippo, elephant, and rhino (largely taken by safari or on license).

‡ Small mammals include all other mammals (impala, kudu, warthog, zebra, etc.) not listed in other list.

**Live and butchered (referred to as carcass yield) weights listed in biomass summaries. Some adjustments were made for immature mammals, differences in average size between male and females, and for carcasses scavenged from predators (when estimates possible).

(Only carcasses witnessed or recorded are listed in kilograms as carcass yield figures.)

Subsequently, I dropped attempts to canvas the community's "harvests" of wildlife and focused instead upon the counts kept by a few individuals, who continued to keep their records independent of scouts (Marks 1994; 1996: Marks and Mipashi Associates 2008). These counts allowed me to assess the changing wildlife dynamics around Nabwalya while I transitioned to a qualitative evaluation (indicators) of what was taking place within its villages. The rest of this chapter describes the impacts of the new wildlife program ADMADE on residents within the Munyamadzi GMA generally by following the events and actors on the developing center stage at Nabwalya, as it was transformed rapidly during the 1990s into the new local designation as "Nabwalya Central."

The Effects of the ADMADE Wildlife Program

When it was announced, the ADMADE program stirred optimism at Nabwalya, promising the potential for the area's long-anticipated "development." After the subauthority solicited suggestions for making the new program relevant, ADMADE rebuffed publicly all local ideas, overruled them in committee, or never acted upon them. As government retrenched its rural services, ADMADE progressively restricted community funds to salaries and predetermined capital items such as schools, teachers' residences and equipment, each a category within a specified percentage of allocated community funds. The program assumed a formal alliance with the "traditional authority" (the government-sanctioned and salaried chief), presumed to represent his community's interests and to implement national policy. Wildlife management never decentralized nor reflected local interests (Marks 2014: chapters 6, 8).

The subauthority and the new CRB institution, begun in 1999, were never certain how much money was in the community bank account at any time as the state wildlife agency controlled the release of funds. Payments were often late and at times never deposited, periodically diverted to meet NPWS or later Zambian Wildlife Authority (ZAWA) deficits. Consequently, local authorities met mainly to redistribute released funds or to obtain credit to pay for previous commitments and contingencies. Their agendas were fixed by outside officials rather than by local representatives. As this dependency grew, so did community debts, with scouts and teachers experiencing delays in their salary up to a year. Some workers went on strike until they received a portion of what was owed. Projects were delayed by valley conditions that constrained the delivery of supplies to the six months of dry season. Securing transport and maintaining vehicles became costly and perpetual bottlenecks.

As clients, some men obtained local employment as temporary workers on community projects, students obtained bursaries for continuing their secondary education from safari donations and community funds, and others served as untrained teachers and temporary wildlife scouts. A few more residents worked for safari operators once the hunting concession's rules changed, thereby improving their chances for employment. The subauthority expected secondary students on loans to return upon completion of their education and help developments in the valley; most saw education as their ticket out of valley poverty and deployed their learning elsewhere. The few employed men brought new and unprecedented wealth into the community; a few stocked local stores with materials such as cloth, salt, soap, canned goods and biscuits that became "essential" markers of prosperity. Yet improvements in welfare and access to the flow of new material resources favored those already employed, with some education, and with those belonging to or having close ties with the Ng'ona clan. Most of this wealth was transient and soon left the community to purchase durable goods and food from elsewhere. The numbers of people settling around the palace blossomed, for it was both a prime site of employment and the point for distribution for relief supplies in times of famine. The palace and surrounding neighborhood contained the main structural initiatives within the GMA, including school buildings, a health center, improved housing for scouts and workers, an administrative block and equipment, wells, grinding machines, and vehicles. Single structures elsewhere in the chiefdom came later, if at all, and were few in comparison to those within the villages around the palace.

Employment and cash became the new basis of local patronage, with capital and purchasing power the currency upon which alliances depended. Insufficient funds and inadequate means and sources of money worried everyone. With very few full-time jobs even in the offering, employment remained seasonal and part time ("piecework"). The scramble for jobs and money forced a sea change in social structures, narrowing focal social relations and rechanneling local largesse in bushmeat distributions (chapter 7). Polygynous marriages became common among employed men; women became increasingly dependent on their spouses and lineage men for economic essentials. A change also occurred in ideology. Feeling the pull and push of forces beyond earlier boundaries, many residents converted wholesale to Pentecostal faiths rather than to the established institutions of Catholic or Protestant denominations.

The Chieftaincy

Despite its many detractors, the ADMADE program benefited some residents. The most important was the chief, who chaired the local subauthority, selected and hired wildlife scouts, made appointments to committees, and commissioned projects, and by direct and indirect means oversaw the flow of ADMADE funds on the ground. The wildlife program greatly enhanced his powers of patronage while the scouts' surveillance enhanced his control over adversaries.

Two events allowed the chief, Blackson Somo, initially to consolidate his power. Although Somo was confirmed in his status as chief in the final months of the Kaunda administration, he found a drastically changing national political arena when the Movement for Multi-Party Democracy (MMD) ousted the President Kaunda's United National Independence Party (UNIP) in the 1991 elections. The incoming president Frederick Chiluba promoted a series of neoliberal economic reforms, including a reduction in state amenities such as health, education, and welfare, with the expectation that rural communities or civil organizations would take on these expensive services particularly in GMAs. Some reforms affected hunting concessions, including the introduction of a longer-term hunting lease and land-use restrictions. As a UNIP supporter, Chief Nabwalya was asked to consider a proposal of a ninety-nine-year lease on his land that was vetted and sanctioned by the new government, which he interpreted as an incipient grab for additional Bisa land. With advice from his national contacts, the chief adamantly refused to accept this deal and temporarily ended this transaction and used the occasion to rally his subjects behind him to save the homeland. With his dream of a long term lease somewhat diminished, the entrepreneur obtained a concession for half of the Munyamadzi GMA during the formal government-tendering process and gradually overcame the initial local misapprehensions to become a staple contributor to community development and conservation programs. Unlike other private concession operators, this entrepreneur did not depend on making a profit from the concession as he could selectively tailor his culls of animals to yearly conditions and pay the state for the allocated but unused animal quota. As a proprietor of a concession, he also managed so his friends could actualize their sport hunting wishes to shoot African big game with modern bows and arrows, a means prohibited earlier by the colonials. As a goodwill gesture to the community, he paid for the delivery of famine relief by aircraft during the drought of 1992 and petitioned successfully for ADMADE to return his expenses in 1993.[29]

The second event was almost tragic. The first thunderstorm of the 1994 rainy season produced a brilliant bolt of lightning that completely obliterated everything within the chief's temporary quarters while his palace was being built. Fortunately, the chief and his wife had sought refuge from the hot, sultry weather and were staying in their reception hut in front of the palace. Learning of the chief's newfound destitution, his friends showered him with gifts and contributed funds for building a new and more substantial palace.[30] The construction of the new palace with burnt bricks and a metal roof took three years to complete. The chief's friends and wellwishers also stocked his palace with new furniture, a refrigerator and conventional equipment to facilitate his business and farming enterprises. Later in the decade, the two safari concessionaires gave the chief a used Land Cruiser.

Chief Nabwalya also experienced the downside of visibility and wealth. During the 1997 drought, the police arrested him on suspicion of stealing money from a visiting government delegation authenticating registration documents. This source turned out to have been false and the chief was cleared eventually of all charges. Yet, both he and his wife were beaten severely, suffered the indignities of an arrest, and spent time in prison. His wife died the following year supposedly from internal injuries suffered during this ordeal.

Somo appointed a steady flow of confidants to positions within the subauthority and its successor in 1999, the Community Resources Board (CRB). Some of these appointed men became transients as they took the blame during crises and paid the consequences for reports and calculations beyond their training and grasp. The chief depended mostly upon two loyal individuals. Lightwel Banda belonged to the chief's clan; Philip Njovu did not. Njovu was a long-time friend from their years on the Copperbelt and helped Somo during his ascension to become a chief. Both individuals were crucial for developing and sanctioning new alignments of power within this chiefdom.

Born in 1947, Lightwel had a secondary education, which included work in bookkeeping and accounting. He returned to the valley in 1979 at the behest of Kabuswe Mbuluma to assist with education and development. In his mother's village of Kalimba, Lightwel established a self-help school that became a showpiece for the district. He also was elected the political chairman of his local ward. Upon the retirement of Julius Mwale in 1989, Blackson Somo appointed him as a court justice, a position that brought Lightwel to the center stage of politics at Nabwalya. Besides this court position, Lightwel became a part of the inner circle, serving as chairman of the local cooperative, vice chairman of the local Catholic congregation, chair and main spokesman at most public gatherings, and chief's representative in regional and district meetings. Most importantly, he became an active member of the Munyamadzi Wildlife Sub-Authority.

Given his background and work experiences, Lightwel's activities affected local hunters profoundly. Armed with the chief's blessing of plausible amnesty and as a ward chair, Lightwel encouraged many illegal gun owners to turn in their unregistered weapons. This 1992 initiative netted forty-seven unregistered firearms, with promises of cash rewards in addition to pardons. Unfortunately, these guns were taken away in an undercover ruse by another wildlife command from Eastern Province and became part of its booty in a government anti-poaching initiative there. Residents yielding their guns received nothing (see chapter 3). In 1993, Lightwel sought his late grandfather's muzzle-loading gun. Owning a weapon was a way he could tap into the perks of subauthority membership.[31] As a member, he had first dibs on the meager quota of local NPWS game licenses allocated annually for community members. Once in hand, these licenses allowed their holders a legal windfall. The license owner could sell his licenses to safari operators at an inflated price, kill the animal(s), and sell the meat on the plateau, or use the meat for local patronage. Prominent individuals monopolized these buffalo licenses and turned them into substantial sums of money and favors.

Lightwel's high-profile activities touched the lives and sentiments of many residents and outsiders. He was fully aware of the jealous detractors within his lineage that were threatened by his energy and prospects. Fearing that he might take over an important leadership position within the subauthority, a lineage member confronted, threatened, and "forced" Lightwel to withdraw his candidacy. Again, after receiving an invitation to a regional conference, he was "forced" to stay behind by similar hostility.[32] After successfully attending a national ADMADE meeting in 1995, he fell sick and almost died. As a deputy to the 1996 ADMADE convention, he fell sick again with an inflamed eye. A Kitwe surgeon operated on his eye, but Lightwel never recovered his normal vision, and his health deteriorated even as the subauthority and others contributed large sums of money toward his consultations with specialists.

In September 1997, Lightwel served as master of ceremonies at the chief's Malaila ceremony, fell ill from exhaustion, and died the next day. In death, he was given a hero's burial in the royal cemetery along Bemba Stream. At the gravesite, orator Philip Njovu listed Lightwel's accomplishments and pleaded with the assembled crowd, "Today Lightwel has left a very big gap because the community will no longer benefit from his many contributions. We will really miss him and every Bisa must know this for I am urging the Ng'ona people that they must stop this dangerous trend of jealousy." Everyone knew this reference was a euphemism for suspected witchcraft. People speculated on who was responsible. Some assumed the cause was bad relations among his kin on the judiciary. Others suggested those on the subauthority were to blame, while still others asserted that his death was brought about through competition over a woman. The next meeting of the subauthority featured an eloquent eulogy by the unit leader, and the assembly allocated to Lightwel's estate two hippo licenses from its quota. Ng'ona clansmen appropriated the cash from these meat sales as well as all of Lightwel's material possessions. His two wives and children remained destitute, hesitant to return to their home villages.

Like Lightwel, Philip Njovu was born in the Luangwa Valley, leaving in 1945 "so he could find something to wear" to express to me what his options were at that time. He held several jobs in town before becoming a supervisor in the Kitwe mines. He was a cousin to Blackson Somo and remained a close confident. As an impressive and assertive individual, Njovu became the prime mover and spokesman behind the new order. In local parlance, he became the "second chief" and orchestrated many strategic plans. His appointments covered the local power grid as a member of the subauthority, chair of the local school association, chair of the cooperative, and treasurer of the local Catholic church, and he was also in charge of most monetary transactions and community projects. People both feared and respected him.

Because Njovu acted so effectively and authoritatively, he suspected that others wished him ill. In 1997, while acting in the chief's absence, he received a report that an elephant had been killed nearby. He immediately realized his options were to cover all bases until the chief's return; otherwise he might be accountable and the chief vulnerable. His impromptu decisions impinged upon other residents' activities, which upon the elephant's demise becoming known by outside authorities might implicate local people in plausible criminal acts (introductory chapter). Njovu's death in 2000 coincided with another "unexpected" elephant's death, whose carcass provided a windfall feast for the locals and government representatives participating in the funerary protocols. After losing his two lieutenants, Lightwel and Njovu, the chief became dispirited and personally devastated. It took him several years to recover momentum and regain authoritative traction.

The Remaining Local Hunters

Many of the standard-bearers within the earlier lineage husbandry of wildlife, who were visible in earlier decades, were dead or soon would be. Julius Mwale, headman, renowned hunter, and long-term court justice, suffered from tuberculosis and died on 21

May 1991. Ackim Mbuluma passed away on 12 June 1996, as did Lowlenti that same year. When his second wife died, Lowlenti returned to Paison Village where he faced tough times. There his first wife refused him, claiming that he had inappropriately "rested" his late wife's spirit, and, given his reputation and history, no one trusted him with their firearms. He struggled to support himself by fishing and became a mere shadow of his former active and defiant self until his death.

Mathew Sokwe's wife was attacked by a crocodile late one afternoon and died in transit by hammock to the hospital. As she belonged to the same lineage as Hapi, rumors spread that she was a victim of a "sent crocodile" and suspicion centered on her husband as the wizard (see chapter 5). Kunda Jacobi justified his savage attack with an iron bar on Sokwe's head and torso because his elder had been labeled a "witch." Sokwe made a quick exit to the plateau for treatment for his head injuries and filed an assault report with the Mpika police. Upon his return months later, suspicions about Sokwe continued. Later, he and a son were summoned before the chief's court and charged with showing disrespect for not visiting the chief and by "being bad and selfish persons." In court, he recognized his relatedness to the chief and suggested that circumstances had prevented him from making courtesy calls to the palace as he feared accusations as an "informer" (a pejorative colonial label) similar to what others had claimed about the late Chibale. Perhaps more relevant was that he and his accusers belonged to contesting national political parties during an election year.

In the process of solidifying their local reputations, Malenga Morgan and Kunda Jacobi had the most to lose from the expanding presence of the ADMADE scouts. The two continued to press and expand their separate social strategies for several years, with necessary periodic deference to scouts and their detractors. Malenga's predicament was more precarious: not only did he belong to a lineage whose predecessors were often perceived as being at odds with the local Ng'ona lineage but his residence was also proximate to both the palace and the scout camp. Kunda was caught up in the more factious individualism of the time and, as part of the breakup of Chibale Village, eventually created his own homestead with fields away from the rest of his kin. He kept respectful rela-

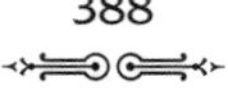

tions with the chief, retained links with powerful outsiders, and left his isolated settlement and family for long stretches of time. Later in the decade as his health declined, he accused an uncle of witchcraft and beat him with an iron bar. Still trying to find a resolution for his declining powers in his malignant social relationships, he visited a witch diviner across the Luangwa River and died while there in December 2004.

The Wildlife Scouts

With their primary education completed, some local youths gained employment as wildlife scouts. Recruitment as scouts depended upon their education, local connections, and luck for their employment. One of these fortunate ones, Richard, the grandson of a local headman, finished his primary education just in time to join the first class of wildlife scouts.[33] Stationed at Kanele Wildlife Camp, he became the main scout at the culling station. Richard distinguished himself as a conscientious scout, a fearless hunter, and as a community servant, but his job was never easy.

Once, while temporarily assigned to another wildlife camp, Richard heard gunshots. He gathered other scouts and went to investigate. They found a vehicle belonging to a development agency in the possession of his unit leader and other uniformed wildlife scouts. These men had just killed an antelope and were butchering it before loading it in a vehicle. Despite his demands to see their license, Richard received only silent stares, convincing him that his duties were not welcomed. Richard had witnessed an increasingly normal exchange, one in which outsiders combined their vehicles and contacts with the firearms and authority of wildlife staff to cement clandestine deals, even as such infractions were known to infuriate residents.

Scout activities included protecting wildlife and safeguarding government property and the safari camps. During the persistent famines, such assets became tempting and the thieves mean-spirited. In July 1995, two thieves, Benson and Musonda, raided a safari camp while safari clients were out hunting and stole a large amount of cash and ammunition. Richard apprehended Benson on his own the next morning, and, while the two walked to the safari camp, they came upon Musonda, who was unknown to

Richard as the other robber. Musonda got the jump on Richard, clobbering him on the head and torso with an axe and leaving him near death and covered in blood. Both thieves disappeared and fled to town. When the other scouts searched and found Richard, he was hardly recognizable with his swollen head and bloody uniform. They took him to the clinic where he slowly recovered. This ordeal troubled Richard, and he never showed the same enthusiasm for subsequent assignments. He was often sick with headaches for which he visited the witch finder, and he was later transferred to another ADMADE unit elsewhere in the Luangwa Valley.

The wildlife scout program's initial decade featured ups as well as downs. Scouts managed to discourage some hunting, particularly with muzzle-loading guns, and served as a deterrent to the most visible local infractions. They targeted, arrested, and even employed some "notorious poachers," channeling their skills to keep them and others out of the meat "business." The wildlife scouts and teachers became the most discernible presence of state authority and power. Safari clients and civil servants supported the wildlife program, as its staff improved their access and protection on a wildlife "frontier"; scouts could minimize their costs in hunting or securing bushmeat; and, given the appropriate circumstances and connections, these outsiders could reciprocate with uncommon experience and monetary benefits. ADMADE's imposed organizational structures locally camouflaged or deflected the heavy dictates of these outside influences. In particular, safari operators poured millions of Zambian currency into supplying, supporting, and advising scouts while lending their transport and giving rewards for anti-poaching arrests. The government publically promoted scout activities and arrests in the media even as they provided access to wildlife for political cronies and party functions (ministerial game licenses). Wildlife revenues generated and dispersed under the ADMADE program were suspect because of claims of aberrant transfers at all levels. The tragedy was that no program promoter cared to learn of its downsides; it was simpler to believe in the stated mission and to expect that time would lessen any detrimental aspects that surfaced within any outsider's brief exposures. Moreover, the careers of project staff required promoting successful outcomes, even if the snapshots of observational dots were discursive, unrepresentative, and even anecdotal impressions.

Despite heavy outside investment, the wildlife agency and its local agents lost ground on both the local and regional "poaching" fronts. Commercial hunters with their long strings of carriers descended the escarpment and annually took buffalo, elephant, and hippo by the hundreds. Besides providing some immunity against capture and prosecution, the manpower and firepower of these gangs swamped that of scouts arrayed against them. Many impoverished plateau youths entered this tragic fray to stay even against an economic tide that, even in their home villages, was not in their favor (Marks 2001; Brown and Marks 2007).

Scouts filled their reports with numbers of arrests and patrols, but also with their local skirmishes over beer, women, and fears of public reactions to prescribed roles. Subauthority minutes and local records included specific cases of scout indiscipline, adultery, rumors, wizardry, failure to perform duties, and fights. With little national oversight and supervision, unit leaders didn't last long within this GMA. As outside recruits, they usually lost credibility through their wanton behavior or failed to carry out the orders of local authorities. As the government responded to donor insistence that it reduced its numbers of civil employees, all scouts and guards eventually lost their employment during the transition to ZAWA in 1999. Very few were rehired as ZAWA passed on most enforcement responsibilities and expenses to the local community resource boards. Former scouts received large payout packages, which they frivolously spent on luxuries (e.g. expensive radios, videos, and vehicles). They terminated their town marriages or liaisons and prepared for reentry, if they still could, into a dismal village life, which most feared as well as detested.

The century ended in famine and drought. In December 1999, the chief went out to Mpika to petition the vice president for assistance. There he was told to expect relief in February. When the first relief flight arrived on 21 February 2000, the helicopter was booked as a media event and contained only a few maize bags of famine assistance. On board, the Member of Parliament representing the district deplaned for a photo opportunity and appeared more concerned about her reelection then than managing the surrounding destitution. School had been adjourned through mid-March so children could assist their parents, who were collecting *lupunga* (wild rice) and "harvesting" other wild products (bushmeat) to survive.

On the Demise of Diversity: Out of Many a Coercive and Corrosive Dependency

Development is a cipher for political processes contingent upon a government's interests and activities, just its "lack" (or underdevelopment) is supposedly a result of governmental neglect. Given this label northern nations have proposed and implemented various types of imaginative schemes around the world in "less developed" regions. Through a callous and calculating jargon of selected ideas and "technical" ways to implement their economic models, these elsewhere places appear exceptional locales for new policy pronouncements and as targets for development planning. Booked as "a-political" official texts, planning documents of development undertakings lack pertinent political and institutional analysis, as well as relevant cultural and environmental evaluations, and suppose that the national arenas and situational fields where their projects are applied are receptive realms for their prosaic fantasies. When presented as "win-win" development agendas for the donor and the client state, these plans supposedly incorporate essential improvements for the latter's subjects and their wishes for modernization and enhanced welfare. From the initiators' perspective, the Zambian "community-based" wildlife program, ADMADE and its successor under ZAWA, might be considered as a well-intended program of integrated wildlife conservation and improved human welfare within the Luangwa Valley. Yet, from the perspective of an incisive half-century case study, this wildlife program has succeeded so far only to increase the morbidity and traumas for most the Munyamadzi GMA's human residents and through its anti-poaching wildlife policies and slanted regulations decreased the viability and variability of some wildlife populations.

Buffalo bushmeat was a stock component in the identity and destiny of Valley Bisa lineage hunters. This cultural good assumed its symbolic significance through its complementarity with agricultural products, the stock of women's work. Buffalo and bushmeat consumption served as a meaningful spiritual reading of the ancestral lineage spirits evaluation of their living descendants' compliance with the corporate (community) moral norms within a sheltered "subsistence economy." The procurement of bushmeat was a generational-honored and decades-honed way of becoming a man, an identity as a protector of lineage hegemony and welfare. A metonym for good health, well-being, and "the good life," bushmeat retained its value as a component in ordinary meals and assumed local significance greater than its nutritional and monetary values.

For lineage hunters, buffalo meat and guns were metonyms for this provincial system of power, privilege, and status, the bastions of a provincial masculinity and its cultural emblems. As a social currency and as a perishable commodity, buffalo meat had its limits. As a local product, bushmeat was bulky and visible with a comparatively short life under blistering conditions. Hunters used their knowledge, conceptual expertise, and adopted means (muzzle-loaders) to exploit this source enmeshed within their lineage's overall strategy to improve wealth and welfare. As a productive stratagem, hunting values were decipherable by kith and kin sharing responsibilities from the beginning to the ending of its procedures. People understood the glitches, uncertainties, and problems in these processes, which were subjected to local negotiations as well as reparations. Dreams and petitions of close relatives sometimes stimulated these forays just as surely as lineage quandaries and decisions over the culpability of its members might end them (Marks 1979a). Lineage squabbles, persistent grudges, and wizardry might halt a hunter's production of meat just as the breaking of a ritual norm was thought to expose a hunter's kin to physical or mental harm.

All small and insulated cultural bastions depend on the persistent and cohesive strengths of their members against both internal and external conspiracies. Cultural convictions may become targets for those excluded as well as its members' progeny's dispositions awaiting their moments of generational ascendancy. In the Valley Bisa case, decreasing isolation through changing national and local economies of scale, imposed structures and new laws, through education and the exposure to new ideas, increasing poverty and epidemic diseases as well as gender and generational conflicts, and an expanding youthful population all worked synergistically to transform these earlier provincial strongholds. Ideas and behaviors, like Trojan horses, can take root and cultivate change, or take shape and appear benign, when accepted and embraced. In this case, the many scattered lights of an earlier tangible patronage based upon lineage elders, bushmeat, ancestral cults, face-to-face accountability, and protection faded during the late1980s, to become replaced by a single source centered on a chief's patronage within more distant and uncertain national sanctions, expanding communities of hopeful, global parishioners, and the fungibility of money.

The epigraph of this chapter, "In killing a buffalo, the game guard likens it to his mother," is a riddle I recorded in my notes during the initial ADMADE program years. This idiom offers a fitting way to end this chapter. By juxtaposing a negative image or constraint (game guard)[34] between two positive icons (buffalo, mother), the saying plays upon residual cultural values by showing that the guards' subversive approach threatens the stability of the older cultural order. The identity for a Valley Bisa man allegedly begins at birth and is embedded in the welfare of his mother's lineage. Within this paradigm, the proper destiny of buffalo is to feed, protect and support one's matrilineage rather than to contest that fate. This meat meant for "respectful" disposition and consumption among descent group members was not for sale, similar to the guns producing it, except in times of extreme circumstances. To become an adult man, a youth marries and produces children, yet the identities of these children belong to his wife's lineage. Upon marriage, a man mentors and contributes to his sister's children, for they stand to inherit his status, possessions, and legacy. This indigenous epigram cautions wildlife scouts that in assuming their employment as protectors (of wildlife) and procurers of local wealth (employment, bushmeat), they should not withdraw their protection from or jeopardize (demean, kill) their mothers (local birthrights) by judiciously following the shorter-term objectives of foreign bosses. Rather, young scouts should temper their exuberance and temporary ascendancy by realizing that their heritage and security remains in the hands, hearths and memories of local kin. Thus the injunctive proverb appearing in chapter 1, "An important traveller [foreigner] may become your mother," as an individual opportunity during an earlier era had become transposed into a threat to the current descent groups' security and environmental welfare.

Notes

1. *Gameguard nyama wapaya nyina; wako wapaya mboo* (recorded 1989).
2. Although my field research began in 1966, the stories and information collected during this initial year were mainly those of personalities and events from the earlier decade that continued to be influential during the 1960s and 1970s.
3. Many of these headmen claimed to have killed one or more elephant, eland, and "many buffalo" (Marks 1976: table 36, p. 200). The taking of buffalo was expressed then with verbal gusto and hand gestures suggesting overwhelming numbers.
4. One headman and the resident district tax collector had a shotgun, each acquired after long service and official support. The chief, by virtue of his status, owned a .375-caliber rifle, two shotguns, and had several muzzle-loaders in his possession.
5 Eustice Poles, a Game Ranger stationed for several years beginning in 1946 made foot patrols several times a year into the Luangwa Valley. He kept detailed observations of his daily observations on wildlife during his foot patrols as well as recorded his impressions of residents' and elephant control guards' actual and inferred activities upon wildlife. See particularly his earlier reports (Poles 1947, 1949, 1951a,b, 1953), in author's personal collection.
6 These game mammals were listed on a different schedule and required the purchase of a separate license. Lions were still listed as "vermin" as were most of the wild predators and could be shot on sight. Only later were they listed as game animals requiring additional licenses and safari fees.
7. My records of all the hunting potions and ritual prescriptions for pacifying the carcasses of elephant, lion, and eland show the imprint of this chief as their prescriptions remained the same, unlike those for other species, which varied and represented more local lineage formulae. After his initial kill of a rhino, even Jackson sought the patronage and protection of this chief.
8. Local rumors abounded about the "genetic" madness, back-biting, and endemic wizardry among members of the chief's lineage. This madness was evident in some by their strange and unpredictable

behaviors. Untoward assaults and claims among Ng'ona members were longstanding and enduring; neither time nor distance seemed to resolve them. Kabuswe Mbuluma was one of eight siblings "of the same womb" (four sisters and four brothers). By the beginning of the 1960s, the chief had outlived all his siblings, including their children, with the exception of Goliat.

9. Hapi told me this story in 1993. It underscored his earlier relationship with his beleaguered father (before becoming his client) and the context and significance for Luben's quest for a buffalo at that time.
10. When Kalaile died in 1975, some suspected him of being a "wizard" and of using witchcraft. According to these accusers, he had "constructed crocodiles" to prey on his relatives and neighbors. These claimants mentioned three young girls, including the daughter of a game guard, all taken by crocodiles during a short interval. Responding to these tragedies, the guards snared the crocodile and cut it into pieces. Coincidentally, Kalaile went to the Kanele Wildlife Camp and, upon hearing that "his crocodile" was dismembered, reportedly died on the spot. Local lore was that the person responsible for such a "constructed animal" would expire simultaneously with his product. As most of Kalaile's relatives were in town and showed no inclination to hunt, his relatives sold his shotgun to a policeman; a nephew inherited the muzzle-loader.
11. It was rumored that Goliat's spirit had possessed a nephew, who two decades later was found unsuitable as a candidate for chief.
12. Actually this requirement began with the district councils, whose territories included GMAs. This bushmeat was a source of patronage for district politicians and officials. When councils failed to provide staff and vehicles to transport this bounty from the Munyamadzi GMA to Mpika, safari operators began dumping the carcasses at the chief's palace.
13. The delighted DS paid Hapi fifty pounds sterling (a sizeable sum) and returned to Mpika with his vehicle overloaded with two buffalo carcasses. The chief remained with the other two; Hapi got a large share.
14. There were other dimensions to Luben's dilemmas. He had shifted his village several times, more frequently than normal, while looking for a "peaceful" and productive site at Nabwalya. Confrontations with Lowlenti included the latter's disrespectful comments over beer about Luben and Lowlenti's threats to pull his two sisters out of his uncle's village to establish his own settlement. Luben also had longstanding misunderstandings with the chief about a woman and about not showing appropriate "respect." See also endnote 23 below.

15. Luben initially told me that this new location put him closer to his dry-season employment as a vehicle driver for a safari operator. Yet his desertion of Ng'ona territory and settlement under a Chewa chief, along with his earlier circumstances, had longer-term consequences for his remaining Nabwalya relatives. A single and momentary inquiry usually touches only the surface issues of an individual's major decisions.
16. I was with the late chief when he received the initial message about Kapampa's demise. The search party informed the chief that they had found little of Kapampa's remains and the working presumption was then that he was killed by an elephant late one afternoon while collecting reeds along the Munyamadzi River to make a mat. That evening hyenas had found the body and left few remains. The interpretation that Kapampa was murdered came later in 1988 when I inquired about this death from one of his nephews.
17. Under matrilineal succession, sons neither inherited property nor would succeed to the status of their fathers. The chief's choice of his sons was an indicator of his growing distrust of his living Ng'ona relatives. These problems became visible the next decade during the succession disputes over the chieftaincy.
18. An itinerant honorary wildlife warden happened upon Poulande carrying meat and arrested him, taking him directly to Mpika for trial. At the time, the chief was away so he could not intervene on behalf of his son. Local residents did not protest the arrest when the warden appeared at the palace after the arrest, so the warden took Poulande to Mpika for trial.
19. For example, I estimated that 28 percent of the eleven buffalo killed (representing some 3,650 kilograms of meat) from July to mid-September 1973 was consumed locally. Outsiders killed eight of these buffalo using the expertise of local residents.
20. A detailed description of disassembling these buffalo and other kills for July and August 1973 are in my unpublished field notes.
21. My brief observations then led me to predict the likely demise of the buffalo and zebra at least on the Nabwalya Study Area. This conclusion was a premature judgment in view of subsequent events.
22. An elephant was to kill Edson Nabwalya in July 2006 (see chapter 10).
23. Luben's first wife allegedly was betrothed to Kabuswe Mbuluma before he departed for a long stint as a migrant laborer to Southern Rhodesia. Upon Kabuswe Mbuluma's return years later to Nabwalya, he was summoned and appointed as a chief by the colonial administration. By then, Luben and his wife had produced two children, one of whom later became Mathew Sokwe's wife. Luben's strife

escalated during the late 1950s as the driver for the chief's new Land Rover, which was stolen from the repair shop while being serviced on the Copperbelt. When Luben's spouse died, he was not given a replacement by her lineage. Local "disrespect" and continuing local wrangles were the major reasons behind Luben's decision to shift and settle elsewhere. Mathew Sokwe would have his own trying moments later when he was called to the chief's court to answer for his own shows of "disrespect."

24. Mbewe resided at his well stocked store near Nabwalya palace, yet his home was in Bunyolo Village, some distance away. He used his guns to take game occasionally on his bird license. He kept these guns mainly for his status as a businessman and was not a "hunter." During the 1970s, Mbewe owned a small second-hand van to transport supplies from Chipata to his store.
25. The reader might recall Kameko Tembo (KT) and Mulenga Morgan (MM) from their life histories (chapter 5) and Doris Kanyunga (DK), Rose Chiombo (RC), Joseph Farmwell (JF) and Mumbi Chalwe (MC) from their activities (chapter 2). All these individuals and the others in this table were participants in the social and activity surveys in 1988–89.
26. The assumption of equal distribution of meat during 1966–67 is appropriate given that 40 percent of the recorded meat tallied in relish dishes could be attributed to the activities of the elephant control guards, who killed six elephants during those months, five to protect cultivations , one for killing a returning migrant. These elephant kills were "common property," as meat available to all attending the butcheries. The more focused and detailed time and motion studies in 1988–89 were representative of the cultural and wealth stratification that year. On a randomly selected weekday, each participant household was observed for a whole day, including the relish dish consumed.
27. The NPWS neglected to allocate a quota of game licenses for area residents, describing it as "a technical oversight," meaning that all wildlife harvests by local residents during 1989 and most of 1990 were technically "illegal." This admission occurred because the Chief had contacted [letter dated August 22, 1990] the Northern Provincial Member of UNIP's Central Committee about this issue, who wrote the Director of NPWS on September 18, 1990 [REF MCC Northern Province to Director NPWS Allocation of Resident Hunting Licenses for Nabwalya included a minute MCCNP 9/4/1990 to the Warden at Mpika] as follows: "This office supports Chief Nabwalya's observations in not considering residents for hunting quotas. It is true this

policy is not doing the country any good because residents have resorted in killing animals. As a result the Nation's efforts to control poaching will be fruitless as this will be perpetuated om a large scale by residents who are denied their hunting quotas. I urge you to grant hunting licenses to residents as well as a way of controlling poaching which is on a large scale in some areas of the country." The Director of NPWS [REF: NPWS/WLM 13/1 NPWS Director to Chief Nabwalya 9/18/90] confessed to the chief as follows: "It is regrettable that [this] issue was not resolved this year. I would like to assure you that we will by every means facilitate the full participation of the people in the planning, management, and use of wildlife." The Director set the year's quota of animals as "Buffalo (15), Impala (20), Puku (20), Warthog (20) total seventy-five animals." He suggested that the subauthority hold a raffle to apportion the quota among hunters possessing firearm licenses. The letter arrived later and after most local hunting was over for the year and people were busy cultivating fields. Subsequent quotas for buffalo were never this high for residents again. The 1991 residents' quota for eighty-eight animals included baboons (ten) and hippo (five), both species not preferred by residents, and only eight buffalo.

28. These claims are based on observations and interviews with the culling station manager and staff, including reviews of local and subauthority documents during my visit in 1993. As the station's quota arrived late in 1991, the scouts recorded killing few animals. In 1992, the culling supervisor reported that the safari operator took nineteen animals (37 percent of licensed animals including six buffalo and one hippo) while the culling station took thirty-three head of game (including sixteen buffalo and four hippo). Outsiders often obtained animals at local prices, and many transactions were on credit with much money going unreported and unpaid. ADMADE's unit leader was responsible for sales, and he was often alone during these transactions. (Culling supervisor, 1992 Annual Report Munyamadzi GMA; minutes of 1993 Wildlife Management Authority; Wildlife Management Sub-Authority Committee minutes and personal interviews, July 1993).
29. Minutes of the Munyamadzi Wildlife Management Authority, Mpika, 21 May 1993. All hunting leases in Zambia expired in 2012 and were subjected to another round of tendering processes. This foreign fiancier withdrew his bid from reconsideration after the tendering process became "political" after the death in office of President Michael Sata.

30. The chief was quoted in the *Zambia Daily Mail* (8 October 1994) as saying, "I lost all my valuable property in the fire, kindly publicise [*sic*] this tragedy and let my subjects and other well-wishers assist me out. He [the chief] further said that he would appreciate any form of assistance that may be rendered."
31. Lightwel knew more about modern weapons than he did about muzzle-loaders. While he was in town he joined the Home Guard and was certified as a marksman. Ownership of any weapon, particularly the possession of a certificate of gun ownership (termed a blue book), was essential to obtain a game permit and a local resident's license. Lightwel had yet to cultivate the contacts necessary to obtain either a shotgun or rifle.
32. I have placed "forced" in quotes to contextualize the syntax of the language and the seriousness of these allegations. Threats of witchcraft are implied if the individual does not comply with the curser's wish.
33. Richard's lineage had produced a long line of distinguished local hunters. When he finished his primary schooling, he worked briefly with me in 1988 on initiating the social profiles and the time-motion survey. When he expressed a wish to become a wildlife scout, I wrote him a letter of recommendation, as it seemed to be in his long-term interest. He was trained in the first class of local wildlife scouts.
34. As a colonial loan term, "game guard" persists even as designations have evolved to "wildlife scout," "community scout," and recently "wildlife police officer." In 2006, many residents still referred to all these recruits as "game guards."

Section III

The Challenges of Decreasing Entitlements

Chapter 10

On Coping within a Cornucopia of Change

Bwanga Leya (sixty-two-year-old woman): *This European [Stuart Marks] is mostly about this place. In the future you will find that this place is claimed by Europeans.*

Doris Kanyunga (sixty-year-old woman): *No, staying here is not what he wants. He wants to understand the problems we face in our daily activities and living.*

Bwanga Leya: *What will he do when he finds it? Is it that he will bring us a good road? Still the game guards are troubling us for the meat we consume. We are now eating wild vegetable relishes every day. In the past, we used to eat meat with our porridge and when they killed an elephant, the meat was shared with residents. Now there is nothing and five years have passed without my tasting elephant.*[1]

Conversation recorded on village survey (1989)

Alienation of land for wildlife protection and hunting restrictions favoring state priorities and foreign revenues are not new interventions within the Luangwa Valley or elsewhere.

In 1989, foreign intrusions and losses in land and wildlife were on Bwanga Leya's mind as she engaged Doris Kanyunga in the conversation presented in the epigraph. Kneeling beside a flat stone grinding sorghum seeds into flour, Kanyunga was a participant in an activity survey I supervised during 1988–89. Her activities represented those of a particular demographic, an elderly resident married to a local wage earner. A local recorder also timed her enterprises as he inscribed her relationships, activities, and occasional conversations during a randomly selected day each week during that agricultural year. Within local memories, the presence of visible strangers and their undertakings were often harbingers of impending losses in local land and liberties. Despite Doris Kanyunga's attempt to reassure her friend, Bwanga Leya remained unconvinced. Her response became a personal challenge to convert my perceptions, conversations, and observations into meaningful translations of time and place.

The Challenges of Decreasing Entitlements

Any exercise in resource planning is either innovative or conventional; it either builds toward a new social order or strives to reinforce an existing one. Programs in resource development require explorations in new social forms, whereas those in resource conservation seek to strengthen the status quo (Firey 1960). These days the negotiations over what to include within such regimes are rarely local concerns as they take place at all levels between the global and specific sites where they are implemented. Yet the ideas and procedures that situate resource regimes along the novice to commonplace continuum begin primarily at the global or national levels; individuals and organizations whose funds support planning exercises influence at all levels (Buscher 2013). Despite the contrary claims of those involved, there was little originality behind the "community-based" wildlife programs sponsored on the Zambian scene during the 1980s. Planners and prime actors were either professional wildlife expats or

Zambian civil servants whose funding, ideas, and materials came from donor and international agencies, which despite stated intentions remained largely fixed on naive ideas about the relative ease of changing village lives. I examine policy documents and events leading to the inception of the Zambian program in the 1980s before reviewing two village incidents in 2006 that indicate the reverberations on persistent unresolved land and wildlife issues.

Two "community-based" initiatives sprung from the Lupande Development Workshop in 1983: the Luangwa Integrated Resource Development Programme (LIRDP) and the Administrative Management Design for Game Management Areas (ADMADE). Planners laid their conceptual bricks and mortar upon the colonial foundations of centralized power and its extractive privileges as they extended earlier pejoratives about village (in)capabilities as the need for "universal" expertise to save "their game," reflecting professional as well as personal blinders. These limitations become apparent in a close examination of national policy documents and events leading to the inception of LIRDP and ADMADE. Both programs were products of their time, focusing primarily on wildlife as a "resource" within a management system backed by violence and economic incentives in the state's and foreign interests. LIRDP's objectives were to integrate all natural resources (fish, timber, wildlife, agriculture) within a defined geographic area and to gain hegemony over a national park and the adjacent GMA as an experimental territory in sustainability. NPWS's objectives were to increase law enforcement and manpower within a small pilot area to enable it to gather short-term numerical information for demonstrating its effectiveness as a conservation agency for donors. Both the Norwegian and American embassies allegedly prompted the initiating documents prepared by foreign and Zambian nationals, who were given assurance that funds were available under certain conditions. Both programs were accepted, but neither was as innovative as its promoters claimed.

Political Skirmishes: Two Wildlife Initiatives during the 1980s

We are honest people, who are keepers of the wildlife. We do not like poaching and we have been keeping the animals here a long time for the government, but we receive no benefit for this service. . . . Teach us how to manage wildlife ourselves and we will protect and keep wildlife here always.
—Address of Hon. Chief G. Malama at Lupande Development Workshop, 1983.[2]

During the 1980s, economic plans and resource managers backed by international financiers changed the incentives within Zambia's wildlife institutions. Concerned over the depressing declines of wildlife within and around its protected areas, a small group of wildlife biologists, conservationists, government resource officials, and donors convened in September 1983 on the land of Chief Malama. This small chieftaincy abutted the South Luangwa National Park and the occasion was the Lupande Development Workshop.[3] As the sole local representative, the chief addressed the foreigners and reportedly made the following four major points. First, despite residents reserving wildlife for tourists, safaris, and government, they had received no benefit. Second, there were local needs for a clinic as well as good roads to retain their people at home. Third, jobs were scarce, so employment in the conservation sector would help. Fourth, since safaris employed no local residents, they should hunt elsewhere.[4] Chief Malama's presence was significant in other ways as this workshop's proceedings and resolutions shaped subsequent national wildlife policies and structures for decades (Dalal-Clayton 1984).

The workshop's participants resolved that circumstances demanded a new policy, a working model that showed effective management and efficient use, and that conditions required international funds and expertise to counter the massive scale of wildlife poaching. All agreed that local residents within the Luangwa Valley's GMAs should participate in the development and management of natural resources and receive benefits. A strong faction representing European agencies and local conservationists disagreed with the NPWS contingent's proposal. They sought instead to integrate all natural resources used by villagers, a developmental space for their project, and a new institutional structure free from Zambian politics into a larger management model. The meddling of politicians and corrupt bureaucrats (including some within the NPWS contingent) had confounded this group's earlier ambitions. After the workshop, they sought hegemony over the South Luangwa National Park (SLNP) and the adjacent Lupande GMA, external donor funding, and the support of President Kaunda to bring their plan to fruition (Larsen, Lungu, and Vedeld 1985). Under the auspices of the National Commission for Development Planning, this group filed its proposal for a feasibility study to the Norwegian Agency for International Development (NORAD). NORAD accepted their proposal and the National Commission appointed two biologists as consultants, who submitted their plan in September 1985. NORAD later funded this proposal as the LIRDP.[5]

With President Kaunda's backing and through his chairmanship of its steering committee, LIRDP obtained authority over the SLNP and the Lupande GMA. The president also secured the services of a European wildlife ecologist from Malawi to serve with a Zambian as the project's co-directors. Begun in 1986 as a response to the failure of earlier donor projects and to the poor performance of NPWS, LIRDP's stated goals were mainly wildlife conservation and improving community welfare through the sustainable uses of all the GMA's natural resources. Its target was financial viability with eventual national benefits. Its three initiating ecologists remained influential until the 1991 national elections, when the Kaunda government was removed from office. Although LIRDP continued to receive funding from NORAD, the incoming Zambian president brought a new agenda, replaced the program's directors, and later integrated LIRDP with NPWS under the Ministry of Tourism.[6]

The Beginnings of ADMADE

As the legislated national wildlife agency and instrumental in convening the Lupande Workshop, the NPWS officers recognized their liabilities in their rival's (LIRDP's) plans. Unilaterally, they judged their wildlife mandate sufficient to meet all GMA residents' needs and conceived their own separate pilot project to secure donor funding. Other Zambian officials at the workshop supported this program and its concerns.

Within the GMAs, residents are allowed to live and grow crops, but the wildlife remains state property. Residents could legally access wildlife by purchasing game licenses from NPWS, which progressively increased in price as quota numbers of species on licenses decreased. NPWS officials assumed that the residents' lack of demand for these legal licenses meant that they persisted in making illegal kills and cooperated with outside poaching gangs. Because of its visible wildlife, the Luangwa GMAs were prime tourist hunting locales whose concessionaires generated sizeable capital for commercial profits and state taxes. NPWS biologists favored safari concessions and safari hunting clients as both shared similar wildlife values and their enterprise generating revenues, which enabled NPWS to employ them and increase departmental control over rural areas. In addition, biologists required powerful and wealthy allies to protect their donor funds, which they pursued outside the channels constraining other governmental agencies.

To gain international donor assistance, NPWS knew that they needed an immediate project that could provide quantitative information showing that their ideas worked effectively. They selected Chief Malama's area of two hundred residents and a "buffer zone" slice (5 percent) of a GMA for their pilot project. The adult population was sufficiently small to demonstrate the short-term ripple effects of employment and the rapid infusions of cash into its local economy. As anticipated in a small, poor, rural community, economic gains and promises showed a temporary large shift in local approval of NPWS's project. The depiction of this initial research site was anecdotally crafted to indicate positive results.

Before the pilot project began, the safari operator operating within the concession reportedly earned some US$ 350,000 while residents received less than 1 percent from employment and gratuities. Five government and six village scouts lived in a single enforcement camp and ineffectively policed their assigned zone. Rather than undertake prescribed patrols, these scouts responded only upon hearing gunshots. Poaching, particularly of elephant and rhino, was depicted as "rampant." Residents actively participated in these illegal activities, and they resented government scouts for arresting them while they enthusiastically supported foreign hunters rather than enhanced local welfare (Dalal-Clayton

1964; Lewis, Mwenya, and Kaweche 1991). Such abysmal and intransigent conditions appeared to demand immediate funding, technical assistance and developmental remedies that a donor might find irresistible. Donor funding could provide the capital and expertise, and the gross revenues generated annually by the current safari firm appeared as insurance for the project's sustainability. Part of the experimental design was for the NPWS to show its research competency as well as to enhance its own legal mechanism (aka the Wildlife Conservation Revolving Fund) for retaining wildlife revenues generated within the project area to further its purposes. NPWS community scouts were to remain as year-round "local custodians of their wildlife resources," supervised and monitored by a civil servant. Given these perspectives, the proposal was an inevitable "win-win" program for everyone (Lewis, Kaweche, and Mwenya 1988).

Two revenue streams, the auctioning of hunting rights to safari companies and the processing of hippo products, went into the NPWS's revolving fund to support their pilot project. This fund immediately enabled the department to expand its staff ("beyond its government-approved quota of civil servants") and to infuse large amounts of cash into the local economy.[7] The project emphasized wildlife as a "resource" defined in numerical terms as revenues from a set quota of species, the numbers of patrols mounted and arrests made to protect wildlife, and impromptu surveys administered by scouts to assess changes in local attitudes toward departmental management. The designers assumed that residents, as they benefitted from employment as game scouts and as workers, would act as economically rational consumers. Economic language and concepts (efficiency, positivism, rates of increase) were the project's focus and the basis for its short-term assessments.[8]

Apprehensive about their rivals' progress in LIRDP, NPWS promptly internalized their project's basic organizational plan as the program frame for ADMADE. Like its pilot project, the NPWS's new program, when underwritten by donor funding would boost resources (cash, equipment, staff) while shielding it from meddling politicians. NPWS knew that donor funding could expand ADMADE into a national program if its hard data demonstrated progress in curbing the slaughter of elephant and rhino and if they could show that GMA residents responded favorably

to its management on communal lands. Their pilot gave them the data points for constructing their case for donor assistance. Some funding from the World Wildlife Fund came in 1988 and allowed the department to expand into the Munyamadzi and five other GMAs. A year later, a $3 million (US) grant from USAID and additional international funding secured ADMADE's tenure on the national wildlife landscape (Marks 2014: 238–48).[9]

Higher Political Dynamics

In his study of the political economy of Zambian wildlife policies under President Kaunda (1964–91), Clark Gibson (1999) shows that individuals (the president, various project managers, chiefs) and groups (donors, professional hunters, wildlife officials, scouts) shaped wildlife procedures through their decisions and activities. His institutional approach illustrates how organizational incentives influenced both individual and group choices. Under the one-party Zambian state in the 1980s, the two "community-based" initiatives (ADMADE and LIRDP) sought to reinforce their respective authority through external funding while seeking to limit politicians' options and manipulations. Project managers' decisions about budgets, staff deployments, and developments were geared more toward protecting their projects from national political uncertainties than toward promoting decentralized initiatives that the branding rhetoric ("community-based") might suggest. The Norwegian scheme (LIRDP) sought President Kaunda's sponsorship while hoping to contain his guidance. Its directors sought a legal foundation to buffer the president's influence, which would require eventual national funding. Similarly, NPWS sought external resources (mainly USAID), wishing to expand its authority and territory while minimizing possible conflicts with other state agencies. Gibson and Marks (1995) describe ADMADE's influences upon rural villagers and scouts regarding wildlife selection and protection. In this section, I present evidence that subsequent Zambian politicians learned from these initial skirmishes and molded the newer wildlife policies in the Zambia Wildlife Act of 1998 to promote their interests. In addition, politicians excluded local voices and sought ways to limit wildlife experts' and staff prerogatives (Manspeizer 2004).[10]

Politics and money are conspicuous assets in planning and for executing projects, yet are often missed, dismissed or redirected as these exchanges often occur unobserved, off-stage, or in secrecy in the higher corridors of power. After the 1991 Zambian elections, public commentaries about foreign influences within the national conservation initiatives became more prominent within the Zambian press. Justified in terms of economics, the new slogan became "to survive, wildlife must pay its own way," as conservation projects were to demonstrate wildlife's competitive advantage in land use values over "marginal" agriculture. Neoliberal policies demanded a shift from public to private ownership together within a broader spectrum of Northern Hemisphere ideals such as "good governance," "democracy," and "transparency." Concerned about governmental corruption, donors sought to initiate private entrepreneurial opportunities by supporting civil society (such as NGOs, representative organizations) and select businesses to work as partners with established state authorities. The presumption was that donor assistance to nongovernmental partners (particularly NGOs) would enhance the expansion and diversification of interests that would eventually become expressed in democratic processes. The resulting diversity in private national interests would reduce government monopolies over economic activities and manifest good governance (Reed 1995; 2001).

Subsequent to USAID's funding of ADMADE and NORAD's support for LIRDP, backing from the European Union (EU) in the 1990s aimed to "modernize" the Zambian wildlife sector. The EU expected the new authority to become an autonomous, reliable, fiscally sound structure insulated from Zambian politics. The restructuring of NPWS to become the wildlife authority began in 1992 and involved five different consultancy teams and more external wildlife experts. Each of these teams operated in Zambia for specific purposes during different times through 2001. After years of contentious negotiations and considerable outside pressure, the National Assembly passed the Zambia Wildlife Act in 1998. This act effectively eliminated the authority of earlier administrative staff over future wildlife decisions.

NPWS leaders and actors faced repeated crises throughout the frequent quibbles between donors and national politicians before a weak, underfunded Zambia Wildlife Authority (ZAWA) emerged as a national political prize. Beginning in October 1999, the ZAWA board chair was replaced four times in the first eight months of the institution's existence. In addition, President Chiluba dismissed four ministers of tourism in five months beginning in November 2000, while appointing a suitable director general took more than two years.[11] Rivals jockeyed for political and party leadership as it became clear that the president lacked support for a third term in office.[12] As political trophies, ZAWA staff appointments gave access to significant foreign exchange and to generous campaign contributions from safari concessionaires wishing institutional backing for their claims to lucrative hunting tracts, one of which was the Munyamadzi GMA.[13]

During an address to the Zambian Parliament in 2001, President Frederick Chiluba unexpectedly canceled the tender proceedings for renewing leases to hunting concessions. The public context was that Zambia's wildlife needed protection to recover from heavy hunting, poaching and bureaucratic corruption during the agency's transition. The presidential decree originally banned all hunting; however, after a public outcry, only safari (overseas) hunting remained prohibited. This interdiction lasted two years, and as there were no safari revenues for funding wildlife scouts during this time, the EU and safari companies, anticipating another "wildlife holocaust," poured millions of Kwacha (the Zambian currency) into specific GMAs to support anti-poaching units. During this moratorium and to keep itself funded, ZAWA unilaterally retained earlier revenues from safari access on communal lands, which belonged legally to rural communities. The new Community Resource Boards (CRBs) took ZAWA to court to win back their revenue claims, which ZAWA eventually paid later in inflated currency. While many accusations of predatory behavior emerged during the presidency of Frederick Chiluba, Jan Kees van Donge (2009) cautions that important international and national cultural influences work in these political confrontations and are worth more detailed analyses.

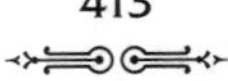

Political Presumptions and Meager Incentives on the Ground

Under the ADMADE program, wildlife planners made erroneous assumptions about the significance and symbolism of wildlife for rural societies and about the dispositions of chiefs. In addition, they assumed that all GMA residents profited from unlicensed hunting and suspected that every rural man "poached" or had the disposition and access to equipment to do so. Under the intense surveillance of locally employed scouts, managers imagined that residents would quickly learn that unlicensed hunting was costly and would relinquish it for the "community channels" and benefits under their new wildlife protocols. They expected that the programs' successes in reducing extralegal and local hunting would increase wildlife numbers as targets for safari hunters, whose license and permit purchases in turn would boost state revenues and augment cash incentives for developments within rural communities. This chain of economic infusion and supposition was intended to convert the "doubting Thomases" everywhere (Lewis 1993).

ADMADE's community incentives were neither appropriate nor sufficient to change the behavior or attitudes of many individuals at village level. Assuming their ADMADE roles, chiefs obtained new powers and controlled incoming revenues, effecting their positions as wealthy patrons. The influences of chiefs were apparent in the program's distribution of benefits and in the placement of structures. Permanent buildings, such as schools, clinics, and storerooms, along with water wells, grinding machines, and other signs of wealth appeared around their palaces. Moreover, the chiefs decided whom to employ as community workers and scouts, thereby enhancing and affecting a sizeable clientage. Community funds contributed to palaces and religious centers as well as toward travel and the "sitting fees" for appointed subauthority members. The wildlife warden retained the final word on "community" projects, authorizing only those that he considered commensurate with his agency's understanding for "conserving" wildlife and perpetuating the status quo. Wardens also provided little positive supervision on the ground as they had no applicable developmental training and their institution was ill-equipped to facilitate such a role. Consequently,

rather than reconnecting rural Zambians to their wildlife, as the institutional rhetoric initiating ADMADE claimed, the program progressively sought to restrict local uses of wildlife, land, fish, and timber while privileging its access by outsiders, legally and selectively. Under donor pressure, wildlife officials eventually acknowledged their more visible problems, which were published as "lessons learned papers."[14]

Structural flaws and misdirected objectives eventually led to donor demands for reform. The eventual shift from NPWS to ZAWA was more than a name change as the authority's structural designs were contested; donors and Zambian officials promised or reneged on revenues and proposed staff as the revisions got caught in the political wrangles during the waning years of the Chiluba presidency. ZAWA emerged from these conflicts as a weak, underfunded wildlife agency with businessmen, rather than biologists, staffing its prominent echelons. Its mission was to acquire sustainable revenues supporting its authority.

The 1998 Zambia Wildlife Act incorporated many of ADMADE's suppositions about the economic values of wildlife, its presumptive control over human activities within the GMAs, and the donors' neoliberal agenda, which demanded a reduction in state power and monopoly, the removal of some subsidies, and the creation of free markets within which enterprise, good governance, and democracy might flourish. The Act's new policies endorsed cooperative and "devolved" partnerships between businesses (safaris) and an imposed "representative" local institution, the CRB that replaced the truculent Wildlife Subcommittee, to serve under the "professional guidance" of the new, struggling wildlife authority. The local chief became a patron of the CRB and received a stipend from community revenues while monitoring its activities through his appointed delegates within the CRB. The guidelines of the Zambia Wildlife Act specified that CRB members were to act on behalf of their respective village groups. In conjunction with ZAWA, the CRB was to negotiate "beneficial co-management agreements" with safari and tourist operators. The act also stipulated that the CRB would dispense its revenues equitably in projects, assume the main costs for anti-poaching and local wildlife management, together with some social services, even as it received less funding than its predecessor. The CRB was responsible for reassuring its partners that GMA residents did not engage in wildlife transgressions.

Intended as an autonomous body with wide-ranging functions, ZAWA's main predicaments remained (through 2011): meager revenue streams and dependency upon foreign subsidies, ineffective leadership, and a lack of comprehensive vision. Internal staff traumas produced unanticipated and perpetual crises. ZAWA imagined itself as the overlord, provided minimal guidance and training for communities, and placed unilaterally its demands on the CRBs while providing them less revenue. The point is that rural communities were never intended as the main beneficiaries under either the ADMADE program or ZAWA. Donor support and centralized management did not achieve stability in any institutional sense. Neither did they help alleviate rural poverty or lead to sustainable wildlife uses. The National Resources Consultative Forum report (2008) showed habitats degenerating and wildlife decreasing at high rates, and it documented that all GMA communities were less developed and poorer than most non-GMA rural communities. These and other dilemmas, controversies, and associated problems were repeatedly documented in a stream of consultancies and published reports, yet government took little, if any, tangible action.

On the fortieth anniversary of my studies within the Munyamadzi GMA in 2006, I was hoping to find some representative evidence for improved community welfare, especially after eighteen years under this "community-based" program and after millions of US dollars were allegedly distributed to improve the Munyamadzi GMA. Instead, most of the 525 interviewees related a continuing decline in their household welfare, including reduced food and personal security, and complaints of insufficient and skewed employment opportunities. In addition, over a third (38 percent) volunteered details that either they or a close relative had been arrested, beaten, or jailed for a recent wildlife violation. Tragic incidents with wildlife continued frequently with most incidents unrecorded (table 3.1, p. 134). Even the smaller species of wildlife together with the other bush products upon which residents' well being depended showed noticeable declines in both quality and quantity (figure 3.1, p. 128). The supposed "democratically elected" wildlife institution met occasionally, but local CRB representatives, elected under spurious arrangements, remained aloof of constituents' needs. Many respondents agreed with one of ADMADE's guiding principles, a willingness to exchange wildlife for development revenues, but overwhelmingly qualified this acceptance with the need for more equitable sharing of these benefits among all residents. Women generally received more indirect benefits (clinics, children's education) from this program as wildlife and hunting were masculine domains. Overwhelmingly, residents considered the building of an all-weather road as their major need, as they had since 1978 (Marks 2014).

What was anticipated as good news in the language of the 1998 Wildlife Act for GMA communities got strangled in the national and international wrangles over the financing and structuring of the new wildlife authority (ZAWA) and in the inevitable squabbles with safari concessionaires over the tenure procedures. These disputes prolonged the emergence of ZAWA, which materialized as an underfunded agency with little sense of direction. In 2006, I sensed that an informal consensus among ZAWA officials and safari operators was against constructing a serviceable road connecting Nabwalya to the plateau. These groups showed no empathy for equity issues or for the protection of local properties or lives from "problem animals" unless they, individually or corporately, could prosper from these episodes. Yet some district officials seemed to realize that time and change ran inevitably against this privileged status quo. Indeed, the Patriot Front government, elected in 2011, signed a contract with a Chinese firm in 2014 to construct an all-weather road from the western plateau connecting to roads within Eastern Province. This road was to pass through Nabwalya and the firm was to build a bridge across the Luangwa River. Finally, in March 2015, "the Zambia Wildlife Act will be repealed and replaced with the government agency that will be created" (Lusaka Times 13 March 2015, http://www.lusakatimes.com/2015/03/13/zawa-abolished/).

Two Persistent Dilemmas: Land and Wildlife

"Wildlife is our main resource, poverty and hunger our greatest problem."
Interview with a Nabwalya resident, 2001

The Valley Bisa face numerous challenges in their daily struggles to extract a meaningful life from their parsimonious valley environment. Over the decades, their efforts have become more protracted as their growing numbers, particularly among the young, bring escalating demands upon local resources, social services, and relationships. Local welfare progressively depends on appropriate rainfall, access to fertile but scattered pockets of soil, and unresolved conflicts with abundant wildlife, as well as scarce employment and ineffective social services. In recent years, the first source appears increasingly erratic while the latter four remain within the uncertain capacities of a distant government and charities. Government agents purportedly respond only if it suits their respective interests, rather than react to local needs. Fiscal retrenchments narrow employment opportunities and social services, with residents increasingly dependent on the whims of private donors and the benevolence of religious groups.

Facing this same advancing constellation of uncertain and demeaning circumstances for at least a century, the Valley Bisa have shrunk noticeably in confidence as they get drawn into dependencies with uncertain outside agents and offices. Residents live within the shadows of an adverse, ineffective government as well as that of increasing numbers of large, capricious mammals. Both entities remain beyond their control, yet each can deliver systematic and devastating attacks on whatever residents find valuable, even life itself. Such circumstances constantly corrode most human identities and energies as they erode cultural confidence within the muck of the most demeaning poverties—impotency and despair. Even under these dismal circumstances, a very few individuals, noticeably at times, imperceptibly at others, appear to flourish.

During 2006, the occurrence of two events reminded me of similar episodes witnessed in earlier years and how their persistence enduring as local conundrums without official recognition had worn on the humanity and developmental character of most valley residents. I begin with the description of a meeting held by Chief Nabwalya with his headmen in July of that year. My notes and the meeting's transcript allow me to recall the "voices" of various participants and thereby portray the seriousness of local sentiments about land and how they felt about the state's subtle and encroaching boundaries and nooses. Notes of an earlier meeting between government officials and residents in 1967, the latter fearful of facing dispossession of their land as well as their livelihoods, capture some of the earlier emotions and fervor on these issues. The second episode chronicles my learning of Edson Nabwalya's death by an elephant later that same July in 2006. These formal meetings over land and events surrounding an untimely death reflect unresolved cultural challenges and outside pressures that continuously afflict the lives and welfare of valley residents. Such episodes also indicate some ways in which residents maintain their own spaces in the face of seemingly overwhelming adversities.

The Issues of Land and Boundaries (2006)

On 1 July 2006, the chief called a meeting to discuss a concern that had remained unresolved for years. The mood was somber as the headmen and their representatives gathered under a grove of trees in front of the palace. Eighty-five men and four women sat in a semicircle in the shade while the chief, his councilors, and guests sat in chairs facing their audience. After the chief's constable and the head teacher introduced the agenda, the meeting opened with a prayer and with introductions.[15] The head nurse mentioned the problems he faced at the clinic. Since the government had cut back on rural services, the health clinic depended on donations from missionaries and safari operators. The clinic had no mattresses, so the nurse advised patients to bring their own sleeping mats. In turning over the meeting to the head teacher, he mentioned his imminent departure for a midwifery course that would last a year. Whether or not he returned would depend on his outside sponsor.

The head teacher spoke about education as the key to success; "With education, people are comfortable," was his theme. He instructed local parents to encourage their daughters to attend school rather than to marry older men. The local school had enrolled 670 students that year, the largest attendance in its history. Many students had returned following the devastating famines of the previous years. Three trained teachers, seven untrained teachers, and a woman provided supervision at the school.

The chief indicated that the land issue remained unresolved and was going badly. "What I have to say, you have to contribute as headmen," he began, "as I cannot do everything alone. When you have no land, it means you are buried forever. This issue of boundary is very critical and that's why people have to be educated as the head teacher said in his speech. Nabwalya is isolated, detached from the rest of the world and has no light." He recalled that missionaries helped residents during colonial days, so he had invited the Baptist missionaries to establish a church and the professor had returned. "When I came to the throne, many people said they didn't want me because I didn't smoke or drink as the former chief. But knowledge is not based on drunkenness and smoking." He continued, "I am able to understand issues affecting development, but there are people who do not know anything even after getting independence. Where there is no research, there is no development."

According to the chief's account, outside scouts from the SLNP arrested some boys fishing in the Mupamadzi River several years ago. This incident had rekindled the issue of the boundary between the Munyamadzi GMA (his chiefdom) and the national park. A district magistrate had dismissed the trespassing case as the scouts presented no evidence that the boys were inside the territory of the park, most of which was former Bisa land. More recently, the chief alleged that ZAWA had extended the park's boundary further into the GMA and placed permanent beacons. This act was done without officially informing the "owner of the land." He angrily emphasized that "People must be informed of all changes. The problem is that most ZAWA people come from areas where there are no animals and they are jealous of us getting the benefits from our animals."

In the chief's view, this problem started when the Movement for Multi-Party Democracy (MMD) came to power because they had "politicized the whole program of tourism instead of giving power to the people to care for their resources and to benefit from them. ZAWA must change its attitude, or else should quit."[16] The MMD politicians had "scrapped off" the little development the GMA had from Irish Aid and left residents in poverty. "This government has no intention to help us," the chief continued. "It's better for us to take care of ourselves and maintain our resources. If ZAWA cannot remove their animals from here, then let them remain for us and the Nation. . . . Why should the people of Nabwalya be chased from their land? Who are those [people] eligible to enjoy wildlife benefits from here? To every resource, those who reside in the area are the ones to use them the most."

Then the chief instructed the nurse to read and display two articles that had appeared recently in national newspapers. In the earlier article, the headline read in bold letters, "CHIEF NABWALYA TELLS ZAWA TO LEAVE HIS AREA."[17] In it, the chief alleged that ZAWA had incorporated a portion of his land into the national park and that "nothing had been done about the matter despite several appeals to government." He continued, "ZAWA is harassing people that go to the area to fish and get firewood. I don't want them [ZAWA] in my area, they should leave. . . . They should remove the beacons, if they refuse to leave that would be the beginning of war [*nkondo*]. We shall look after our own animals."

After reading and displaying the newspaper, several headmen made comments. One noted that the boundary issue was not for the chief alone, but that the assembled headmen should remove the beacons. He punctuated his remark by reciting a common courtesy, "When one is given a place to build a house by the "Owner of the Land," he is not granted permission to do anything on the land without additional consent." All headmen agreed that no boundary could be established in the absence of the stakeholders and, if ZAWA wouldn't compromise, "we should fight them." A councilor to the chief was concerned about survival: "There is no place left for our activities like collecting firewood, poles, and honey. This area is hunger-prone and our survival depends on the forest from where we get our food. The best way is to remove the beacons so that ZAWA knows that the Bisa community is angry." Perhaps feeling the momentum of the moment, the single ZAWA representative noted that "as ZAWA workers, we cannot do anything but this is entirely up to you to react. The map we have shows the old boundary, but this has been exaggerated by the "uppers."[18] The problem is with the exposure of land to investors, because there must be something private that they want from this land, such as precious stones."

After further discussion, the assembly reached a consensus to break the beacons. The chief's constable announced that everyone should assemble the following morning for this mission. The constable then announced that food would be supplied to participants.[19] The chief had the final word:

> This meeting should be written in the newspaper so that others will know that the government should give people choices to decide about their resources. Decentralization is just on paper, as Valley Bisa are not happy with what has happened here to us. As a democratic country, Zambia the government doesn't want to give rural people choices. Tourism should be decentralized as well because it is our local industry while government should develop its artificial industries in towns. Politicians have their own industries as the backbone of this country. Northern Province is less developed because our resources and labor are taken out to other provinces. God gave man what they should enjoy and government should not dictate what the community should feel.

Early the next morning, a consultant arrived unannounced at the palace. He was hired to gather information for a pending court case against a safari firm whose contract and hunting concession ZAWA wished to terminate. This unexpected visit delayed the beacon removal to the next day. After the consultant's departure, participants returning from that beacon exercise said the nearby beacons were destroyed.

In late 2007, a resident wrote me that both ZAWA and the community had remained silent about the beacons. The community, he noted, was satisfied with what it had accomplished and was ready whenever ZAWA wished to discuss the boundary issue. Whereas the local community felt disadvantaged with the little revenues they received under the government program, this confrontation was more serious because their land was at issue.

ZAWA proceeded against the safari firm in court, basing its case on charges of illegally obtaining the concession, overshooting its quota by one animal, and not meeting its expected schedule of payments. Prominent members of the CRB entered the court case on the side of the safari operator and testified against ZAWA's claims.[20] The injunction placed by the firm was upheld by the

High Court, but ZAWA appealed the case to the Supreme Court and won on a "technical issue." There remained many unsettled questions about these procedures, but their impacts on the chief, the CRB, and the foreign safari industry were unmistakably devastating. The concession was given to the previous concessionaire reportedly as a political favor for his support in the previous election (Marks 2014: 254–64).

An Earlier Confrontation over Land (1967)

The rhetoric of the 2006 meeting over the beacons was tame in comparison to a 1967 standoff over government proposals for Valley Bisa resettlement (Marks and Marks, unpublished manuscript).[21] On this occasion, the newly independent government announced its plan to meet with residents about their resettlement (chapter 3). Learning of this impending delegation, residents became so enraged that the chief, Kabuswe Mbuluma, sent word to Mpika that a large delegation might provoke a fight. When officials failed to appear as scheduled, residents held their own meeting instead.

Speakers on this occasion were deeply angered by what they had heard of government intentions coming so soon after their joint fight for political independence. Orators accused their ministers of forgetting what freedom was like and compared them to Europeans, who loved dogs and placed them in the front seats of their vehicles while their accompanying Africans sat uncomfortably on the floorboards in the back. They accused these "vulnerable" ministers, who had inherited the good life from the colonials, of now seeking Valley Bisa land for farms, forcing residents to leave the land of their ancestors "to become like other people instead of letting us be ourselves." One speaker spoke metaphorically of residents' distaste for "cutting trees and making cassava gardens," references to the agricultural practices of the Bemba and Bisa on the poorer soils of the plateau. An elderly spokesman told of Europeans finding them with so many animals that "the elephants [rather than people] made their roads" through the escarpment. He accused officials of being like "a rabbit beating a drum with its beats intoning meaning *'pano pesu, pano pesu'*—this is our land," as if the repeated rhythmic beat would assure the territory as their inheritance.

All speakers adamantly refused to consider resettlement under any circumstance. One elder asked rhetorically, since residents never interfered with government officials hunting on their land, "Why should [officials] stop us from living here? In the past," he continued, "locals had no government guards for we drove the animals from our fields ourselves. We took care of ourselves then and we can do it now. Why do we need game guards to kill raiding animals when they never have enough shells to kill animals that are destroying our gardens? . . . Our ancestors came here and did not leave this land. We will not leave either."

Kabuswe Mbuluma was not surprised at what he heard and said that he would repeat their messages to the government representatives, if they appeared. "When the first chief came here from the Congo, he did not come alone; they settled in areas which suited his clan." He continued,

> The chief that settled here did so because the land was fertile and his people would not be hungry. I have told the government that the early chiefs and their followers are buried here and we cannot leave them and go anywhere else. If the early colonials had loved this area, they would have made our ancestors leave and Europeans would have settled here. But they did not and we remain. The "great fence" [the escarpment] has kept us and our animals here. I will not let you be taken to any other land to suffer. This meeting is not mine, but was called by those from Mpika, even though they have not appeared. If the officials come another day and want to meet with you, they must find you where you are.

When the Senior Bisa Chief Kopa and two junior officials arrived a day later from Mpika, they called a meeting and met a very angry crowd. The senior chief opened the gathering by reminding attendees that he had come to listen and asked the two officials to speak. Both spokesmen complained about their arduous travel and the terrible condition of the road. They said the escarpment made developments in the valley impossible, and it constrained government's abilities to resolve their problems with wildlife. If the audience agreed to move, they would not face these same problems and would be given jobs and money, together with good schools, a dispensary, and even cars. The government wished for residents to move peacefully and for their own benefit.

Those gathered as the audience demanded to know if the government they had fought for was the same as that asking them now to leave their homeland. "You must represent an alien state for you talk like Europeans! How can our government turn suddenly against us and favor the elephants? We had no development under the Europeans, yet we did not expect the same treatment from our fellow Africans. How is it that we are being left behind unless we agree to leave our homes?" This speaker suspected that someone was trying to obtain Bisa land for money. A second speaker was more abusive. He declared that residents had heard many times what the strangers now proposed. As always, the response was a resilient rejection of these plans, and he declared the meeting adjourned immediately.

This act insulted Senior Bisa Chief Kopa, who sought to restore order. He recited two proverbs to remind the audience of his mission. "If someone is sent to ask for tobacco, would you insult him? Likewise don't insult me, just say what you feel and I will tell that to the government." The second adage was that "when a person is sick, visitors come to call, not to mourn. The same applies to us here. I am just paying you a visit as a sick ['undeveloped'] client; I am not here to mourn as if you were dead. This is no cause for a fight, as you have the right to refuse or to agree."

Subsequent speakers noted that the proposals were refused earlier and that no minds were changed. Chief Kabuswe Mbuluma closed the meeting by reminding his subjects that "even during the previous [colonial] government, I was fighting like I am now. Now you hate me even as you hated them. . . . I cannot sell this area." What really bothered him was that the money from access and game licenses generated in his area no longer brought schools, hospitals, and other developments to the Bisa.[22] Instead the government used the money for developments elsewhere. He closed with a plea: "Trust me. Don't hate or kill me. I will tell the government at the next meeting exactly what you have said. Now release me from these struggles today." The assembly agreed and the meeting adjourned.

After 1967, the newly independent Zambian government largely left the issue of Valley Bisa resettlement alone while it struggled with more compelling concerns elsewhere. Moreover, it contended with an international five-year United Nations Development Program consultancy and review of wildlife and land-use survey in the Luangwa Valley (Dodds and Patton 1968). Among the consultancy's many recommendations were restructuring the colonial game department, resituating and modernizing its title as the National Parks and Wildlife Service, and including the Munyamazi's Chifungwe Plain in the SLNP. In 1972, and without consulting with the Valley Bisa chief and residents, this plain was unilaterally incorporated through the legislative act of promulgating the new national park boundaries (Astle 1999). The eastern border of the plain was left indeterminately marked on the ground within the GMA perhaps because its boundary was close to Chief Kabuswe Mbuluma's palace. Local memories proved stronger than government protocols and became the "bones of contention" and speculations during the 2006 meeting. From the standpoint of the chief and many residents, land boundaries and the management of viable land assets remain unresolved as of 2014, yet these issues are critical to their survival and for identities that would provide them a meaningful future.

The Problem of Living with Large, Dangerous Wild Animals

Longer-term studies allow opportunities to continue conversations with individuals at different stages of their lives, to know them as persons, as well as to check at intervals on their well-being. I first knew Edson Nabwalya while he was a student in 1966–67 attending Nabwalya Primary School where my wife Martha had volunteered to teach English. Later when Edson settled in his wife's village fifteen kilometers away, I met him less frequently. His mother was one of Chief Kabuswe Mbuluma's younger wives, and her daily tasks of preparing food and beer kept the polity of the palace running. She had no formal education and never traveled much beyond her natal village and its vicinity.

His mother's marriage to the reigning chief privileged Edson who, as an elder son, assumed important roles. He became an active agent in his father's commercial enterprises and served as the UNIP ward chairman and as a district representative. Edson married a younger woman in the Mupamadzi section of the chiefdom and resided there upon his father's death. As an infrequent hunter, his connections gave him some immunity from the scouts

while his father lived. His arrest for firearm and wildlife violations occurred shortly after his father died (chapter 9). He was arrested again in 2004 when scouts found dried buffalo meat in his house. They beat, arrested, and took him to Mpika where he waited in jail for three months before his case came to trial. The magistrate released him on a two-year suspended sentence as the buffalo meat in Edson's possession was salvaged from a carcass resulting from a buffalo trouncing at the river, a death caused by natural (or locally interpreted as supernatural) causes.

On 13 July 2006, a schoolteacher and I spent over an hour with Edson at his homestead. We recollected his recent history, his aspirations, and sentiments about what had happened recently in his world. He and his wife had seven children (from twelve to thirty years old at the time) who still resided in his neighborhood. His welfare had declined considerably after his father's death. He converted to Catholicism, served as a prayer leader, and, unlike his father, was allegedly monogamous. During the severe drought of 2005, a safari manager employed him as a grass-cutter and for other odd jobs. His family had few possessions: a radio for national news, a bicycle for transport, and a muzzle-loading gun inherited from his maternal grandfather. He and his wife grew sorghum, maize, and, in the dry season, vegetables. She brewed beer to purchase household necessities. The homestead possessed a brooding hen and seven pigeons. Edson complained that none of his sons helped him, or "showed respect," even to adult neighbors.

Edson's village area received little development under ADMADE. Some years earlier, a safari operator contributed a grinding mill, but it currently lacked an essential part. The current safari concessionaire was finishing a health clinic nearby as part of his commitment to community development. Edson told us that his CRB representatives did not hold meetings, nor did they inform people of their decisions as "they kept all the benefits for and to themselves." Very few jobs existed; cash was scarce as brewing beer brought little money. "All the problems we face," he told us, "are the result of negligence on the part of government for not delivering on its promises. An all-weather road would bring a very big change making every area easily reached. Some other places have wildlife and roads, why not here?"

A week later and after several hours spent interviewing residents in two isolated villages, I was cycling in the hot afternoon with three of my associates along a bush track returning to Chilima Community School, our temporary encampment. We paused in the shade of a tree adjacent to a seasonal pond to cool off. As we rested and drank water from bottles, we saw a small boy straddling a bicycle proceeding in our direction. Short of stature, the rider peddled from one side, as his feet could not reach the pedals from the seat. From the seriousness with which this youth pursued his journey, my companions immediately suspected that he carried a fateful message. When he saw us, the boy jumped from his precarious position and ran alongside the bike to break its momentum. We complimented the rider on his bravery and for being alone in this desolate space. He informed us that an elephant had killed Edson Nabwalya the previous evening. As Edson's relatives were making arrangements for a wake and burial, the boy did not have many details. Afterward, he quickly mounted the bike as his journey passed through dense game country. He hoped to finish his long circuit ride before nightfall.

That evening in the schoolyard, we learned more about what had happened, about the recovery and disposition of the body, and about the elephant. A group of people using the same road that night came upon Edson's abandoned bicycle. As the night was very dark, they remained unaware of what had happened to the rider and returned the bike to a nearby homestead before resuming their journey. The next morning, young schoolchildren discovered the crushed corpse lying off the road. Frightened, they returned home to tell their parents. Edson's brother-in-law identified the mangled body and covered it with a white cloth.

When the chief heard about the death from the young cyclist, he redirected his tractor driver to collect the body and deliver it to the residence of Edson's mother. On the chief's order, the ZAWA unit leader dispatched community scouts to track the rogue beast, positively identify it, and kill it. The scouts returned, reporting that they followed tracks to a cow and small calf at a lagoon. They saw blood on the cow's tusks, killed her, and watched as the calf ran away.[23]

As prior commitments and time constraints kept us from attending Edson's funeral, we lent a bicycle to the community schoolteacher and requested that he represent us. Beyond his teaching duties, he was the pastor at the local Pentecostal Holiness Church as well as a former member of the subauthority and the initial CRB. In the next days, we learned more about events before and after Edson's death as reports became consistent through reiterative rounds of telling.

After his burial, accounts of Edson's death focused on several events and relationships. On the morning of that fateful day, there was beer at Edson's house. That afternoon, a quarrel broke out between Edson and his sons. Their words turned physical when his wife sided with their sons and insulted her husband. Specifics were not mentioned, just that the sons never got along with their father or "showed respect" to other elders, as Edson had himself recounted to me. Rather than chance a damaging fight, Edson coaxed an intoxicated friend to cycle with him to a distant village for beer. After his friend fell twice from his bicycle, Edson proceeded alone.

When he arrived at the village, Edson informed those present about his circumstances. The headman's wife provided him with a room for the evening and cooked him a meal before he settled into drinking. His friends recalled that at 2200 hours, Edson said that his sons would have passed out by that time and would no longer pose a problem for him if he returned home. Despite his friends' insistence that he stay, Edson mounted his bicycle and began his journey.

Cycling rapidly and alone under a half moon, Edson entered a long stretch of woodland where the deep shadows of the overhead canopy impeded his vision of lurking dangers. He failed to see a cow elephant with a young calf using the same road. Nearby villagers heard an elephant trumpet, but did not know how to interpret its meaning. Beer drinkers following the same road later that night found the bicycle, but not the body lying nearby off the road.

Upon his return from the wake, the community teacher told us that four hundred people attended the burial. He reported that Edson's coffin was "heavy." As a Pentecostal preacher, the community teacher found significance in how the elephant had killed Edson—by breaking both legs, by disemboweling him, and by smashing his skull. For him, this elephant left a figurative statement by marking the differences between humans and animals (upright stance, food consumed, and consciousness) and perhaps between sinners (drunkards) and those truly saved.

The teacher also informed us about the mourners' discussions and what they assumed about this elephant's nature. Some claimed that the elephant was "sent" by someone. They noted that Edson's mother's "womb" had produced seven children; only three now remained alive. Wild animals had killed three of her children (two daughters by crocodile, Edson by elephant). Another son died of tuberculosis. After each death, the mother consulted a witch finder, who affirmed that someone in her lineage had caused these deaths. On the grave of her first son, she was alleged to have put *cilanduzhi* medicine (to kill the person causing the death). She accused a neighboring "brother" of being "this witch," yet his consistent refusals caused the mother eventually to remove this magic from the grave. Most knew that Edson had died simply because he encountered an elephant cow with a young calf, which are known to be dangerous and should be given a large space. But still they sought answers to the "why" questions. Why did this happen? Why Edson? Why now? Many of the mourners refused to consume the flesh of this elephant as that might construe compliance with witchcraft as some had proposed during the wake.

At the burial site, members of Edson's lineage accused his wife of contributing to his untimely death. In family arguments, she always sided with their sons, sometimes even prolonging the quarrels in the process. Disrespectful sons and this visible breech in family etiquette created a space through which detractors, using witchcraft, attacked and successfully killed a vulnerable Edson. After the burial, a witch finder supposedly told the wife's family that the death originated among her in-laws, thus pointing the accusatory finger back toward the lineage of Kabuswe Mbuluma, Edson's father.

Witchcraft accusations are rooted in family struggles over meaning behind fortunes and misfortunes as relatives contend with the perennial problems of assets, poverty, social responsibilities, and legacies. Through the interpretive medium of witchcraft, Edson's death and its social links to other events settled some older scores between the late Kabuswe Mbuluma's relatives and those of the current chief. Such tragedies are contentious arenas in which men and women argue over assets and their respective investments. More problematic for his constituency, Edson was the standing ward candidate for the opposition party in the national elections scheduled that September. His untimely death meant that a plateau candidate would capture the MP seat again and valley interests would continue marginal during the next parliamentary sessions.

Closing Commentary

The current status of Zambia's "community-based" wildlife program underscores several themes within this book. Land and its important assets remain focused assets and central issues in the lives and livelihoods of rural Africans. Wildlife conservation is intimately cultural, political as well as economic, values that can be integrated and bring people together or divide them. All boundaries are cultural never exclusively economic, but always enclosures "not only of geographic space and resources but also of identity, meaning, and power." In turn, these limits become "both multiple and fluid and therefore inherently resistant to uncomplicated clarification" (Harwell 2011: 181). Wildlife conservation in present parlance and through its prospective programs throughout southern Africa has become an abstract northern loan noun that expresses a particular status quo *ante* to sustain a certain atypical lifestyle, which many northerners, as world marketplace consumers or professionals, know about and invest in. Unfortunately, many of these entangled values and practices inculcate most "others" elsewhere, as permanent subordinates supporting commercial and recreational worlds to which they may become attached, but most cannot join in any creative sense. Such eventualities and their unexamined cultural containers put at high risk many expensive conservation arrangements. What appears needed are more open ended exchanges and learning of how to transfer the seeds of this abstruse alien noun (conservation) onto more fertile grounds, where its possibilities can flower proactively as a creative, active, and liberating verb, capable of cultivating the imaginations of other peoples, such as the Valley Bisa, and to culturally challenge all of us into incorporating them within more plausible practical and beneficial global goals. Such exercises can never become one way or imposed learning experiences. So it seems for my perspectives within the Munyamadzi GMA and from the many projects in places I have visited within southern Africa and elsewhere.

Wildlife conservation engages *images* of the world that we, as individuals and members of groups, live in, as well as envision what we would like that world to become. Northerners cherish wildlife aesthetics and its supposed "naturalness" and remember their travels and "adventures" on another's turf. Unfortunately for most Valley Bisa and other groups struggling to subsist and make a living from these same grounds, the cultural meanings within this northern "conservation" ethic are experienced as negative restrictions, injunctions, subordinations, violence, and unrequited conditions. These negative connotations about the worth of "others" are bound in Northern consumptions at home and through travels abroad through their idealized, dichotomized cultural compartments of work/vacations as well as effort/pleasures. The Valley Bisa and others have different experiences, especially with wildlife, and seek to perpetuate these concerns on turf they claim as homeland. At the local level, individuals seek livelihoods as farmers, hunters, merchants, wage earners, cultivators, government workers, as wives and husbands as they seek to define supportive boundaries and social landscapes, rather than culturally suffocate under alien ones.

Residents of the Munyamadzi GMA know that they and "their wildlife" have paid the highest costs and received meager returns from the new wildlife program and its policies. Government deceptions may have worked in the short run for the "instrumental animals" that graze upon these local commons protected by enforced and brutal policies as their accumulating values are funneled off to beneficiaries or interlopers residing elsewhere. Piers Blaikie, (2006: 1953) described how this metaphorical and foreign "Trojan Horse" operates through the interlinked sanctions of a few local elites and entrepreneurs (chiefs, resident businessmen, and civil servants) who channel rural products outward to their patrons in regional (district and provincial offices, tourist and hunting concessionaires), even upward to national centers (ministries, politicians, businesses), and further afield into the global markets of obscure consumers and celebrities (safari hunters, scientists, and tourists). As a result, the imposed wildlife restrictions and land boundary issues remain locally as the unrequited *bête noire*, a French metonym, of experienced vulnerabilities as well as the local "bones of contention" in local contingencies.

Wildlife conservation is also about *identities*—about who we are, about what we do, about how we make our livings, spend our time and are connected, or not, to those who possess the authority to make decisions about common futures. A close reading of the initial Luangwa Valley "community-based" proposals identifies those who were involved (and who weren't). In addition, the documents reveal what decisions were made (maybe, how and why) and who the enablers or detractors were. Project choices were often personal, political, limited, and cultural, differentiating allies from enemies, beneficiaries from contenders, actors from intended subjects. Despite the professed objectives of workshop members and numerous consultants to improve the lives of rural people, most of these inhabitants (as project "targets") became neither "real" participants nor "primary" beneficiaries. Even the managers' audiences, through their arcane reports, were distant strangers, the donors and grantors who had money but not time to care more about the human pulse and cultural dynamics of those inhabiting the Luangwa landscapes. Their documents and proposals have metaphorically morphed into "the elephants between these book covers," a construct whose contours and disposition has taken decades to articulate. The original planners may have possessed good intentions, yet their cultural blindness and boundaries shaped a product and an environment in which readers may become all losers in terms of constructing possibilities for a more inclusive world. Time will tell what was gained or lost as well as by whom and how.

We should examine the pretensions behind the veils of "universal" knowledge and distant "expertise" to expose how these global goals of ecological and human relationships are aligned and managed. Which individuals or groups benefit as well as which pay the costs? At the heart of ADMADE and other community-based programs are claims about the universality of abstract ecological knowledge (outsider's "science") and the need for urbane university-trained experts, who presume to know best how to manage the current technologies and "resources" at all, mostly abstract, levels. Yet sustainable ecological knowledge and practical sustainable efforts depend ultimately upon local groundings in the cultural behaviors and ideas of those gaining long term livelihoods within specific landscapes—the grist of many indigenous and "in place" identities. Resilient human-wildlife relations may be culturally coded, orally learned, expressed otherwise than by "objective reasoning" or in print, but must be valued as if people's lives depended on them.

The local hunters I followed sensed, knew, and understood wildlife and their environments in ways that fell normally outside northern models and disciplinary tracts, including the classrooms, which I was instructed in, and even taught as an "elevated education." The seemingly "untutored" individuals who crafted this local wildlife regime incorporated many strands from elsewhere, underwrote its development through their performances, were accountable to their dependents and clients, and contributed to their group's sustainability with wildlife and other environmental assets for over five decades (1940s–80s). They knew this intricate system, understood, contributed to, benefitted as well as suffered from its processes, were inclined to use it for local goals while creatively mastering its promises and assuming its risks. Not surprisingly, these cultural boundaries unilaterally collapsed as an uncertain world impinged when no one within had the verve, stamina, or inspiration to presevere beyond shortterm prospects.

How a society or individuals respond upon learning that their respective vision is not shared, but maybe an *illusion*, built on myths or fantasies, has everything to do with their possible futures as well as the worlds in which their children and grandchildren might live. One reaction is to become temporarily submissive, that is if the autocratic forces on the outside deploy force, escalate to militarize engagements, or deploy more subtle enclosures. These powers hope, like the colonials before them, eventually to get their subjects' attention or to gain their temporary silences. The current emphasis of the Zambian state together with its private- and public-sector colleagues, including its global allies in wildlife conservation, seems basically of this bent; that is, through "anti-poaching" brutality and coercion to change, convert, and thereby confirm others within a truncated and self-serving vision. At best, the outcome of these activities as of 2014 appears short-term and tragic—a catastrophe of identities and illusions as has happened before on other cultural landscapes worldwide.

Some societies have found ways to diminish their losses and vulnerabilities by coexisting alongside wildlife. Although some others' cultural means and meanings may confront northern sensibilities or approach their limits of language and learning, engagements seeking common ground and dialogue with those, who possess a profound knowledge of history and limits of their places, seem a promising way forward. Moreover, such arrangements might encourage more of us to face the dangerous, destructive "beasts" lurking within and consuming our own souls and environments, which we, as northerners, have yet to confront and tame. Such "elephants" might become more visible through explorations, exchanges, and dialogues with others, whose eyes and experiences have not been so afflicted until recently.

I have not attempted to romanticize or "essentialize" the extensive environmental knowledge of a prominent rural group or of a few individuals within this assembly. Focus and selection are essential to establish a claim, but I have not intended to reduce the obscurities or untidiness of my extended "poaching" forays onto many academic disciplinary fields or force its products to conform to conventional frameworks for analysis. Rather, I have argued against static and erstwhile superior conceptions that some northern professionals or its media, still convey about other peoples' capacities or the plausibility of these others' disparate environmental knowledge. One model never fits all, particularly if its language or mathematics fails to accommodate the subtler imaginations of the human mind or flunks to grasp externalities. Perception of gross difference is always the first glance of any foreigner; to understand difference demands more concerted involvements and time to absorb additional refined perceptions. Moreover, I dispute the rigid separations of human lives, welfare, and cultures from each other and from those of most other nonhuman lives and conditions that have led to the present prognosis about environmental and climate crises. My claim is that appropriate evaluations of the ways in which various human groups have sustained themselves together within their environments and resources over time seems a reasonable as well as a strategic goal for enhancing our collective future.

I have learned about myself, about life, and about wildlife through my associations with the Valley Bisa and others, who have spent generations living sparsely within biodiverse environments. This learning has been reciprocal ventures in ways of thinking with new vocabularies extending in both directions. Spending time to listen and to learn from these extended and reciprocal conversations has been challenging, even sometimes risky. Listening begins with the realization that each site of prolonged human settlement is also a place where active thoughts and activities pulse with history, memories, meanings and changes with more distant contacts, even some not physically present. These exposures take time to absorb and even more to understand. To recognize the nuances of one's natal culture and place takes a lifetime of learning and resilience. Yet, efforts spent studying the ideas and practices of others offer critical comparisons to reflect upon one's own habits and inclinations. Keeping alive the various and different ways in which life is experienced should allow the constructions of more embracing systems of knowledge and agency through avoiding the epistemological snares of privileging one's view over those of all others (Kohn 2013). Northerners may have lost much of these inherent mysteries and obligations of life through their presumed mastery over it.

If I have listened, learned, and transcribed appropriately the gifts that many Bisa have shared with me over these decades about wildlife as well as about human lives, then I may have approached the goal so elegantly articulated for my intents by Doris Kanyunga to her peer in this chapter's epigraph. I initially introduced Doris Kanyunga in 1988–89 (chapter 2) as she went about her chores monitoring the activities of others while showing and being shown "respect." I followed her welfare for several subsequent years (Marks 2014: 78–102) and chronicled her tragic death by an elephant in her field in April 1998 (see chapter 3).

Franz Kafka once wrote to a friend, "A book must be the axe for the frozen sea within us."[24] Such a metaphorical "local axe" is also similar to that suggested in the epigram which began chapter 2, which is a song encouraging a new chief to use his "axe" of affirmation in positive rather than vindictive ways. If my readers have glimpsed something different in other lives, considered possible forays of stepping outside of their comfort zones and lifestyles to witness how wildlife was created, enjoyed, disbursed and locally managed elsewhere in the recent past, to consider that other conversations exist beyond those considered conventionally plausible, to sense the richness of undeserved advantages we think we deserve in our travels and worldwide explorations, then my time recording, reflecting, writing and rewriting this manuscript may serve as an appropriate tribute to the lives, which Doris Kanyunga and many other Valley Bisa have lived and conveyed respectfully through their work, humor and shared through their conversations with me for more than half a century. In doing so, I have not forgotten my acknowledgments to the many others, who have generously and graciously shared their lives, hospitality and even transport to facilitate my passages, stopovers, and chores during these extended hunts and challenges of engaging the "mindful elephants" of conventional wisdom and for empathy with the downs and upsides of such adventures.

Notes

1. **BL:** *Kwena uyu muzungu [Stuart Marks] alikozapo saana pali cino ciaalo. Kuntanzi, mukamona muno ciaalo cikaba cabazungu.*
   DK: *Awe, tekuti kuikala munobafwaya. Bafwaya ukumona utukwata ubwanibombele ne mikalile.*
   BL: *Nomba elyo bakacite shani ngababumona? Bushe elyo bakatuletela umusebo? Nanomba ba game baletucuzha cila buzhiku. Kale twaletobelako inama fwe limbi balikuti nga baipaya inzovu leelo takwaba. Ine nomba papita ifyaka fizano na fitatu ukwabula ukulyapo umunani wa inama ya nzovu.*
2. Address reprinted in Dalal-Clayton (1984) and Larsen et al. (1985). Chief Malama's words were likely written by or succinctly abbreviated by one or more of the European authors to "indigenize" support for the project's objectives rather than those expressed by the chief. The chief's plea has yet to be achieved after more than twenty years.
3. Thirty-eight individuals attended this workshop; nine were NPWS officers and twenty were from various government ministries whose responsibilities impinged on wildlife issues. Fifteen were Europeans. Notwithstanding the stated intents of the workshop to represent rural Zambian interests in the proceedings, only the local chief was invited.

4. I doubt that Chief Malama articulated these telling points in such a succinct way. This text was used strategically to build consensus around the plans and ideas put forward at the workshop for both management models that emerged from it. During the development of community plans under ADMADE and LIRDP, chiefs might be advised ("consulted," told what to expect), while the ideas of other rural inhabitants were neither sought nor brought into the conversation. NPWS had experienced problems with GMA residents, and they hoped the benefits granted chiefs would give them incentives to bring their subjects around to accepting the new regime. NPWS overlooked colonial legacies and how their field agents significantly had changed chiefs' roles.
5. As a Senior Fulbright Research Professor in Zambia and resident at Nabwalya during 1988–89, Dr. Thor Larson asked me to comment "as a devil's advocate" on his draft. My main concern was that while local perceptions, uses, and access to wildlife were given as the project's focus, these issues were not specifically addressed. I wrote a letter suggesting that monitoring local resources, culling destructive wildlife, and distributing meat were functions earlier performed by GMA residents and they could be encouraged ("empowered" is the development term) to do so again. Local hunters could use and develop their own techniques and knowledge for these tasks. NPWS might be given a monitoring role in these activities, but I cautioned that most residents in the Munyamadzi GMA considered that "NPWS personnel represented the most corrupt, brutal, inhuman, and untrustworthy branch of the national government." Further, the normative use of the term "poaching" to cover all extralegal hunting in GMAs did not fully capture what was going on at the village level. I wrote that "to ensure local participation in planning, decision-making, and implementation" of the project (another project goal), much more basic research was required. I suggested that his team might examine the ideas of and learn from what happened to Norman Carr and others in creating Nsefu Game Reserve, a project from the 1950s managed by a local chief, who received most of the tourist revenues. I also thought that programs for women needed to be developed within the project. (Author's letter to Larsen, 27 February 1989.)
6. LIRDP produced a number of weighty documents, both in draft and printed form, while implementing its program during the first year. Among them are: "Proposals for the Phase 2 Programme June 1987" (LIRDP Project Document 3), "The Luangwa Resource Development Project," "The Phase 2 Programme" (LIRDP Project Document 4), and a paper by Richard H. V. Bell (the co-director, technical), "An

Introduction to the Luangwa Integrated Resource Development Project" (thirteen mimeographed pages). As a member of LIRDP's first International Union for the Conservation of Nature's (IUCN) review mission, I had access to and read its main documents during September 1989.

7. These safari auction revenues were proportionately distributed to the chief (40 percent) and recycled (60 percent) back into the GMA for wildlife management, mainly for anti-poaching and supplies. The proceeds from the harvest of hippos, a practice purposely kept labor intensive to maximize local employment, were equally shared between NPWS goals and those expected for the local community.
8. Some measurements used by this project's managers, particularly for declines in elephant and rhino poaching and for assessing local attitudes, were criticized as anecdotal by social researchers (Clark Gibson, 1999: footnote 44)
9. Other documents examined include: World Wildlife Fund, "Program Report on the Matching Grant for a Program in Wildlands and Human Needs" (awarded by USAID, 9 March 1987); "Project Grant Agreement between the Republic of Zambia and the United States of America for Regional Natural Resources Management" (AID Project Number 690-0251, August 1989). As a Fulbright Scholar, I arrived at Nabwalya in September 1988 and followed village-level meetings and interactions with NPWS staff in the central Luangwa as ADMADE was implemented.
10. My information and inferences come from tracking events and from several trips to Zambia while employed by Safari Club International Foundation (1995–2002) to remain informed of national political controversies. I also draw upon the insights of Ilyssa Manspeizer, an anthropology graduate student, who followed these issues while she resided in Zambia during the late 1990s. I served on her PhD dissertation committee.
11. The emergence of ZAWA was complicated by internal MMD maneuvers, including those by the vice president and prominent party and ministerial rivals who had financial stakes in safari companies and sought to appoint key ZAWA staff. The four ministers of tourism were fired for many reasons, including their alleged affiliation with groups opposing Chiluba's third-term prospects as well as their failures to give special licenses to political clients or to sanction game permits in support of political functions.
12. Both the Zambian and MMD constitutions dictated that the president could serve a maximum two terms at five years per term.

13. Hunting concessions bestow legal access to safari operators and are the means by which they generate profit and state revenue. Concessions are taxable accounts and come attached with a safari firm's commitments to wildlife management and for community development. ZAWA receives much of its operating revenues through these GMA concessions and licenses, while off-loading responsibilities for management and development onto the safari companies and CRBs.
14. Donors had previously supported a number of consultancy reviews of ADMADE, including one by Mano Consultancy Services Ltd, "An Evaluation of the ADMADE Programme with Special Reference to the 'Strengthening Phase 1995–1997'" (report written for USAID/Lusaka, January–February 1998) and by Ernest and Young Chartered Accountants, "Report on the Wild Life Conservation Revolving Fund (WCRF) Financial Management Capacity" (report for USAID/Lusaka, February 1998). In their papers, ADMADE officers identified some problems, including that they did not transfer any real authority for wildlife management to communities, that they failed to pay communities promptly and fairly, and that they collected heavy appropriations for wildlife protection. These officers also doubted the capacity of rural communities to meet management goals, expressed their conflicts with local chiefs, bemoaned the fickle competence of safari operators to market hunts, and indicated a number of ecological and physical factors that affected the program. These confessions came after critical reviews and were never effectively or politically acted on or implemented, thereafter to my knowledge.
15. I was just completing the third week of a four-month stay during which I had recruited and trained several local associates to administer a detailed questionnaire to all GMA village groups. A young German missionary couple and I sat with the councilors; we were formally introduced and asked to respond to our welcomes.
16. National food shortages and progressive economic declines led to the formation of the MMD in 1990 in opposition to UNIP's governance since independence. From 1991 to 2011, the MMD was the dominant party until it was defeated narrowly by the Patriotic Front (PF).
17. The articles were printed in the *Zambia Weekly Post*, dated 10 June and 16 June 2006. In the former article, the chief complained about the road, which he claimed was last graded in 1978. He noted that the poor road contributed to poor health, to the lack of medicines in the clinic, and that "patients were forced to seek medicines from witchdoctors." The latter article is also referenced in chapter 3.

18. This reference to "uppers" concerns this officer's superiors, who had information that was not known to the "lowers" (residents).
19. Headmen were fed the meat from an "unauthorized" buffalo shot the previous day by a community scout. This meat was cured in a pit next to the community center, and it had already fed the assembled crowd as it would for the next two days. Kanele Wildlife police officers had conspired with a community scout to kill this buffalo for the occasion. The scout shot the buffalo in a field next to his village, in full view of villagers (his relatives). When the CRB vehicle came to collect the kill, villagers surrounded the van and demanded meat. Officials initially refused the request, but the continuing protests demanded a public pronouncement and apology. Consequently, the scout was placed on disciplinary leave without pay, yet appeared later the same month when his services were required (see EN 23).
20. A rumor in the valley and in town was that this safari company had received the concession "unexpectedly" because of the political infighting within the MMD party prior to the election. The safari tendering process, which granted concessions for a ten-year period, was stopped by President Chiluba. Another rumor was that the company, which earlier "owned" the concession, was a large contributor to the MMD party, and the owner wished to get his prized concession back. See "Michelson's Detention Saddens Kabimba," *Sunday Post* (4 March 2007); also, personal conversations while I was in Zambia and communications subsequent to my departure in September 2006.
21. The quotations for this 1967 meeting are from an unpublished manuscript (Marks and Marks, nd).
22. Under indirect rule, most fees paid by outside safari hunters went to the native authorities (chiefs), who were largely responsible for managing the controlled hunting areas.
23. The community scout who killed this cow elephant was the same individual who shot the buffalo that fed the assembly of headmen (EN 19) and was then on disciplinary leave. He was a "notorious poacher" who was employed as a community scout to "occupy him from illegal activities." Such hunters were valued for their bush skills and courageous "spirit"; they were commanded often to do dangerous chores by ZAWA Wildlife police officers. The expectation was that their experiences and pay would "convert" them to become icons for conservation. His name was Kazembe (see p. 15).

24. Franz Kafka was an Austrian (born in Czechoslovakia) writer (1883–1924), and this quote appears in a letter he wrote to Oskar Pollak, dated 27 January 1904 (www.themodernword.com/Kafka/Kafka_quote.html, under letters). I thank Bert Kaplan for bringing this quote to my attention after I presented a paper to a faculty seminar on psychoanalytic readings at UNC-Chapel Hill in 2010. My talk was on the lives and struggles of seven Valley Bisa men, whom I had known and sporadically followed over six decades. We compared their life histories with a study on a cohort of Harvard University men that had graduated during the twentieth century.

Afterword

Readings "Out Loud" about Land and Wildlife as Properties

Without a doubt, British colonial and Zambian postcolonial wildlife policies and programs have profoundly altered the physical and cultural landscapes within the central Luangwa Valley. The colonial and subsequent independent state appropriated large sections of formerly sparsely occupied Valley Bisa landscape by establishing no-trespass boundaries (game reserves, national parks) and criminalizing many customary practices, such as taking wildlife and other natural products for sustenance and for protecting lives. Administrative decisions to provide meager health and education facilities, as well as official reluctance to improve road and communications infrastructure within this GMA, have diminished residents' welfare and competitiveness with other groups within Zambia. As they have increased in number and continue to struggle with subsistence, the Valley Bisa have also greatly affected this topography. While a few individuals have prospered temporarily from casual employment or through dependency relationships with outsiders, the plight of most has remained precarious, more laborious, and even more uncertain as futures are constructed daily and somewhat ambiguously.

State Management of Land and Wildlife as "Property"

Risks, both intended and unintended, associated with development programs and contracts include social, political, economic, and cultural processes, yet none of these relationships and their elasticities are well known. Recipient individuals and groups simultaneously and selectively absorb, ignore, or reject components of activities, entities, and ideas offered by foreigners. Current relationships between small communities and their adjacent resources are ultimately not local, but global, particularly given the changing international connections of production and destruction. As smaller social and marginal groups are increasingly incorporated into national states, they face incessant transformative demands. Paraphrasing Folke, Colding, and Berkes's (2003) commentary on these inevitable skirmishes, individuals within marginal communities must learn to build resilience, enabling them to cope with uncertainties as well as to nurture diversity and an openness to assimilate different kinds of knowledge and experiences. These small group adaptive and adopted transformations may enhance their self-directed searches to retain higher levels of autonomy in their searches for appropriate futures (2003: 383). As observed by an eminent African scholar, "History has shown that people who live by borrowed ideas are never respected even by the people from whom they borrow" (Ogot 2001: 31). Such proverbial wisdom seems appropriate for everyone, everywhere.

Dimensions of State Controls over "Common Lands"

The creation of state demarcations, such as provinces, chieftaincies, protected areas, and GMAs, inscribe governance, surveillance, and control over land and the development of new subjectivities. Along with these inscriptions, governments codify legislation and establish standardized procedures for creating new claims by which some people access land and valuable products while denying these assets to others, even those living among these assets. Behind such promulgations are usually the interests, claims, and aspirations of some well-placed people, who, through their wealth, awareness, location or status, are more likely than others, presently subsisting within these designated spaces, to legally benefit from the new proscriptions and regulated claims.

Some controls over land and resources have been around for a long time; newer ways are complex and often insidious. Influenced by global commercial or international conservation interests far removed from the people and sites they affect, developing states employ variegated combinations of "enclosure, territorialization, and legalization" backed by force, or threats of violence, to implement and secure programs affecting their land and resources. These recent procedures are different, however, in their frequencies, creativities, complex alliances, and often the subtle and subversive ways that they are implemented in space or through time (Peluso and Lund 2011). These schemes evolve from negotiations between unprecedented alliances of actors, funding, and institutions, which form working partnerships within the narrow "consensus" confines of the current neoliberal political-economy (Harvey 2007). Estienne Rodary (2009) argues that by attaching conservation initiatives so thoroughly to the ideology and power of world neoliberal capitalism, current conservationists' initiatives within Africa are different from those in the colonial past. These new programs are "biopolitical" procedures that profoundly affect the daily lives, resource decisions, and welfare of rural villagers. Such penetrating policies and practices aim to progressively erode local institutions and social boundaries, thereby placing their subjects together with their livelihoods onto the escalator of continuous uncertainty ("social mobilization") powered by events and capital from elsewhere. I find Rodary's thesis compelling, as similar processes of "governmentality" and commercial finance likewise affect most of us living in the northern world.[1] We are all becoming subordinated subjects to those possessing both capital and political power (Piketty 2014, 2015). As world citizens living in a universe lessened in former freedoms and earlier values, our environments now appear reduced in both humanness and diminishing visual biodiversity, progressive processes we barely are aware of or understand. Given this premise, my thesis is that we can share important lessons about life and wildlife with rural people and learn from them, as they may have spent generations cultivating their living environments differently from our ethnocentric ideas and practices. Unfortunately for us all, much of this local cultural wisdom and practices was affected and diminished by prior generational subjugations by those seeking to change and exploit the people and their environments while exporting its wealth into the northern hemisphere.

Understanding the political-economic motives behind supposed development and resettlement plans for taking large rural areas out of indigenous production and resettling its inhabitants into already burgeoning and competitive cities in order to enhance global biodiversity raises serious questions not only of costs and benefits (directly or indirectly; immediately and subsequently) but also about its proposers as well as those affected. Such resettlement plans and enclosures include the Convention on Biological Diversity's goal of setting 17 percent of the earth's global land surface aside in some sort of protective enclosures. In her historical review of conservation projects involving land taken from indigenous peoples, Alice Kelly (2011) suggests some benchmarks for questioning neoliberal initiatives to control land within their "market driven logics." Her ideas elaborate some subtle ways through which those with capital seek additional assets or incorporate new recruits, as allies or laborers, in expanding their accumulations and managed spaces. Citing examples, she elaborates Rodary's (*op. cit.*) contention that frontiers of cultural confusion are perpetual transformers as residential subordinates struggle under conditions dictated by others with the capital and force to do so. Her nuanced reviews of international projects reveal different interpretations from those picturesque conventional conservation images of "wilderness" and the innately destructive tribesmen lacking discipline and technical knowledge (Huxley 1961). Both of these stereotypical images are flawed as well and serve to perpetuate the wealth and welfare of some over others.

As with many protected and wildlife areas within Africa, those within Zambia's Luangwa Valley were established through unilateral official enclosures, dispossessions, and annulments of resource rights established earlier for communal or "Native Trust Land." Later official actions made possible the subsequent commercial opportunities now claimed for scientists, tourists, safari hunters, allegedly "compliant" government officials, and miners (Neumann 1998; Moore 2006). These national and internationally linked claims became traumatic and irrevocable for residents as portions of their former homelands were enclosed and turned into bewildering "frontiers" to manage wildlife for revenues. Such claims became more invasive and adverse when local customary practices and former rights were criminalized and enforced by the state unilaterally. Beyond losing access to important sources of animal protein, most residents were disarmed and became dependent upon reluctant bureaucrats to protect their properties and lives. Both memories smolder in people's minds, surface in public meetings, and on occasion break into the open as intense confrontations between villagers and enforcement officers (chapter 10).

Frontiers are neither places nor processes, but imagined schemes intending to shape, at least for a time, both spaces and practices within a certain copy. In the promotional tracts of the safari industry, the Munyamadzi GMA becomes "the very heart of prime hunting country," a "wilderness" supporting "one of Africa's richest concentrations of wildlife and birds." Within these commercial materials, residents are largely off-page; if mentioned at all, they are presented as "poachers" or as survivors of a distant and archaic past.[2] These images depict local inhabitants as destroyers of both wildlife and habitats.[3] The few residents, who become known superficially to safari clients in their prescribed status as "chief" or "tracker," may benefit momentarily from these encounters; yet most valley residents fade as undesirable shadows or cunning intruders (Tsing 2003). For safari clients, engagements within these remote regions turn into "adventures" and quests for "trophies," taxidermy and memorabilia to prompt vivid stories about their owner's courage and enterprise within Eden-like vistas. This tradition of trophy taking has an ignoble and lengthy human history (Harrison 2012).

The creation and preservation of enclosures and privatization processes are often violent, accompanied by the progressive transformation of economic, cultural, and environmental relations (Tsing 2005). Through neoliberal conservation practices, remote assets and spaces become construed as abstract capital, such as environmental services, spectacles, and genetic stock, whose values in the global marketplace require cash payments benefitting distant others, not residential inhabitants. Kelley's (2011) expanded review includes all such "enclosures of land, bodies, social structures, and ideas." She shows how some delayed and subtle deals, postponed in time and space, can achieve the same results as schemes with more immediate results. The effects of elusive national wildlife policies, which promote the erosion of local subsistence autonomy and the subordination of subjects as laborers, are not readily apparent during a brief passage on another's landscape. Similarly, the evasive responses of those affected adversely by policies are often difficult to understand or even to interpret. It takes observations and years to establish the links in these extensive and lengthy chains of contacts, connections, and events.

Safari companies compete for multiyear concessions over large land tracts inhabited by dense assemblies of wildlife, paying fees for access and licenses above and sometimes under the tables. Within concessions, companies assume responsibilities for the management of "state properties," including liaison with "local communities." Such entitlements require unconventional and dispersed alliances of players, innovative performances, and global dealings. As the new but temporary "owners of these landscapes," safari management often represents powerful business and state politicians acting in collusion on some distant stage unbeknownst to those affected and living within these conventional boundaries. The perception of wildlife as "property" (construed as "private property") facilitates the prevailing trend in capitalistic commodification, alienation, and specificity by allowing a few people to consolidate gains from their specific knowledge of these processes and how they operate. Furthermore, the perception of "resources" as abstract, discrete units enables state agencies to outsource their management as they track wild products in terms of expected revenue calculations as well as costs in future management and surveillance. Managerial careers depend on fulfilling the boss's demands rather than paying attention to distant, discordant village voices. Dissenting opinions are lost, or frequently "polished," in intermediary translations between inside gainers and outside losers.

When the Northern Rhodesia Department of Game, catering for refined British sporting aesthetics and expecting that Africans might eventually accept colonial values, became the National Parks and Wildlife Service soon after Zambian independence, the structural change represented more than a subtle semantic or spatial shift in political terms. This shift from "game" to a more generic "wildlife" category signaled a broader, more abstract managerial and material change in societal relationships to wild animals. The wild animals became categorical "resources," regulated as minerals and communications, as "generators" (a metaphorical shift to mechanics) of revenues for the fiscal development of the new state. The folding of animal life into resource categories also transformed these living entities as legal "properties" endorsable by new claims. Predators, elephant, and buffalo became more valuable as "licensed targets" for safari hunters, as "photo ops" for foreign tourists, and as "extralegal aims" for sundry gangs of opportunistic entrepreneurs than they did in their relationship to the welfare, development, and protection of GMA "rural communities." The more recent change in the departmental label from "wildlife service" to the Zambia Wildlife Authority further consolidated this top-down dictatorial trend within the structural adjustments of neoliberal economic reforms. Biological "resources" became converted into units serviceable within an expanding global tourist trade as well as the informal, shadowy markets of the local bushmeat traffic. Name changes and title adjustments were sequential harbingers, which progressively alienated assets from previous recipients or "owners," just as they signaled and empowered others to accumulate and profit from them as realigned "properties."[4]

"Abstracting" Wildlife

James Greenberg (2006) perceptively asserts that a disconcerting disconnect between the political-economic and environmental logics within state agencies occur as natural (wild) creatures become fragmented and treated as separate "natural resources." This "gerrymandering of nature" disperses its products (as game, wildlife, fisheries, timber, soils) across bureaucratic domains where each product is subjected to additional treatment as an abstract commodity rather than as an integral part of a whole environment (ecology). An unresolved difficulty is that economic reasoning, markets, and politics are not the same as those regulating the relations among biological entities. Beyond the horizon of the "local," products from nature become split among a hierarchy of territorial (regional, national to international) authorities each demanding its own sphere of jurisdiction. The interactions and relationships within biological communities make human sense only as they are integrated fully into the broader political and economic orders that globally impinge upon them. As Greenberg (2006: 122) contends, the actual tragedies and enduring dilemmas are not with the traditional "commons" itself, but within the subsequent commoditization and regulatory procedures that follow the commercial exploitation of its products.

For Greenberg (2006: 127) this commodification of nature divides conveniently into two processes—a grasp of both is integral for understanding the whole. The first, the "territorializations of nature," includes the practices by which national agencies are assigned rights to certain resource assets and through which they must regulate its production. These recurrences become the "local rules of the game" with sanctions, fees, and punishments, which are often selectively enforced. As shown for the Valley Bisa residents, these bureaucratic rubrics greatly affect their customary livelihoods and present them with continuous challenges as well as persistent handicaps. Resolving these conundrums demands attention to Greenberg's second, less obvious, commodification process. The disconnect between office bureaucrats, who presume to manage the production of "goods," and village residents, who face the realities of divisive mismanagement in place, requires an awareness of how "natural" products, seemingly integral within a local landscape, are uprooted, transported, redefined in terms of rights, repackaged, and relabeled for various consumers within the markets of a globalizing world. Rural residents are at a distinct disadvantage in appropriately interpreting these distant "deals" or in knowing "how to deal" effectively under the circumstances (Peluso and Lund 2011).

Ensuring continuous streams of revenue from annual "crops" of "trophy animals" (measured in length of horn, ivory, density of mane or size), under wild conditions, requires a staggered generational profile for each of the desirable species spread over extensive tracts of land. Most wild species within each of the trophic levels (herbivores, predators, scavengers, decomposers) comprising wild, free-ranging, and biologic communities experience high annual mortality rates. Even many of the larger wildlife species take years to produce the majestic symbols of "wilderness" (horns, size, "cleverness") that tourists admire and trophy hunters seek. Beyond the difficulties of monitoring these wild and mobile populations, they may be exploited with little tangible effects until an ambiguous or unexpected threshold is obtained, whereupon the resiliency of the "wild" population collapses (Thompson 1979; Gunderson, Holling, and Light 1995). Furthermore, international politics at global conventions (such as the Convention on International Trade in Endangered Species [CITES] and the Convention on Biological Diversity [CBD]) may result in withdrawing the products of some species (rhinos, elephants, lions) from legal markets and trades, which may affect national and local management of these wild "targeted" species through reduced revenues and "community" payoffs.

Given the objective to generate consistent revenue surpluses, the role of ZAWA managers and safari operators becomes prosecutorial toward residents, casting them as the villains in shortfalls of expected wild products or revenues. Officials become protective of their assumed "ownership" of wildlife, hence their reluctance responding to residents' appeals for protection from "problem animals." Local appeals may be self-serving just as scouts' reluctance appears (to save the ammo for their own purposes), yet their officials cater largely for higher-paying foreign hunters. Within GMAs, residents experience repeated interferences with their daily activities and restrictions to earlier entitled products from their land and lagoons. The critical answers to resolve these resource dilemmas are simultaneously and continuously "geographical, historical, anthropological, political, economic, and sociological—all at once, a matter of culture/power/history/nature" (Biersack 2006: 27).

Firm borders around porous, mobile living entities that rationalize a certain type of resource management reflect a northern notion of "property" and a proprietary interest in their physical state as abstract "resources." Such a disposition predisposes the existence of other narratives and claims, which state management forcibly displaces or silences (or stigmatizes) by declaring them illegal. Yet, state agents and their foreign beneficiaries count on the asymmetrical relationships among human groups as they unilaterally force their claims, just as they disregard local livelihoods and identities in implementing their agendas. Some local identities, based in farming and on natural resources, may be more intimately connected to animals, both domestic and wild, through different cultural idioms or rationales of memory, definitions, and meanings than are those of state agents. As Emily Harwell (2011: 181) notes, livelihoods are never exclusively economic but are imbedded in cultural, social politics and histories at all levels and times. Revealing the prior existence of an officially "silenced" wildlife management system and the engagements of local residents in creatively contributing and operating within its guidelines has been a major goal of this book. This regime was a local creation as well as production; it was one that residents understood, influenced, and promoted. The regime was not static, but continued to change in response to circumstances on the ground and elsewhere until its demise. An inflating national economy, the death of its chief architect, and a weak centralized government disciplined by foreign aid ideals of "structural adjustments" finally overwhelmed local initiatives and ownership.

After more than two decades under a supposedly integrated development and conservation program, Luangwa Valley residents and ZAWA still remain at loggerheads over the legitimacy of the state's "exclusive" rights over wildlife. Promises and expectations given by agents on both sides have never been fulfilled. Representations of village needs were quickly silenced within the ranks of those "appointed" to the original subauthority or later "elected" to the CRBs. The more traumatic events, violence by the scouts toward suspected residents, meager incentives, and imprisonments, have yet to achieve compliance with either conservation or development goals. Under the alleged "community-based" wildlife management, wildlife diversities and numbers of smaller game have decreased around settlements, while increases in the elephant and buffalo numbers around village fields have intensified the morbidity and dangers to villagers and local properties. Further, armed and allegedly vigilant anti-poaching patrols have yet to stop the massive valley poaching from the Zambian plateau or the persistent and assumed privileged access of some government officials and outsiders (Marks 2014).

Because claims to rights are inherently inseparable from broader social contracts, new questions emerge about relations between holders and nonpossessors of particular properties. We should inquire what factors support or stimulate resistance to newly created cultural boundaries. Under what conditions can individuals or groups enforce their claims and obligations (Harwell 2011: 181)? The perpetual high arrest rates of Valley Bisa residents together with the alleged "poaching" and convictions among government officials, wildlife scouts, and many others explicitly show the lack of a durable contract among these interested parties. Persistent poachers have subtle strategies for evading these inquiries, and scandalous violations of existing game laws occur among all groups of hunters as well as "poachers." Rosaleen Duffy (2010) questions the uncertainties within the conventional wisdom surrounding these "poaching wars," including the differences between legitimate and extralegal hunting and between the nature of subsistence and commercial wildlife uses, and asks questions of who benefits and how. I found very few conservation converts or supporters for ZAWA's management performance among residents in 2006 (Marks 2014). Persistent violence by state agents against marginal groups also suggests a basic weakness as it indicates that the "targeted people" implicate a project as "inefficient, damaging, or simply irrelevant" for them (Hirschman 1970; Marks 1999; Li 2007: 16: Songorwa 1999).

Perspectives on Development, Properties, and Neoliberal Economics

There will always be downsides and unpredictable consequences of any development program, yet these costs and outcomes are particularly deplorable when there is neither the political nor financial will at the national level to monitor or adjudicate demeaning human processes. New relationships of governance, land, and resource control surge across former homelands, rupturing past relations and practices, yet these tsunamis also offer challenges, new opportunities, and even unprecedented freedoms. Indeed, the rapid rise and miscalculations of new local leaders along with the ebbs and flows of coercive influences and commercial markets open spaces for depressive as well as creative responses (Marks 2014: 236–76). Despite the present dismal prospects, how the futures of some, a majority of, or most Valley Bisa residents will emerge from these circumstances remains an open yet inchoate question.

Privatization is a common and divisive subject among those involved in development, inequalities, and social justice, as the remaining common lands appear destined for progressive transformation into private property. Privatization and property relations are particularly relevant for understanding neoliberal economic rationales. These projects effectively alter society, nullifying nature relationships through intensified enclosures of land and resources that sustained previous generations of households in rural communities. Neoliberal agendas typically include deregulation, commodification, marketization, and liberalization of trade. Yet, it is privatization and its obscurity within the complexities of new property rights together with other unseen schemes (the disciplinary work of the state, market functions, and the objectives of project planners) that are shaping these neoliberal concepts. By acknowledging the complexity of property relations, by reaffirming that the "ownership" of any property is always a social relationship of interdependence, by suggesting the existence of choices as well as of other property regimes, and by indicating that the practice of privatization yields mixed and often contradictory results, Becky Mansfield (2008a: 2) contributes to an understanding of how these neoliberal processes affect our lives and those of others.

Mansfield (2008a, b) reminds that privatization is not necessarily the inevitable and evolutionary process implied through the neoliberal doctrines of "free markets," for governments are the creators, definers, and ultimate defenders of the properties involved in these transactions. Further, unilateral state transactions are neither as straightforward nor as simple as enclosures of land, definitions of resources, and new proprietary regimes as each of these processes entails the redesigning of environments, livelihoods, and identities altogether and simultaneously. As the possession of "exclusive rights as private property" by an individual or a group often implies the "dispossession" of others' previous rights, this perception forces both the new possessors as well as those dispossessed to become different "market subjects" (owners and nonowners respectively). Both are expected subsequently to discipline themselves according to their respective status within the newly imposed social order. Moreover, this working definition of neoliberal exclusive property fails to consider the inequalities created and the existence, and the preference, of others for perhaps different property regimes.

"Property" may appear superficially about only "tangible things," but it is more profoundly about "social arrangements," which allow an individual or group to acquire specified rights in particular objects (Cohen 1927). Yet what really counts in these transactions is for the "others," the previous users "legally" alienated from these assets, to acknowledge and acquiesce to the recently acquired possessor's claims. As this acceptance is never definite but disputed, such contestations mean that new claims must be defended ultimately by the state, even violently at times, for the state (perhaps encouraged by other organizations) is the architect, definer, and defender of such assets. Privatization is more than a mere shift in institutional structure, from public regulation to private hands, for it implies values for these new entities as properties within global markets. As the key to and initial process in creating new possessions (in this case, innovative commodities from nature and its products), privatization creates new channels, enhancing capital flows through which values accumulate and circulate (Polanyi 1944).

Neoliberalism also employs property in ways that silence and simplify the complexities of life. In its ideology, property becomes "private property" with supposedly "exclusive ownership"; it is this simplified definition that legitimizes developmental projects linking capital growth with individual "freedom" and "free markets." Missing from this definition ("property") are questions of whose freedoms are enhanced and how inequalities were created during the processes of closure and dispossession. Also absent from this neoliberal construction are other types of properties (particularly communal or common properties), which, if mentioned at all, are considered part of the problem that the state must overcome. The development of the common property literature (Berkes, Feeny, McCay, and Acheson 1989) and the numerous contributions of the 2009 Nobel Laureate in Economics, the late Elinor Ostrom (1990, 1995), and her network, show that common property is not the same thing as "open" resources that necessarily end up as Garrett Hardin's (1968) "tragedy of the commons." This common property literature examines how people have collaborated and organized themselves to manage natural resources (wildlife, fisheries, and forests) without government involvement, thereby debunking much of the conventional wisdom about the need for government intervention and regulations. Many of the common property scholars depict the high costs of government and privately sponsored interventions by reworking people's basic relationships among themselves and with others, by dismantling their livelihoods as well as their relationships with local resources that create powerful oppositions to and distrust of state intrusions (Mansfield 2008a, b). By their very natures, the profundities inherent in the intended neoliberal transformations in "nature"—human relations together with the resistances to these processes, are so complex and convoluted that what these transformations and their antitheses may mean still remain open questions. By provoking transformations in multiple spaces and ways, both neoliberal proponents and their project audiences forge and rework new interconnections into alternatives that continually threaten and change the intentions and disciplines of their creators. These shaped and refashioned alternatives may become what Mansfield (2008a: 11) suggests should give us hope for more sustainable futures.

Examples of Unintended or Unforeseen Consequences

When viewed over wide vistas of time and space, enclosures and repressive means often have unexpected consequences for those groups effecting them, for those affected, as well as for still others, who eventually may benefit even if it takes a generation or so. During the eighteenth and nineteenth centuries, the English gentry's imposition of progressively restrictive land and game laws against the lower classes of rural neighbors eventually forced the latter to migrate to the industrializing cities as wage laborers to work for a rising merchant class. According to Thompson (1975), these conflicts over land and privilege, between country landowners and urban wealth, were really over "power and property rights" as the gentry fought to preserve their way of life, to

continue their political predominance in country life over that of the cites, and to perpetuate their own status as "gentlemen." The gentlemen, as rural elites, preferred their land and the "status quo" over the money then in the hands of parvenus and urban merchants.

Eventually status, wealth, means, and much of the earlier landscape were to shift dramatically. The mounted gentry with their regimented rides and colorful rituals were left with only the inedible fox to pursue as the larger mammals were eliminated through the broader landscape transformations promoted by the wealth of the expanding merchant class. This ascendant group's wealth was acquired through trade and from appropriations of the labor of those recently displaced from the land. The merchants' taste in field sports was on the ground with shotguns after birds ("feathers rather than fur"), activities enhanced through landscape changes as well as by incubators and gamekeepers, all of which made their large battues possible. A close and intergraded reading of Thompson (1975), Longrigg (1977), Vanderell and Coles (1980), Munsche (1981), and Deucher (1988) makes these historical inferences plausible.

Similar impositions and events occurred in the southern United States after the Emancipation Proclamation and the conclusion of the American Civil War in 1865. Africans had been violently wrenched from their societies and transported across the Atlantic Ocean to be sold as slaves (private properties) for work on the southern plantations of their white owners. As slaves, these men and women surreptitiously kept their forging techniques and former identities alive for generations. For both whites and blacks, hunting exploits on the sporting fields turned into important markers of class and race. As did their British and European counterparts, prosperous southern whites took to their sporting fields as displays of wealth unburdened by the necessity of hunting that drove other poorer whites and blacks in their subsistence practices. For African Americans, hunting and fishing pursuits remained a form of labor, a struggle for survival as well as a symbol of their cultural, economic, and spatial separation and subordination to whites. When freed en masse before and after the conclusion of the Civil War, many African Americans dispersed into the neighboring open and woodland areas where, as did many poorer whites, they subsisted largely as autonomous small groups of households. Within these sites, they relied upon wild products through mixed strategies of hunting, fishing, and foraging combined with subsistence agriculture as well as occasional work for hire or barter. Eventually, these widely dispersed subsistence settlements and the common commercial trades in wild meats diminished most wildlife and extirpated the larger forms from vast regions within the US South.

Between Emancipation and the early twentieth century, the rural US South became a "frontier" as its citizens strove to rebuild from the ravages and losses of a major civil conflict. The large numbers of freed and independent-minded African Americans challenged the earlier racial hierarchy as they appeared for many whites to threaten the region's future prosperity. Adding to these racial tensions, hunting and fishing tourism in the South became the focus of a major industry promoted to rebuild the economic and cultural reunification between former enemies. This development suited large southern landowners, as it offered profitable prospects for their land as sporting fields and new allies endorsing a mythical "Dixie Land," a genuine "southern experience" on an antebellum plantation complete with subordinate and supportive blacks (Marks 1991: 39–61). To build a political coalition with white landowners, businessmen and city sportsmen took decades to convince poorer rural whites to join them by subordinating their customary foraging rights to new legislation and professional wildlife management. A larger part of their argument was that their restoration of control over African Americans was the key to the region's white cultural, economic, and sporting future. Southern whites accomplished this political goal during the dreadful Jim Crow decades when the civil rights of many African Americans were violently abrogated, including their hunting and fishing entitlements, which made them progressively dependent on a white-dominated economy. During the early twentieth century, a majority of influential white sportsmen and legislators passed laws to sell game licenses in support of game enforcement as well as wildlife management and federally funded restoration projects (Giltner 2008). Not until the national Civil Rights era beginning in the 1960s were these civil and sporting fields again legally opened to an expanding African American

middle class. Similar tragic stories are found in Africa as well as elsewhere (Neumann 1998; Ramutsindela 2003; Jacoby 2003; Greenough 2003; Greenough and Tsing 2003).

In *Transforming the Frontier,* Bram Buscher (2013) describes some of the challenges within the latest neoliberal fashions of the Trans-frontier Peace Parks initiatives extending beyond the international borders of South Africa. On these frontiers, he describes the problematic contradictory riddles and how conservationists and their neoliberal partners joined ranks. His book provides a critical and constructive reading of neoliberalism's politics and decentralized governance as well as the ways promoters stimulate and mask their paradoxes. Buscher offers a constructive way to evaluate some conservation contradictions, especially those conflicts between epistemological realities and their representational promotions in combining conservation and development objectives (Buscher 2013: 226). His analysis addresses the three neoliberal modes of political conduct by which the Trans-frontier Peace Park group reached consensus: by 1) employing the language of management to capture separate powerful audiences while retaining dominance; through 2) espousing *antipolitics* (the purging of deviant options and opinions to reach an often predetermined consensus); and by 3) the use of *marketing techniques* (abstract concepts and visual images to gain competitive advantages over rival's proposals) in their frontier skirmishes over the constitution of the reality and representation(s) in human-nature relationships. To confront these wrangles, Buscher posits four strategies: (1) incorporating political science models to clarify political processes in anthropological and geographical disciplines; (2) becoming more politicized in offering constructive solutions to get beyond specifying perverse problems; (3) confronting the legitimacy of marketplace implications and representations that preclude more inclusive conservation outcomes supporting cultural differences; and (4) broadening the frontier spaces of reality and representation to include indigenous worldviews and epistemologies that expose the domination of neoliberalism economics in conservation schemes (Buscher 2013: 229–32).[5]

Notes

1. Besides these two articles cited, I referenced additional manuscripts as background reading before writing this section. These publications include: Arun Agrawal (2005); Bram Buscher (2013); D. R. Brockington, R. Duffy, and J. Igoe (2008); H. de Soto (2000); M. Foucault (1991); David Harvey (2007); Anna L. Tsing (2005); Tanya M. Li (2007); Donald Moore (2006); and James Scott (1990). As this book was going to press, I was reading James Ferguson (2015) and Thomas Piketty (2015) about the new politics and possibilities for redistribution.
2. The words in quotes are from conversations with safari operators or taken from various advertisements or brochures featuring the Luangwa Valley. Similar quotes for the Zambian Ministry of Tourism, Environment, and Natural Resources (MTENR) and the Zambia Wildlife Authority (ZAWA) are found in a brochure provided to safari clients, "Safari Hunting in Zambia" published by ZAWA in 2002.
3. In 1966, a senior wildlife official described to me the people within the Munyamadzi Corridor, a space recently opened to risky seasonal transport and before my initial visit, as those "who habitually torched, pillaged, and raped the environment!"
4. The chronological litany of name changes begins with the Northern Rhodesia Department of Game and Tsetse Control in the 1940s and 1950s and its replacement by the Department of Game and Fisheries under the Ministry of Native Affairs prior to Zambian independence. The Department of Game and Fisheries was reorganized under the Zambian Ministry of Lands and later placed under the Ministry of Natural Resources and Tourism. Its name was changed to the Department of National Parks and Wildlife Services (NPWS) while hosting the five-year United Nations Development Project, which ended in 1973. The department's name and functions changed to Zambia Wildlife Authority (ZAWA) in 1999 as part of the neoliberal structural adjustments under the MMD party (Marks 2012).
5. See also Paige West (2006) and Sian Sullivan (2009) for further critiques of conventional community-based wildlife management.

References

Achino-Loeb, Maria-Luisa (ed.). 2006. *Silences: The Currency of Power.* Oxford: Berghahn Books.

Adams, William M. 2005. *Against Extinction: The Story of Conservation.* London: Earthscan.

Agrawal, Arun. 2005. *Environmentality: Technologies for Government and the Making of Subjects.* Durham, NC: Duke University Press.

Alpers, E. 1975. *Ivory and Slaves in East Central Africa: Changing Patterns of International Trade to the Later Nineteenth Century.* Berkeley: University of California Press, 1975.

Anderson, David and Richard Grove (eds.). 1989. *Conservation in Africa: People, Politics, and Practice.* New York: Cambridge University Press.

Anderson, Virginia D. 2004. *Creatures of Empire: How Domestic Animals Transformed Early America.* New York: Oxford University Press.

Anker, Peder. 2001. *Imperial Ecology: Environmental Order in the British Empire, 1895–1945.* Cambridge, MA: Harvard University Press.

Antrobus, Sally. 1983. 'If You Shoot to Kill, They Only Learn to Run.' *Safari Magazine* 9(4): 38–42.

Astle, William L. 1999. *A History of Wildlife Conservation and Management in the Luangwa Valley, Zambia.* Bristol, UK: British Empire and Commonwealth Museum. Research Paper Number 3.

Atran, Scott and Douglas Medin. 2008. *The Native Mind and the Cultural Construction of Nature.* Cambridge, MA: MIT Press.

Barnett, Rob. 2000. *Food for Thought: The Utilization of Wild Meat in Eastern and Southern Africa.* Nairobi, Kenya: TRAFFIC. Chapter 5 on Zambia, 113–40.

Bayart, Jean-Francois, Stephen Ellis, and Beatice Hibou. 1999. *The Criminalization of the State in Africa.* Oxford: James Currey.

Benedetto, Robert (ed.), Winifred K. Vass (trans.). 1996. *Presbyterian Reformers in Central Africa: A Documentary Account of the American Presbyterian Congo Mission and the Human Rights Struggle in the Congo, 1890–1918.* Amsterdam: E. J. Brill Academic Publishers.

Bender, Barbara. 1993a. 'Stonehenge—Contested Landscapes.' In Barbara Bender (ed.), *Landscape: Politics and Perspectives,* 245–79. Oxford: Berg,

——. 1993b. 'Introduction: Landscape—Meaning and Action.' In Barbara Bender (ed.), *Landscape: Politics and Perspectives,* 1–17. Oxford: Berg.

——. 1999. 'Subverting the Western Gaze: Mapping Alternative Worlds.' In P. J. Ucko and R. Layton (eds.), *The Archaeology and Anthropology of Landscape.* London: Routledge.

Berkes, F., D. Feeny, B. J. McCay, and J. M. Acheson. 1989. 'The Benefits of the Commons,' *Nature* 340: 91–93.

Biersack, Aletta. 2006. 'Reimagining Political Ecology: Culture/Power/History/Nature.' In Aletta Biersack and James B. Greenberg (eds.), *Reimagining Political Ecology,* 3–40. Durham, NC: Duke University Press.

Birmingham, David. 1976. 'The Forest and Savannah of Central Africa.' In John E. Flint (ed.), *The Cambridge History of Africa,* 5: 300. Cambridge: Cambridge University Press.

Blaikie, Piers. 2006. 'Is Small Really Beautiful? Community-Based Natural Resources Management in Malawi and Botswana,' *World Development* 34: 1942–57.

Bonner, Raymond. 1993. *At the Hand of Man: Peril and Hope for Africa's Wildlife.* New York: Alfred A. Knopf.

Bordieu, Pierre. 1977. *Outline of a Theory of Practice.* Cambridge: Cambridge University Press.

——. 1980. *The Logic of Practice.* Cambridge, UK: Polity Press.

Borgmann, Albert. 1992. *Crossing the Postmodern Divide.* Chicago: University of Chicago Press, 119–20.

Boyd, John Morton (ed.). 1992. *Fraser Darling in Africa: A Rhino in the Whistling Thorn.* Edinburgh: Edinburgh University Press.

Bowden, Mark. 1982. 'Rhino, the Last Look?' Philadelphia, PA: *The Philadelphia Inquirer* [Reprinted articles for February 7–10, 1982 including article on Zambia's Luangwa Valley, p. 11–19 with photographs by Chuck Isaacs].

Brelsford, W. V. 1956. *The Tribes of Northern Rhodesia.* Lusaka, Zambia: The Government Printer.

Brockington Dan. 2009. *Celebrity and the Environment: Fame, Wealth, and Power in Conservation.* London: Zed Books.

Brockington, D., R. Duffy, and J. Igoe. 2008. *Nature Unbounded: Conservation, Capitalism, and the Future of Protected Areas.* London: Earthscan.

Brosius, J. Peter, Anna L. Tsing, and Charles Zerner (eds.). 2005. *Communities and Conservation: Histories and Politics of Community-Based Natural Resource Management*. Walnut Creek, CA: AltaMira Press.

Brown, Taylor and Stuart Marks. 2007. 'Livelihoods, Hunting, and the Game Meat Trade in Northern Zambia.' In Glyn Davies and David Brown (eds.), *Bushmeat and Livelihoods: Wildlife Management and Poverty Reduction*, 92–105. Oxford: Blackwell Publishing.

Burton, Richard (trans.). 1973. *The Lands of Kazembe: Lacerda's Journey to Kazembe in 1798*. London: John Murray.

Buscher, Bram. 2013. *Transforming the Frontier: Peace Parks and the Politics of Neoliberal Conservation in Southern Africa*. Durham, NC: Duke University Press.

Cancel, Robert. 2013. *Storytelling in Northern Zambia: Theory, Method, Practice and the Necessary Fiction*. Cambridge: Open Book Publishers (www.openbookpublishers.com)

Chabal, Patrick and Jean-Pascal Daloz. 1999. *Africa Works: Disorder as Political Instrument*. Oxford: James Currey.

Chanock, Martin. 1998. *Law, Custom, and Social Order: The Colonial Experience in Malawi and Zambia*. Portsmouth, NH: Heinemann.

Chipungu, Samuel N. 1992. *Guardians in Their Time: Experiences of Zambians under Colonial Rule, 1890–1964*. London: The Macmillan Press, Ltd.

Cioc, M. 2009. *The Game of Conservation: International Treaties to Protect the World's Migratory Animals*. Athens: Ohio University Press.

Cohen, M. R. 1927. 'Property and Sovereignty,' *Cornell Law Quarterly* 18: 8–30.

Colson, Elizabeth. 1971. 'The Impact of the Colonial Period on the Definition of Land Rights." In Victor Turner (ed.), *Colonialism in Africa: 1870–1960*, chapter 3. Cambridge: Cambridge University Press.

——. 2006. *Tonga Religious Life in the Twentieth Century*. Lusaka, Zambia: Bookworld Publishers.

Comaroff, Jean and John Comaroff. 1992. *Ethnography and the Historical Imagination*. Boulder, CO: Westview Press.

——. 2012. *Theory from the South: Or, How Euro-America is Evolving toward Africa*. Boulder, CO: Paradigm Publishers.

Connolly, William E. 2013. *The Fragility of Things: Self-Organizing Processes, Neoliberal Fantasies, and Democratic Activism*. Durham, NC: Duke University Press.

Croll, Elizabeth and David Parkin (eds.). 1992. *Bush Base: Forest Farm; Culture, Environment and Development*. London: Routledge.

——. 1992. 'Cultural Understanding of the Environment.' In Elizabeth Croll and David Parkin (eds.), *Bush Base: Forest Farm*. London: Routledge, 11–36.

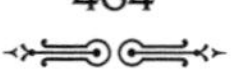

Crehan, Kate. 1997. *The Fractured Community: Landscapes of Power and Gender in Rural Zambia*. Berkeley: University of California Press.

Cronon, William. 1996. 'Introduction: In Search of Nature.' In William Cronon (ed.), *Uncommon Ground*. New York: W. W. Norton and Company, 23–56.

Cumming, R. Gordon. 1857. *A Hunter's Life in South Africa*. Alberton, South Africa: Galago Publishing (Pty) Ltd. Reprinted in 1986; facsimile of 1857 edition.

Cunnison, Ian. 1959. *The Luapula Peoples of Northern Rhodesia: Custom and History in Tribal Politics*. Manchester, UK: Manchester University Press.

Dalal-Clayton, D. B. (ed.). 1984. *Proceedings of the Lupande Development Workshop*. Lusaka, Zambia: Government Printer.

Das, Veena, Michael Jackson, Arthur Kleinman, and Bhrigupat Singh (eds.). 2014. *The Ground Between: Anthropologists Engage Philosophy*. Durham, NC: Duke University Press.

de Luna, Kathryn. 2012. 'Hunting Reputations: Talent, Individuals, and Community in Precolonial South Central Africa,' *Journal of African History* 53: 279–99.

de Soto, Hernado. 1989. *The Other Path: The Invisible Revolution in the Third World*. New York: Harper and Row.

——. 2000. *The Mystery of Capitalism: Why Capitalism Triumphs in the West and Fails Everywhere Else*. New York: Basic Books.

Darling, Frank Fraser. 1960. *Wildlife in an African Territory*. London: Oxford University Press.

Deuchar, Stephen. 1988. *Sporting Art in Eighteenth-Century England: A Social and Political History*. New Haven: Yale University Press.

Dodds, Donald G. and David Patton. 1968. *Report to the Government of the Republic of Zambia on Wildlife and Land-Use Survey of the Luangwa Valley*. Rome: United Nations Development Programme/Food and Agriculture Organization of the United Nations. Report no. TA2591.

Dove, Michael R., Percy E. Sajise, and Amity A. Doolittle (eds.). 2011. *Beyond the Sacred Forest: Complicating Conservation in Southeast Asia*. Durham, NC: Duke University Press.

Dowie, Mark. 2009. *Conservation Refugees: The Hundred-Year Conflict between Global Conservation and Native Peoples*. Cambridge, MA: The MIT Press.

Duhaylungsod, Levita. 2011. 'Interpreting Indigenous Peoples and Sustainable Resource Use.' In Michael R. Dove, Percy E. Sajise, and Amity A. Doolittle (eds.), *Beyond the Sacred Forest: Complicating Conservation in Southeast Asia*, 180–215. Durham, NC: Duke University Press.

Duffy, Rosaleen. 2010. *Nature Crimes: How We're Getting Conservation Wrong*. New Haven, CT: Yale University Press.

Easterly, William. 2007. *The White Man's Burden: Why the West's Efforts to Aid the Rest Have Done So Much Ill and So Little Good.* New York: Penguin Books.

——. 2010. 'Foreign Aid for Scoundrels,' *The New York Review of Books* 57 (18): 37–38.

Ellis, Stephen and Gerrie Ter Haar. 2004. *Worlds of Power: Religious Thought and Political Practice in Africa.* New York: Oxford University Press.

Evans-Pritchard, Edward E. 1937. *Witchcraft, Oracles, and Magic among the Azande.* New York: Oxford University Press.

——. 1956. *Nuer Religion.* New York: Oxford University Press.

Feinburg, Harvey M. and Joseph R. Solodow. 2002. 'Out of Africa,' *Journal of African History* 43 (2): 255–61.

Ferguson, James. 1994. *The Anti-Politics Machine: "Development," Depoliti-zation, and Bureaucratic Power in Losotho.* Minneapolis: University of Minnesota Press.

——. 2015. *Give a Man a Fish: Reflections on the New Politics of Distribution.* Durham, NC: Duke University Press.

Firey, Walter I. 1960. *Man, Mind and Land: A Theory of Resource Use.* Glencoe, IL: The Free Press of Glencoe.

Folke, Carl, Johan Colding, and Fikret Berkes. 2003. 'Synthesis Building Resilience and Adaptive Capacity in Socio-Ecological Systems.' In Fikret Berkes, Johan Colding, and Carl Folke (eds.), *Navigating Social-Ecological Systems: Building Resilience for Complexity and Change,* 352–87. Cambridge: Cambridge University Press.

Foucault, M. 1991. *Discipline and Punish: The Birth of the Prison.* New York: Pantheon Books.

Fraser, Donald. 1923. *African Idyll: Portraits and Impressions of Life on a Central African Mission Station.* New York: Negro Universities Press. Reprinted in 1969.

Freehling, Joel and Stuart Marks. 1998. 'A Century of Change in the Central Luangwa Valley of Zambia.' In E. J. Milner-Gulland and Ruth Mace (eds.), *Conservation of Biological Resources,* 261–78. Oxford: Black-well Science.

Gamitto, A. C. P. 1960. *King Kazembe and the Marave, Chewa, Bisa, Bemba, Lunda and Other Peoples of Southern Africa.* 2 vols. Trans. I. Cunnison. Lisbon: Estudios de ciencias politicas e socias, Nos. 42, 43.

Gann, L. H. 1969. *A History of Northern Rhodesia: Early Days to 1953.* New York: Humanities Press.

Garland, F. 2008. 'The Elephant in the Room: Confronting the Colonial Character of Wildlife Conservation in Africa,' *African Studies Review* 51: 51–74.

Gibson, Clark and Stuart Marks. 1995. 'Transforming Rural Hunters into Conservationists: An Assessment of Community-Based Wildlife Management Programs in Africa,' *World Development* 23(6): 941–57.

Gibson, Clark C. 1999. *Politicians and Poachers: The Political Economy of Wildlife Policy in Africa.* New York: Cambridge University Press.

Giltner, Scott. 2008. *Hunting and Fishing in the New South: Black Labor and White Leisure after the Civil War.* Baltimore: The Johns Hopkins University Press.

Greenberg, James B. 2006. 'The Political Ecology of Fisheries in the Upper Gulf of California.' In Aletta Biersack and James B. Greenberg (eds.), *Reimagining Political Ecology,* 121–48. Durham, NC: Duke University Press.

Greenough, Paul. 2003. "Pathogens, Pugmarks, and Political 'Emergency': the 1970s South Asian Debate on Nature." In Paul Greenough and Anna L. Tsing (eds.), *Nature in the Global South: Environmental Projects in South and Southeast Asia,* 201–30. Durham, NC: Duke University Press.

Greenough, Paul and Anna Lowenhaupt Tsing (eds.). 2003. 'Introduction.' In Paul Greenough and Anna L. Tsing (eds.), *Nature in the Global South: Environmental Projects in South and Southeast Asia,* 1–23. Durham, NC: Duke University Press.

Gunderson, L. H., C. S. Holling, and S. S. Light. 1995. *Barriers and Bridges to the Renewal of Ecosystems and Institutions.* New York: Columbia University Press.

Guyer, Jane I. 1995. 'Wealth in People—Introduction,' *Journal of African History* 36(1): 83–90.

——. 2004. *Marginal Gains: Monetary Transactions in Atlantic Africa.* Chicago: University of Chicago Press.

Guyer, Jane I. and Samuel M. Eno Belinga. 1995. 'Wealth in People as Wealth in Knowledge: Accumulation and Composition in Equatorial Africa,' *Journal of African History* 36(1): 91–120.

Haggard, H. Rider. 1885. *King Solomon's Mines.* Parklands, South Africa: AD. Donker (Pty) Ltd. 1985 edition.

Hall, P. E. 1910. 'Notes on the Movements of *Glossina morsitans* in the Lundazi District, North-Eastern Rhodesia,' *Bulletin of Entomological Research* 1(3): 183–84.

——. 1950. 'Memories of Abandoned Bomas, No 2: Nawalia,' *Northern Rhodesia Journal* 1(5): 55–57.

Hammond, Dorothy and Alta Jablow. 1970. *The Africa That Never Was: Four Centuries of British Writing about Africa.* New York: Twayne Publishers, Inc.

Hanna, A. J. 1956. *The Beginnings of Nyasaland and North-Eastern Rhodesia, 1859–95.* Oxford: Clarendon Press.

Hardin, Garrett. 1968. 'The Tragedy of the Commons.' *Science* vol. 162, issue 3859: 1243–1248.

Harris, W. Cornwallis. 1852. *The Wild Sports of Africa: Being the Narrative of a Hunting Expedition from the Cape of Good Hope, Through the Territories of the Chief Moselekatse to the Tropic of Capricorn*. Cape Town, South Africa: L. C. Struik (Pty) Ltd. Reprinted in 1987; facsimile of the 5th ed.

Harrison, Simon. 2012. *Dark Trophies: Hunting and the Enemy Body in Modern War*. New York and Oxford: Berghahn Books.

Harvey, David. 2007. *A Brief History of Neoliberalism*. Oxford: Oxford University Press.

Harwell, Emily. 2011. 'The Social Life of Boundaries: Competing Territorial Claims and Conservation Planning in the Danau Sentarium Wildlife Reserve, West Kalimantan, Indonesia.' In Michael R. Dove, Percy E. Sajise, and Amity A. Doolittle (eds.), *Beyond the Sacred Forest: Complicating Conservation in Southeast Asia*, 180–215. Durham, NC: Duke University Press.

Hayden, Sherman Strong. 1942. *The International Protection of Wild Life*. New York: Columbia University Press.

Hirschman, A. 1970. *Exit, Voice, Loyalty: Consumer Responses to Declines in Firms, Organizations, and the State*. Cambridge, MA: Harvard University Press.

Hochschild, Adam. 1999. *King Leopold's Ghost*. New York: Houghton Mifflin, Co.

Holling, C. S. 1987. 'Simplifying the Complex: The Paradigms of Ecological Function and Structure,' *European Journal of Operations Research* 30: 139–46.

Hughes, David M. 2006. *From Enslavement to Environmentalism: Politics on a Southern African Frontier*. Seattle: University of Washington Press.

Huxley, Julian S. 1961. *The Conservation of Wildlife and Natural Habitats in Central and East Africa*. Paris: UNESCO.

Isaacman, Allen F. 1972. 'The Origin, Formation, and Early History of the Chikunda of South-Central Africa,' *Journal of African History* 13: 443–61.

Isaacman, Allen and Barbara S. Isaacman. 2004. *Slavery and Beyond: The Making of Men and Chikunda Ethnic Identities in the Unstable World of South-Central Africa, 1750–1920*. Portsmouth, NH: Heinemann.

Isaacman, Allen and Anton Rosenthal. 1984. 'War, Slaves, and Economy: The Late Nineteenth-Century Chikunda Diaspora,' *Cultures et Development* 16: 639.

Jackson M. and I. Karp (eds.). 1990. *Personhood and Agency: The Experience of Self and Other in African Cultures*. Stockholm: Almqvist & Wiksell.

Jacoby, Karl. 2003. *Crimes against Nature: Squatters, Poachers, Thieves, and the Hidden History of American Conservation.* Berkeley: University of California Press.

Johnson, Mark. 1987. *The Body and the Mind.* Chicago: University of Chicago Press.

Johnston, Harry W. 1897. *British Central Africa.* New York: Negro University Press [1969 reprint of 1897 edition].

Jones, William O. 1957. 'Manioc: An Example of Innovation in African Economics,' *Economic Development and Cultural Change* 5: 97–117.

Kapferer, Bruce. 1967. *Cooperation, Leadership, and Village Structure.* Lusaka, Zambia: Institute for Social Research, Zambian Papers No. 1.

Karp, Ivan. 1990. 'Power and Capacity in Iteso Rituals of Possession.' In M. Jackson and I. Karp (eds.), *Personhood and Agency: the Experience of Self and Other in African Cultures,* 79–93. Stockholm: Almqvist & Wiksell.

Kelley, Alice B. 2011. 'Conservation Practices as Primitive Accumulation,' *Journal of Peasant Studies* 36(4): 683–701.

Kennedy, Pagan. 2002. *Black Livingstone: A True Tale of Adventure in the Nineteenth-Century Congo.* New York: Viking.

Kingsley, Judith. 1980. 'Pre-colonial Society and Economy in a Bisa Chiefdom in Northern Zambia.' Unpublished PhD dissertation in anthropology. Ann Arbor: University of Michigan.

Kingsolver, Barbara. 1998. *The Poisonwood Bible: A Novel.* New York: HarperCollins.

Klein, Julie T. 1996. *Crossing Boundaries: Knowledge, Disciplinarities, and Interdisciplinarities.* Charlottesville: University Press of Virginia.

Kohn, Eduardo. 2013. *How Forests Think: Toward an Anthropology beyond the Human.* Berkeley: University of California Press.

Lane-Poole, E. H. 1938. *The Native Tribes of the Eastern Province of Northern Rhodesia.* Lusaka, Zambia: The Government Printer.

Langworthy, Harry W. 1972. *Zambia before 1890: Aspects of Pre-colonial History.* London: Longman.

Larsen, Thor, Fidelis B. Lungu, and Trond Vedeld. 1985. 'Preparation Report on the Luangwa Integrated Resource Development Project (LIRDP).' Lusaka, Zambia (September 1985).

Leach, Melissa and Robin Mearns (eds.). 1996. *The Lie of the Land: Challenging Received Wisdom on the African Environment.* Oxford: James Currey.

Leach, M., R. Mearns, and I. Scoones. 1999. 'Environmental Entitlements: Dynamics and Institutions in Community-Based Natural Resource Management,' *World Development* 27(2): 25–247.

Leader-Williams, N. and S. D. Albon. 1988. 'Allocation of Resources for Conservation.' *Nature* (vol. 6199): 533–535 (8 December 1988).

Leader-Williams, N. and S. D. Albon, and P. S. M. Berry. 1990. 'Illegal Exploitation of Black Rhinoceros and Elephant Populations: Patterns of Decline, Law Enforcement, and Patrol Effort in the Luangwa Valley, Zambia.' *Journal of Applied Ecology* 27: 1055–87.

Leader-Williams, N. and E. J. Milner-Gulland. 1993. 'Policies for the Enforcement of Wildlife Laws: The Balance between Detection and Penalties in the Luangwa Valley, Zambia,' *Conservation Biology* 7(3): 611–17.

Letcher, Owen. 1911. *Big Game Hunting in North-Eastern Rhodesia.* London: John Long.

——. 1913. *The Bonds of Africa.* London: John Long, Ltd.

Levi-Strauss, Claude. 1966. *The Savage Mind.* London: Wiedenfeld and Nicolson.

Lewis, Dale, G. B. Kaweche, and A. Mwenya. 1988. 'Wildlife Conservation outside Protected Areas: Lessons from an Experiment in Zambia,' Nyamaluma Conservation Camp: Zambian National Parks and Wildlife Service, mimeo.

Lewis, Dale, A. Mwenya, and G. B. Kaweche. 1991. 'African Solutions to Wildlife Problems in Africa: Insights From a Community-Based Project in Zambia,' *International Journal on Nature Conservation* 7: 10–23.

——. 1993. 'The Zambian Way to Africanize Conservation.' In D. Lewis and N. Carter (eds.), *Voices From Africa: Local Perspectives on Conservation,* 79–98. Baltimore: World Wildlife Fund Publications.

Lewis, Dale and Andrew Phiri. 1998. 'Wildlife Snaring: An Indicator of Community Response to a Community-Based Conservation Project.' *Oryx* 32 (April): 111–21.

Li, Tanya M. 2007. *The Will to Improve: Governmentality, Development, and the Practice of Politics.* Durham, NC: Duke University Press.

Livingstone, David. 1874. *The Last Journal of David Livingstone.* Edited by H. Waller. London: John Murray.

Longrigg, Roger. 1977. *The English Squire and His Sport.* New York: St. Martin's Press.

Louis, Wm. Roger. 1966. 'Sir Percy Andersons's Grand African Strategy, 1883–1896.' *The English Historical Review* 81 (no. 319): 297–314.

Lyell, Denis D. 1910. *Hunting Trips in Northern Rhodesia.* London: Horace Cox.

——. 1924. *The African Elephant and Its Hunters.* London: Heath Cranton Ltd.

MacGaffey, Janet. 1991. *The Real Economy of Zaire: The Contribution of Smuggling and Other Unofficial Activities to National Wealth.* Philadelphia: University of Pennsylvania Press.

MacKenzie, John. M. 1988. *The Empire of Nature: Hunting, Conservation and British Imperialism.* Manchester, UK: Manchester University Press.

Madan, A. C. 1906. *Wisa Handbook: A Short Introduction to the Wisa Dialect of North-East Rhodesia*. Oxford: Clarendon Press.
Mansfield, Becky. 2008a. 'Introduction: Property and the Remaking of Nature-Society Relations.' In Becky Mansfield (ed.), *Privatization: Property and the Remaking of Nature-Society Relations*, 1–13. Malden, MA: Blackwell Publishing.
——. 2008b. 'Property, Markets, and Dispossession: The Western Alaska Community Development Quota as Neolibralism, Social Justice, Both, or Neither.' In Becky Mansfield (ed.), *Privatization: Property and the Remaking of Nature-Society Relations*, 86–105. Malden, MA: Blackwell Publishing.
Manspeizer, Ilyssa 2004. 'Considering Wildlife Conservation in Zambia at the Turn of the Millennium.' Unpublished PhD dissertation. Binghamton University, State University of New York.
Marks, Martha S. and Stuart A. Marks. N.d. "Who Owns the Animals? Profile of a Zambian Community in Transition." Unpublished manuscript.
Marks, Stuart A. 1973a. 'Settlement History and Population of the Valley Bisa in Zambia,' *Zambia Museums Journal* 4: 43–56.
——. 1973b. 'Prey Selection and Annual Harvest of Game in a Rural Zambian Community,' *East African Wildlife Journal* 11: 113–28.
——. 1976. *Large Mammals and a Brave People: Subsistence Hunters in Zambia*. Seattle: University of Washington Press.
——. 1977a. 'Hunting Behavior and Strategies of the Valley Bisa of Zambia,' *Human Ecology* 5: 1–36.
——. 1977b. 'Buffalo Movements and Accessibility to a Community of Hunters in Zambia,' *East African Wildlife Journal* 15: 251–61.
——. 1979a. 'Profile and Process: Subsistence Hunters in a Zambian Community,' *Africa* 49: 53–67.
——. 1979b. 'An Integrated Social-Environmental Analysis of the Luangwa Valley in Zambia,' *Case Studies in Development Assistance 7*. Washington, DC: Development Studies Program, USAID.
——. 1982. 'Arguing from the Present to the Past: A Contemporary Case Study of Human Predation on African Buffalo.' In D. Hopkins et al. (eds.), *Paleoecology of Beringia*, 409–23. New York: Academic Press.
——. 1984. *The Imperial Lion: Human Dimensions of Wildlife Management in Central Africa*. Boulder, CO: Westview Press.
——. 1991. *Southern Hunting in Black and White: Nature, History, and Ritual in a Carolina Community*, Princeton, NJ: Princeton University Press.
——. 1994a. 'Local Hunters and Wildlife Surveys: A Design to Enhance Participation,' *Journal of African Ecology* 32: 233–54.
——. 1994b. 'Managerial Ecology and Lineage Husbandry: Environmental Dilemmas in Zambia's Luangwa Valley.' In Mary Hufford (ed.),

Conserving Culture: New Discourse on Heritage, 111–22. Urbana, IL: University of Illinois Press.

——. 1996. 'Local Hunters and Wildlife Surveys: An Assessment and Comparisons of Counts During 1989, 1990, and 1993,' *Journal of African Ecology* 34: 237–57.

——. 1997. 'Snaring Wildlife: A Hidden Transcript: of Livelihoods and Ethnicity in Zambia's Luangwa Valley.' Unpublished paper, delivered at the International Association for the Study of Common Property, Berkeley, California.

——. 1999. 'Contextual Factors Influencing a Rural Community and the Development of a Wildlife Management Regime in Zambia (1987–1997),' *Journal of Policy and Planning* 1: 235–46.

——. 2000. 'Combining Wildlife Conservation with Community Development: A Case Study and Cautionary Assessment from Zambia.' In Moses K. Tesi (ed.), *The Environment and Development in Africa*, 187–201. Lanham, MD: Lexington Books.

——. 2001. 'Back to the Future: Some Unintended Consequences of Zambia's Community-Based Wildlife Program (ADMADE),' *Africa Today* 48 (Spring): 120–41.

——. 2002. 'Creating Colonial Wildlife Policies in the 1930s: On Cake as Metaphor for Elephant Control in Northern Rhodesia.' Unpublished paper, presented at University of Northern British Columbia seminar, 1 November 2002.

——. 2003. 'A Culling of Hippos Failed to Resurrect a Drought-Stricken Community: A Riddle in Community-Based Wildlife Management.' Paper presented at a teleconference symposium at Carnegie Mellon University, 25 April 2003; online at http://hdgc.eppcmu.edu/misc/SMA%20cull%20of20Hippos.doc.

——. 2004. 'Reconfiguring a Political Landscape: Transforming Slaves and Chiefs in Colonial Northern Rhodesia,' *Journal of Colonialism and Colonial History* 3: 13 pp.

——. 2005. *Large Mammals and a Brave People: Subsistence Hunters in Zambia*. New Brunswick, NJ: Transaction Publishers [Second Edition with a New Introduction and Afterword].

——. 2009. 'Rural People and Wildlife in Zambia's Central Luangwa Valley: Precautionary Advice from a Long-Term Study.' In Rolf D. Baldus, *A Practical Summary of Experiences after Three Decades of Community-Based Wildlife Conservation in Africa "What are the Lessons Learnt?* 52–71. Joint publication of FAO [Food and Agriculture Organization of the United Nations] and CIC [International Council for Game and Wildlife Conservation] Budapest, Hungary.

Marks, Stuart A. 2012. 'The Name of the Game: Community Cost of Zambian Wildlife Policies within a Game Management Area (1966–2012)." Unpublished paper, presented at "Old Lands: New Practices?" Conference, held at Rhodes University, Grahamstown, South Africa, 14 September 2012.

——. 2014. *Discordant Village Voices: A Zambian Community-Based Wildlife Programme*. Pretoria, South Africa: UNISA Press.

——. 2016. 'Disclosure on a Wildlife Enclosure: Transition on a Zambian Frontier 1988–92. Invited presentation to Gwendolen M. Carter Conference 'Tropics of Discipline: Crime and Punishment in Africa', University of Florida, Gainesville, FL [1–2 April 2016].

Marks, Stuart and Mipashi Associates. 2008. *On the Ground and in the Villages: A Cacophony of Voices Assessing a 'Community-Based' Wildlife Program after 18 Years—The Munyamadzi Game Management Area, Central Luangwa Valley*. Privately published under grant for distribution (100 copies).

Marnham, Patrick. 1987. 'Counting All Elephants' and 'A Manhunt in the Game Park.' In *Fantastic Invasion: Dispatches from Africa*, chap. 1 (pp. 21–36) and chap. 2 (pp. 37–53). London: Penguin Books.

Mauss, Marcel. 1954. *The Gift: Forms and Functions of Exchange in Archaic Societies*. Trans. I. Cunnison. London: Cohen and West.

Miller, Joseph C. 1988. *Way of Death: Merchant Capitalism and the Angolan Slave Trade, 1730–1830*. Chap. 2. Madison: University of Wisconsin Press.

Milner-Gulland, E. J. and N. Leader-Williams. 1992. 'A Model of Incentives for the Illegal Exploitation of Black Rhino and Elephant: Poaching Pays in the Luangwa Valley, Zambia,' *Journal of Applied Ecology* 29: 337–39.

Mloszewski, M. J. 1983. *The Behavior and Ecology of the African Buffalo*. Cambridge: Cambridge University Press.

Moore, Donald. 2006. *Suffering for Territory: Race, Place, and Power in Zimbabwe*. Durham, NC: Duke University Press.

Moore, Henrietta and Megan Vaughan. 1994. *Cutting Down Trees: Gender, Nutrition, and Agricultural Change in the Northern Province of Zambia*. London: James Currey.

Morris, Brian. 1998. *The Power of Animals: An Ethnography*. New York: Berg.

——. 2000. *Animals and Ancestors: An Ethnography*. Oxford: Berg.

Munsche, P. B. 1981. *Gentlemen and Poachers: The English Game Laws 1671–1831*. Cambridge: Cambridge University Press.

National Resources Consultative Forum. 2008. *The Impact of Wildlife Management Policies on Communities and Conservation in Game Management Areas of Zambia: Message to Policy Makers*. Report by Phylus Simasiku,

Hopeson Simwanza, Gelson Tembo, Sushenjit Bandyopadhyay, and Jean-Michel Pavy, June 2008. Lusaka, Zambia: Natural Resources Consultative Forum; available at www.aec.msu.edu/fs2/Zambia/resources/Final%20_NCRF.pdf.

Neumann, Roderick P. 1998. *Imposing Wilderness: Struggles over Livelihood and Nature Preservation in Africa*. Berkeley: University of California Press.

Northern Rhodesia Government. 1960. *Proceedings of the Sixth British and Central Africa Fauna Conference*. Lusaka, Northern Rhodesia: Government Printer.

Ogot, Bethwell A. 2001. 'The Construction of Luo Identity and History.' In Luise White, Stephan F. Miescher, and David W. Cohen (eds.), *African Words, African Voices: Critical Practices in Oral History*, 31–52. Bloomington: Indiana University Press, 2001.

Orwell, George. 1936. 'Shooting an Elephant,' *New Writing* (Autumn issue).

Ostrum, Elinor. 1990. *Governing the Commons: The Evolution of Institutions for Collective Action*. New York: Cambridge University Press.

Ostrum, Elinor and R. Keohane (eds.). 1995. *Local Commons and Global Interdependence: Heterogeneity and Cooperation in Two Domains*. Cambridge, MA: Harvard University Press.

Parker, Ian and Mohamed Amin. 1983. *Ivory Crisis*. London: Chatto and Windus, the Horgarth Press.

Peters, Pauline E. 2013. 'Conflicts over Land and Threats to Customary Tenure in Africa,' *African Affairs* 112/449: 543–62.

Peluso, Nancy L. and Christian Lund. 2011. 'New Frontiers of Land Control; Introduction', *The Journal of Peasant Studies* 38 (4): 667–681.

Pfeiffer, James. 2005. 'Commodity *Fetichismo*, the Holy Spirit, and the Turn to Pentecostal and African Independent Churches in Central Mozambique', *Culture, Medicine and Psychiatry* 29: 255–83.

Piketty, Thomas, 2014. *Capital in the Twenty-First Century*. Cambridge, MA: Harvard University Press.

——. 2015. "A Practical Vision of a More Equal Society," *New York Review of Books* 62(11) (25 June): 26–29.

Pitman, C. R. S. 1934. *A Report on the Faunal Survey of Northern Rhodesia with Special Reference to Game, Elephant Control and National Parks*. Livingstone, Northern Rhodesia: Government Printer.

Polanyi, Karl. 1944. *The Great Transformation*. Boston: Beacon Press.

Posey, Daryll. 1999. *Cultural and Spiritual Values of Biodiversity*. London: IT Publications and UNEP.

Pratt, Mary L. 1992. *Imperial Eyes: Travel Writing and Transculturation*. London: Routledge.

Ramutsindela, Maano. 2003. 'Land Reform in South Africa's National Parks: A Catalyst for the Human-Nature Nexus,' *Land Use Policy* 20: 41–49.

Reed, David (ed.). 1995. *Structural Adjustment, the Environment, and Sustainable Development*. London: Earthscan Publications Ltd. (published for World Wildlife Fund).

——. 2001. *Economic Change, Governance and Natural Resource Wealth: The Political Economy of Change in Southern Africa*. London: Earthscan Publications Ltd. (published for World Wildlife Fund).

Reed, Robin S. 2012. *Savannas of Our Birth: People, Wildlife and Change in East Africa*. Berkeley: University of California Press.

Richards, Ardrey. 1939. *Land, Labour, and Diet in Northern Rhodesia*. Oxford: Oxford University Press.

Riesman, P. 1990. 'The Formation of Personality in Fulani Ethnopsychology.' In M. Jackson and I. Karp (eds.), *Personhood and Agency: The Experience of Self and Other in African Cultures*, 169–90. Stockholm: Almqvist & Wiksell International.

Riney, Thane. 1967. 'Conservation and Management of African Wildlife.' Rome: Food and Agricultural Organization of the United Nations. Consultancy sponsored by FAO and the International Union for the Conservation of Nature and Natural Resources (IUCN), 35 pp. Photocopy in author's collection.

Riney, Thane and Peter Hill. 1967. 'Conservation and Management of African Wildlife: English-Speaking Country Reports.' Rome: FAO. Consultancy sponsored by FAO and IUCN, 144 pp. Photocopy in author's collection.

Ritvo, Harriet. 1987. *The Animal Estate: The English and Other Creatures in the Victorian Age*. Cambridge, MA: Harvard University Press.

——. 1997. *The Platypus and the Mermaid and Other Figments of the Classifying Imagination*. Cambridge, MA: Harvard University Press.

Roberts, Andrew D. 1970a. 'Pre-colonial Trade in Zamiba,' *African Social Research* 10: 715–45.

——. 1972. *The Lumpa Church of Alice Lenshina*. Lusaka [Zambia]: Oxford University Press. (reprinted from Robert I. Rotberg and A. A. Mazuri (eds.). *Protest and Power in Black Africa*. New York: Oxford University Press.

——. 1973. *History of the Bemba: Political Growth and Change in North-Eastern Zambia before 1900*. Madison: University of Wisconsin Press.

——. 1976. *A History of Zambia*. London: Heinemann.

Rodary, Estienne. 2009. 'Mobilizing for Nature in Southern African Community-Based Conservation Policies, or the Death of the Local,' *Biodiversity Conservation* 18: 2585–2600.

Roe, Emery. 1999. *Except Africa: Rethinking Development, Rethinking Power.* New Brunswick, NJ: Transaction Publishers.

Said, Edward. 1978. *Orientalism.* New York: Pantheon Books.

Schoffeleers, J. M. 1968. "Symbolic and Social Aspects of Spirit Worship among the Mang'anja," PhD thesis. Oxford University.

Scott, James C. 1990. *Domination and the Arts of Resistance: Hidden Transcripts.* New Haven, CT: Yale University Press.

——. 1998. *Seeing Like a State: How Certain Schemes to Improve the Human Condition Have Failed.* New Haven, CT: Yale University Press.

Scudder, Thayer. 1962. *The Ecology of the Gwembe Tonga.* Manchester, UK: Manchester University Press.

Selous, Frederick Courteney. 1881. *A Hunter's Wanderings in Africa: Being a Narrative of Nine Years Spent among the Game in the Far Interior of South Africa.* New York: Macmillan. Reprinted in 1981.

Sider, Gerald M. 2014. *Skin for Skin: Death and Life for Inuit and Innu.* Durham, NC: Duke University Press.

Sinclair, A. R. E. 1977. *The African Buffalo: A Study of Resource Limitation of Population.* Chicago: University of Chicago Press.

Skjonsberg, Else. 1981. *The Kefa Records: Everyday Life among Women and Men in a Zambian Village.* Oslo: University of Oslo, U-Landsseminaret nr.21.

——. 1989. *Change in an African Village: Kefa Speaks.* West Hartford, CT: Kumarian Press, Inc.

Songorwa, Alexander. 1999. 'Community-Based Wildlife Management (CWM) in Tanzania: Are the Communities Interested?' *World Development* 27(12): 2061–79.

Spinage, Clive. 1994. *Elephants.* London: T. and A. D. Poyser Ltd.

——. 2003. *Cattle Plague: A History.* New York: Springer Science and Business Media.

Stefaniszyn, B. 1951. 'The Hunting Songs of the Ambo,' *African Studies* 10: 1–12.

——. 1964. *Social and Ritual Life of the Ambo of Northern Rhodesia.* London: Oxford University Press.

Stigand, C. H. 1913. *Hunting the Elephant in Africa and Other Recollections of Thirteen Years' Wanderings.* New York: The MacMillan Company.

Storrs, A. E. G. 1979. *"Know Your Trees": Some of the Common Trees Found in Zambia.* Ndola, Zambia: The Forest Department.

Strickland, Bradford. 2001. "My Grandfather's Gun was Called 'Field of Children': Ecological History as Indictment of State Development Policy," *Africa Today* 48: 110–19.

Strong, David and Eric Higgs. 2000. 'Borgmann's Philosophy of Technology.' In Eric Higgs, Andrew Light, and David Strong (eds.), *Technology and the Good Life?*, 19–39 Chicago: University of Chicago Press.

Sullivan, Sian 2009. 'Green Capitalism and the Cultural Poverty of Constructing Nature as Service Provider,' *Radical Anthropology* 3: 18–27.
Thomas, F. M. 1958. *Historical Notes on the Bisa Tribe, Northern Rhodesia.* Lusaka, Northern Rhodesia: The Rhodes Livingstone Institute.
Thomas, Keith. 1983. *Man and the Natural World: A History of the Modern Sensibility.* New York: Pantheon Books.
Thompson, E. P. 1975. *Whigs and Hunters.* New York: Pantheon Books.
Thompson, Michael. 1979. *Rubbish Theory: The Creation and Destruction of Value.* New York: Oxford University Press.
Tilley, Helen. 2011. *Africa as a Living Laboratory: Empire, Development, and the Problem of Scientific Knowledge, 1870–1950.* Chicago: University of Chicago Press.
Tsing, Anna L. 2003. 'Agrarian Allegory and Global Futures.' In Paul Greenough and Anna L. Tsing (eds.), *Nature in the Global South: Environmental Projects in South and Southeast Asia,* 124–69. Durham, NC: Duke University Press.
——. 2005. *Friction: An Ethnography of Global Connection.* Princeton, NJ: Princeton University Press.
Turner, Victor. 1967. *The Forest of Symbols: Aspects of Ndembu Ritual.* Ithaca, NY: Cornell University Press.
Uhl, Christopher. 2013. *Developing Ecological Consciousness: The End of Separation.* 2nd ed. New York: Rowman and Littlefield.
Vanderell, Anthony and Charles Coles. 1980. *Game and the English Landscape: The Influence of the Chase on Sporting Art and Scenery.* New York: The Viking Press.
Vail, Leroy. 1977. 'Ecology and History: The Example of Eastern Zambia,' *Journal of Southern African Studies* 3: 129–55.
Van Donge, Jan Kees. 2009. 'The Plundering of Zambian Resources by Frederick Chiluba and His Friends: A Case Study of the Interactions between National Politics and the International Drive towards Good Governance,' *African Affairs* 108/430 (January): 69–90.
van Velsen, Jaap. 1957. *The Politics of Kinship.* Manchester, UK: Manchester University Press.
Vansina, Jan. 1965. *Oral Tradition: A Study in Historical Methodology.* Chicago: Aldine Publishing Company
——. 1966. *Kingdoms of the Savannah.* Madison: University of Wisconsin Press.
Vaughan-Jones, T. G. L. 1938. 'Memorandum on Policy Concerning the Foundation of a Game Department and the Conservation of Fauna in Northern Rhodesia.' Copy of mimeograph in author's possession.

Vitebsky, Piers. 1992. 'Landscape and Self-Determination among the Eveny: The Political Environment of Siberian Reindeer Herders Today.' In Elizabeth Croll and David Parkin (eds.), *Bush Base: Forest Farm, Culture, Environment and Development*, 223–46. London: Routledge.

——. 2006. *The Reindeer People: Living with Animals and Spirits in Siberia*. Boston: Houghton Mifflin Company.

Walker, Ernest P. 1964. *Mammals of the World*, vol. 1. Baltimore: The Johns Hopkins Press.

Walker, John F. 2009. *Ivory's Ghosts: The White Gold of History and the Fate of Elephants*. New York: Atlantic Monthly Press, 2009.

West, Paige. 2006. *Conservation is Our Government Now: The Politics of Ecology in Papua New Guinea*. Durham, NC: Duke University Press.

Western, David and R. Michael Wright (eds.). 1994. *Natural Connections: Perspectives in Community-Based Conservation*. Washington, DC: Island Press.

White, Luise. 2000. *Speaking with Vampires: Rumor and History in Colonial Africa*. Berkeley: University of California Press.

Wiese, Carl. 1983. *Expedition in East-Central Africa, 1888–1891, A Report*. Edited, with introduction and comments by Harry W. Langworthy. Norman, OK: University of Oklahoma Press.

(The) White Fathers' Bemba-English Dictionary. 1954. London: Longmans, Green & Co.; Lusaka, Northern Rhodesia: Northern Rhodesia and Nyasaland Joint Publications Bureau.

Wills, A. J. 1967. *An Introduction to the History of Central Africa*. Oxford: Oxford University Press.

Wolmer, William. 2007. *From Wilderness Vision to Farm Invasions: Conservation and Development in Zimbabwe's South-East Lowveld*. Oxford: James Currey.

Yorke, Edmund. 1990. 'The Spectre of a Second Chilembwe: Government, Missions, and Social Control in a Wartime Northern Rhodesia, 1914–18,' *Journal of African History* 31: 373–91.

Index